Compendium of Plant Genomes

Series Editor

Chittaranjan Kole, Raja Ramanna Fellow, Government of India, ICAR-National Research Center on Plant Biotechnology, Pusa, New Delhi, India

Whole-genome sequencing is at the cutting edge of life sciences in the new millennium. Since the first genome sequencing of the model plant Arabidopsis thaliana in 2000, whole genomes of about 100 plant species have been sequenced and genome sequences of several other plants are in the pipeline. Research publications on these genome initiatives are scattered on dedicated web sites and in journals with all too brief descriptions. The individual volumes elucidate the background history of the national and international genome initiatives; public and private partners involved; strategies and genomic resources and tools utilized; enumeration on the sequences and their assembly; repetitive sequences; gene annotation and genome duplication. In addition, synteny with other sequences, comparison of gene families and most importantly potential of the genome sequence information for gene pool characterization and genetic improvement of crop plants are described.

Interested in editing a volume on a crop or model plant?

Please contact Prof. C. Kole, Series Editor, at ckoleorg@gmail.com

More information about this series at http://www.springer.com/series/11805

Dinesh Adhikary • Michael K. Deyholos •
John P. Délano-Frier
Editors

The Amaranth Genome

Springer

Editors
Dinesh Adhikary
University of Alberta
Edmonton, Alberta, Canada

Michael K. Deyholos
University of British Columbia
Kelowna, British Columbia, Canada

John P. Délano-Frier
Centro de Investigación y de Estudios
Avanzados del IPN (Cinvestav)
Irapuato, Guanajuato, Mexico

ISSN 2199-4781 ISSN 2199-479X (electronic)
Compendium of Plant Genomes
ISBN 978-3-030-72364-4 ISBN 978-3-030-72365-1 (eBook)
https://doi.org/10.1007/978-3-030-72365-1

This Springer imprint is published by the registered company Springer Nature Switzerland AG
The registered company address is: Gewerbestrasse 11, 6330 Cham, Switzerland

This book series is dedicated to my wife Phullara and our children Sourav and Devleena

Chittaranjan Kole

Preface to the Series

Genome sequencing has emerged as the leading discipline in the plant sciences coinciding with the start of the new century. For much of the twentieth century, plant geneticists were only successful in delineating putative chromosomal location, function, and changes in genes indirectly through the use of a number of "markers" physically linked to them. These included visible or morphological, cytological, protein, and molecular or DNA markers. Among them, the first DNA marker, the RFLPs, introduced a revolutionary change in plant genetics and breeding in the mid-1980s, mainly because of their infinite number and thus potential to cover maximum chromosomal regions, phenotypic neutrality, absence of epistasis, and codominant nature. An array of other hybridization-based markers, PCR-based markers, and markers based on both facilitated construction of genetic linkage maps, mapping of genes controlling simply inherited traits, and even gene clusters (QTLs) controlling polygenic traits in a large number of model and crop plants. During this period, a number of new mapping populations beyond F_2 were utilized and a number of computer programs were developed for map construction, mapping of genes, and for mapping of polygenic clusters or QTLs. Molecular markers were also used in the studies of evolution and phylogenetic relationship, genetic diversity, DNA fingerprinting, and map-based cloning. Markers tightly linked to the genes were used in crop improvement employing the so-called marker-assisted selection. These strategies of molecular genetic mapping and molecular breeding made a spectacular impact during the last one and a half decades of the twentieth century. But still they remained "indirect" approaches for elucidation and utilization of plant genomes since much of the chromosomes remained unknown and the complete chemical depiction of them was yet to be unraveled.

Physical mapping of genomes was the obvious consequence that facilitated the development of the "genomic resources" including BAC and YAC libraries to develop physical maps in some plant genomes. Subsequently, integrated genetic–physical maps were also developed in many plants. This led to the concept of structural genomics. Later on, emphasis was laid on EST and transcriptome analysis to decipher the function of the active gene sequences leading to another concept defined as functional genomics. The advent of techniques of bacteriophage gene and DNA sequencing in the 1970s was extended to facilitate sequencing of these genomic resources in the last decade of the twentieth century.

As expected, sequencing of chromosomal regions would have led to too much data to store, characterize, and utilize with the-then available computer software could handle. But the development of information technology made the life of biologists easier by leading to a swift and sweet marriage of biology and informatics, and a new subject was born—bioinformatics.

Thus, the evolution of the concepts, strategies, and tools of sequencing and bioinformatics reinforced the subject of genomics—structural and functional. Today, genome sequencing has traveled much beyond biology and involves biophysics, biochemistry, and bioinformatics!

Thanks to the efforts of both public and private agencies, genome sequencing strategies are evolving very fast, leading to cheaper, quicker, and automated techniques right from clone-by-clone and whole-genome shotgun approaches to a succession of second-generation sequencing methods. The development of software of different generations facilitated this genome sequencing. At the same time, newer concepts and strategies were emerging to handle sequencing of the complex genomes, particularly the polyploids.

It became a reality to chemically—and so directly—define plant genomes, popularly called whole-genome sequencing or simply genome sequencing.

The history of plant genome sequencing will always cite the sequencing of the genome of the model plant Arabidopsis thaliana in 2000 that was followed by sequencing the genome of the crop and model plant rice in 2002. Since then, the number of sequenced genomes of higher plants has been increasing exponentially, mainly due to the development of cheaper and quicker genomic techniques and, most importantly, the development of collaborative platforms such as national and international consortia involving partners from public and/or private agencies.

As I write this preface for the first volume of the new series "Compendium of Plant Genomes," a net search tells me that complete or nearly complete whole-genome sequencing of 45 crop plants, eight crop and model plants, eight model plants, 15 crop progenitors and relatives, and three basal plants is accomplished, the majority of which are in the public domain. This means that we nowadays know many of our model and crop plants chemically, i.e., directly, and we may depict them and utilize them precisely better than ever. Genome sequencing has covered all groups of crop plants. Hence, information on the precise depiction of plant genomes and the scope of their utilization are growing rapidly every day. However, the information is scattered in research articles and review papers in journals and dedicated Web pages of the consortia and databases. There is no compilation of plant genomes and the opportunity of using the information in sequence-assisted breeding or further genomic studies. This is the underlying rationale for starting this book series, with each volume dedicated to a particular plant.

Plant genome science has emerged as an important subject in academia, and the present compendium of plant genomes will be highly useful to both students and teaching faculties. Most importantly, research scientists involved in genomics research will have access to systematic deliberations on the plant genomes of their interest. Elucidation of plant genomes is of interest not only for the geneticists and breeders, but also for practitioners of an array of plant science disciplines, such as taxonomy, evolution, cytology, physiology, pathology, entomology, nematology, crop production, biochemistry, and obviously bioinformatics. It must be mentioned that information regarding each plant genome is ever-growing. The contents of the volumes of this compendium are, therefore, focusing on the basic aspects of the genomes and their utility. They include information on the academic and/or economic importance of the plants, description of their genomes from a molecular genetic and cytogenetic point of view, and the genomic resources developed. Detailed deliberations focus on the background history of the national and international genome initiatives, public and private partners involved, strategies and genomic resources and tools utilized, enumeration on the sequences and their assembly, repetitive sequences, gene annotation, and genome duplication. In addition, synteny with other sequences, comparison of gene families, and, most importantly, the potential of the genome sequence information for gene pool characterization through genotyping by sequencing (GBS) and genetic improvement of crop plants have been described. As expected, there is a lot of variation of these topics in the volumes based on the information available on the crop, model, or reference plants.

I must confess that as the series editor, it has been a daunting task for me to work on such a huge and broad knowledge base that spans so many diverse plant species. However, pioneering scientists with lifetime experience and expertise on the particular crops did excellent jobs editing the respective volumes. I myself have been a small science worker on plant genomes since the mid-1980s and that provided me the opportunity to personally know several stalwarts of plant genomics from all over the globe. Most, if not all, of the volume editors are my longtime friends and colleagues. It has been highly comfortable and enriching for me to work with them on this book series. To be honest, while working on this series I have been and will remain a student first, a science worker second, and a series editor last. And I must express my gratitude to the volume editors and the chapter authors for providing me the opportunity to work with them on this compendium.

I also wish to mention here my thanks and gratitude to the Springer staff, particularly Dr. Christina Eckey and Dr. Jutta Lindenborn for the earlier set of volumes and presently Ing. Zuzana Bernhart for all their timely help and support.

I always had to set aside additional hours to edit books beside my professional and personal commitments—hours I could and should have given to my wife, Phullara, and our kids, Sourav and Devleena. I must mention that they not only allowed me the freedom to take away those hours from them but also offered their support in the editing job itself. I am really not sure whether my dedication of this compendium to them will suffice to do justice to their sacrifices for the interest of science and the science community.

New Delhi, India Chittaranjan Kole

Preface

Amaranth is an ancient C_4 pseudocereal with relatively high protein content, stress tolerance, and environmental adaptability. Although the genus is native to the tropical regions of Central and South America, it is now found around the globe and is cultivated for grain, leafy vegetables, and ornamentals. Some of the amaranth species are economically important weeds, and a few of them have gained resistance to commercial herbicides. Due to its nutritional benefits and agronomic properties, grain amaranth has potential to be one of the high-impact crops for our future generation. The world population is expected to reach 8.54 billion by 2030, available cultivable land is limited and food necessities are mounting. Considering global climate change and its potential impact on the global agriculture system, highly adaptable and superior yielding crops with balanced nutrition profile are in demand. We believe amaranth has a role in supporting global food security. Therefore, it is essential to develop improved breeding tools and resources to integrate genomics and phenomics into amaranth breeding programs and develop elite varieties that could support our future nutritional needs.

In this book, we present an overview of current knowledge on amaranth genetics, genomics, and evolution. The advent of next generation sequencing has accelerated production of molecular information about amaranths, especially in the genome and transcriptome. However, resources for amaranth still lag behind other staple crops (rice, wheat, corn, etc.). Thus, there is a long list of research areas yet to be explored.

This book is composed of 10 chapters that summarize recent advances in the field. All the chapters were prepared by amaranth researchers, for amaranth researchers, and they provide comprehensive information on diverse areas elucidating amaranth diversity, evolution, abiotic and biotic stress response, and the aspects of grain and weedy amaranth that could potentially address global food security.

We would like to thank all our author contributors for taking time and preparing excellent chapters for this book. Without their expertise and involvement, we would not have been able to produce this book.

We would also like to thank the series editor Dr. Chittaranjan Kole and Project coordinator Mr. Prashanth Ravichandran for their guidance during the course of preparation of this book. Thanks, are also due to Springer Nature's publishing team who worked tirelessly to produce this book in this unusual time.

Dinesh Adhikary
University of Alberta
Edmonton, Alberta, Canada

Michael K. Deyholos
University of British Columbia
Kelowna, British Columbia, Canada

John P. Délano- Frier
Centro de Investigación y de Estudios Avanzados del I. P. N.
Irapuato, Guanajuato, Mexico

Contents

Contributors

Sabina Adhikari Department of Biology, Stephen F. Austin State University, Nacogdoches, TX, USA

Dinesh Adhikary Department of Agricultural, Food & Nutritional Sciences, University of Alberta, Edmonton, AB, Canada

Isadora Louise Alves da Costa Ribeiro Quintans Universidade Federal Rural do Semi-árido, Mossoró, Brazil

Rita Andini Faculty of Agriculture, Universitas Syiah Kuala (USK), Kecamatan Syiah Kuala, Provinsi Aceh, Indonesia

Esaú Bojórquez-Velázquez Instituto Potosino de Investigación Científica y Tecnológica A.C, San Luis Potosi, Mexico

Gabriela Cabrales-Orona Departamento de Biotecnología y Bioquímica, Centro de Investigación y de Estudios Avanzados del I.P.N. (Cinvestav), Irapuato, Gto, México

José Luis Cabrera-Ponce Departamento de Biotecnología y Bioquímica, Centro de Investigación y de Estudios Avanzados del I.P.N. (Cinvestav), Irapuato, Gto., México

Kena Casarrubias-Castillo Centro Universitario de Ciencias Biológicas y Agropecuarias, Universidad de Guadalajara, Zapopan, Jalisco, Mexico

Andrea Paola Castellanos-Arévalo Departamento de Biotecnología y Bioquímica, Centro de Investigación y de Estudios Avanzados del I.P.N. (Cinvestav), Irapuato, Gto., México

Ivan Takeshi Cerritos-Castro Instituto Potosino de Investigación Científica y Tecnológica A.C, San Luis Potosi, Mexico

Ana Paulina Barba de la Rosa Instituto Potosino de Investigación Científica y Tecnológica A.C, San Luis Potosi, Mexico

John P. Délano-Frier Departamento de Biotecnología y Bioquímica, Centro de Investigación y de Estudios Avanzados del I.P.N. (Cinvestav), Irapuato, Gto., México

Darshan T. Dharajiya Bio Science Research Centre, Sardarkrushinagar Dantiwada Agricultural University (SDAU), Sardarkrushinagar, Gujarat, India

Daniel J. Fairbanks Department of Biology, Utah Valley University, Orem, UT, USA

Muhammad Kadapi Faculty of Agriculture, Agrotechnology Department, Universitas Padjajaran (UNPAD), Hegarmanah, Kabutapen Sumedang, Indonesia

Upama Khatri-Chhetri Department of Agricultural, Food & Nutritional Sciences, University of Alberta, Edmonton, AB, Canada

Julio A. Massange-Sánchez Unidad de Biotecnología Vegetal, Centro de Investigación y de Asistencia en Tecnología y Diseño del Estado de Jalisco A.C. (CIATEJ), Zapopan, Jalisco, Mexico

Cecilia Nava-Sandoval Escuela Nacional de Ciencias Biológicas del Instituto Politécnico Nacional, Unidad Profesional Lázaro Cárdenas, Prolongación de Carpio y Plan de Ayala s/n, Col. Santo Tómas, Alcaldía Miguel Hidalgo, CDMX, México

Ryo Ohsawa Institute of Life and Environmental Sciences, University of Tsukuba, Tsukuba, Ibaraki Ken, Japan

Erica E. Packard Department of Biology, The University of British Columbia, Kelowna, BC, Canada

Paola A. Palmeros-Suárez Centro Universitario de Ciencias Biológicas y Agropecuarias, Universidad de Guadalajara, Zapopan, Jalisco, Mexico

Sathish K. Ponniah Department of Agriculture, University of Arkansas at Pine Bluff, Pine Bluff, USA

N. N. Prajapati Centre for Crop Improvement, SDAU, Sardarkrushinagar, Gujarat, India

Thiruvarangan Ramaraj School of Computing, College of Computing & Digital Media, DePaul University, Chicago, USA

Anuj Kumar Singh Bio Science Research Centre, Sardarkrushinagar Dantiwada Agricultural University (SDAU), Sardarkrushinagar, Gujarat, India

Muhammad Ikhsan Sulaiman Faculty of Agriculture, Universitas Syiah Kuala (USK), Kecamatan Syiah Kuala, Provinsi Aceh, Indonesia

Fátima Alejandra Talamantes-Herrera Instituto Potosino de Investigación Científica y Tecnológica A.C, San Luis Potosi, Mexico

Kapil K. Tiwari Bio Science Research Centre, Sardarkrushinagar Dantiwada Agricultural University (SDAU), Sardarkrushinagar, Gujarat, India

Selma Valles Instituto Potosino de Investigación Científica y Tecnológica A.C, San Luis Potosi, Mexico

Abbreviations

AFLP	Amplified Fragment Length Polymorphism
APMC	Agricultural Produce Market Committee
ARS	Agricultural Research Service
AUJ	Agriculture University, Jodhpur
AVGRIS	AVRDC Vegetable Genetic Resources Information System
DBSKKV	Dr. Balasaheb Sawant Konkan Krishi Vidyapeeth
GBPUAT	Govind Ballabh Pant University of Agriculture and Technology
IARI	Indian Agricultural Research Institute
ICAR	Indian Council of Agricultural Research
IIHR	Indian Institute of Horticultural Research
INR	Indian Rupee
ISSR	Inter-Simple Sequence Repeat
ITS	Internal Transcribed Spacer
KAU	Kerala Agricultural University
NBPGR	National Bureau of Plant Genetic Resources
OUAT	Odisha University of Agriculture and Technology
PCA	Principal Component Analysis
RAPD	Random Amplification of Polymorphic DNA
RFLP	Restriction Fragment Length Polymorphism
SDAU	Sardarkrushinagar Dantiwada Agricultural University
SNP	Single Nucleotide Polymorphism
SSR	Simple Sequence Repeats
TNAU	Tamil Nadu Agricultural University
UAS	University of Agricultural Sciences
USDA	United States Department of Agriculture
VPKAS	Vivekananda Parvatiya Krishi Anusandhan Sansthan

1 Genome and Transcriptome of Amaranth Species

Sabina Adhikari, Upama Khatri-Chhetri, and Dinesh Adhikary

Abstract

After decades of abandonment, amaranth crop was re-domesticated and brought to the attention of farmers and researchers. In recent years, the nutritionally rich and agronomically important C4 plant has become the preferred commercial grain and economically important weedy species around the globe. Improved knowledge of its genome, transcriptome, and inherent genetic imprints will facilitate crop improvement and develop molecular approaches to control some of its weedy relatives. While much research exists in the molecular markers, nutritional facts, ethnobotany, taxonomy, and morphology of amaranths, there is limited omics information for this high-value crop. However, with the advent of sequencing technology, high throughput sequencing cost has decreased, and omics research is slowly progressing on the crop. This chapter will, therefore, focus on the innovation that has been made in the field of amaranth genome and transcriptome in both grain and weedy species.

S. Adhikari
Department of Biology, Stephen F. Austin State University, Nacogdoches, TX, USA

U. Khatri-Chhetri · D. Adhikary (✉)
Department of Agricultural, Food & Nutritional Sciences, University of Alberta, Edmonton, AB, Canada

1.1 Introduction

Grain amaranth (*Amaranthus hypochondriacus* L., *Amaranthus caudatus* L., *Amaranthus cruentus* L.) is an ancient crop that originated in Central and South America (Sauer 1949). The protein-rich C_4 pseudocereal is highly tolerant to drought and salinity, thereby adapting to a wide range of environmental conditions (Kong et al. 2009). In addition to desirable agricultural traits, amaranths have beneficial nutritional properties (Huerta-ocampo et al. 2014). Seeds are used as a grain, and young leaves are used as leafy vegetables (Tandang-Silvas et al. 2012; Venskutonis and Kraujalis 2013). Amaranth grain supplies easily digestible protein with all the essential amino acids, including a relatively high content of lysine (Pedersen et al. 1987; Silva-Sanchez et al. 2008; Vecchi and Añón 2009). Amaranth seeds have also been proposed as an alternative source of squalene, which was traditionally harvested from marine mammals (He and Corke 2003). Earlier studies have shown that squalene affects cholesterol biosynthesis and decreases blood cholesterol levels (Qureshi et al. 1996; Berger et al. 2003; Martirosyan et al. 2007). Finally, compared with many other commonly used food crops, amaranth grain has a higher content of dietary fiber, lipids, calcium, magnesium, iron, potassium, zinc, vitamin E, and B complex vitamins (Betschart et al. 1981; Tosi et al. 2001; Alvarez-Jubete et al. 2009).

D. Adhikary et al. (eds.), *The Amaranth Genome*, Compendium of Plant Genomes,
https://doi.org/10.1007/978-3-030-72365-1_1

While much research exists in the molecular markers, nutritional facts, ethnobotany, taxonomy, and morphology of amaranths, there is a dearth of reliable research data available on the genomic and transcriptome dynamics. One of many constraints has been the limited funding resources for research on the crop, this is even though the crop has high nutritional value and agronomic potential. However, with the advent of sequencing technology, high throughput sequencing cost has decreased and that has helped researchers to access the omics information of the crop. This chapter will focus on the innovation that has been made in the field of amaranth genome and transcriptome dynamics in recent years.

1.2 Amaranth Draft Genome

A. hypochondriacus is a diploid species with 32 chromosomes (n = 16) (Sunil et al. 2014) and an estimated genome size of 466 Mb per haploid nucleus (Maughan et al. 2008; Bennett and Smith 1976). In the last 20 years, the use of advanced molecular tools to study the genomics of grain amaranths has been increasing. Genetic resources have been built to improve the crop and to understand the species distribution of both grain and weedy species, including genetic molecular markers SSRs (Mallory et al. 2008; Kietlinski et al. 2014), SNPs (Maughan et al. 2008), and Bacterial Artificial Chromosome (BAC) libraries (Maughan et al. 2008). Moreover, recent advances in sequencing technologies have helped generate transcriptome information to study biotic and abiotic responses (Délano-Frier et al. 2011), herbicide resistance genes (Riggins et al. 2010), and also a draft genome and transcriptome (Sunil et al. 2014; Clouse et al. 2016). The earlier draft sequence information was based on *A. hypochondriacus* landrace (Rajgira), which was collected from Karnataka, India (Sunil et al. 2014). The assembly consisted of 367,441 scaffolds, with an N50 of 35,089 bp. The total length of this assembly was 40% larger than the expected size of the amaranth genome (Bennett and Smith 1976), suggesting significant disassembly. Thus, additional sequencing work was needed to build a quality reference genome and transcriptome. Clouse et al. (2016) published a high-quality draft genome sequence of the commercially important Plainsman variety of *A. hypochondriacus* (hereinafter written as Plainsman). The genome assembly contained 377 Mb of the total estimated genome size (466 Mb per haploid nucleus) as 3518 scaffolds with an N_{50} value of 371 kb (Bennett and Smith 1976; Clouse et al. 2016). The study reported an improved version of the physical map based on the BioNano Genomics (BNG) mapping technology, and the genome annotation relied on deeply sequenced transcriptome information from eight different stages of development along with abiotically stressed tissues. Furthermore, the study provided a comparative analysis of four amaranths species, *A. hypochondriacus*, one accession of each of *A. cruentus, A. caudatus*, and *A. hybridus*.

In 2017, Lightfoot and researchers applied chromatin conformation capture technology (Hi-C) and developed an improved contiguous chromosome assembly of grain amaranth (*A. hypochondriacus*) with an N50 and contig value of 24.4 and 1.25 Mb, respectively. Scaffolding chromosome sequences from a fragmented assembly is a difficult task, using Pacific Biosciences (PacBio), the researchers have utilized single molecule for sequencing and closing the missing information in the earlier assemblies (Sunil et al. 2014; Clouse et al. 2016). The use of the in vivo Hi-C was the first report in *Amaranthus* species assigning 98% of the assembly into the chromosome (Lightfoot et al. 2017).

Some *Amaranthus* species (*A. tuberculatus, A. hybridus,* and *A. palmeri)* are agronomically important weeds. Understanding the genome evolution within these wild varieties has been a prime focus in recent years. Combining PacBio long-read sequencing and chromatin contact mapping to assemble sequences, Montgomery et al. (2020) presented draft assemblies of the three weedy species; *A. palmeri* was scaffolded to near chromosome level resolution with the N50 value of 20.1 Mb. Additionally, to address the potential issues of heterozygosity, the study produced haplotype assemblies of the remaining two species, *A. tuberculatus* and *A. hybridus*, with the contig N50 of 2.58 and 2.26 Mb, respectively (Montgomery et al. 2020). Although a robust genetic and physical map in amaranth are still missing, having access to the high-quality genome assemblies in both weedy and cultivated varieties will help accelerate genomic and molecular studies revealing the evolution of agronomic traits and assisting in the fast-track molecular breeding of the crop. Additionally, the development of such genomic resources will contribute to the global effort to develop molecular tools and resources to improve the cultivated crop and to understand the noxious weeds at the molecular level; hopefully, one day the weeds can be controlled.

1.3 Amaranth Transcriptome

The transcriptome is a panoply of RNA transcripts, which includes rRNA, mRNA, tRNA, and various other types of regulatory, structural, and non-coding RNA, that is present in a living cell at a given time in response to a specific environmental stimulus. Functional gene annotation, as an understanding of where, when, and under what conditions these transcripts are expressed, is an integral step in the gene discovery (Liu et al. 2014; Clouse et al. 2016; Conesa et al. 2016), genetic marker designing (Okada et al. 2018) and, eventually, marker-assisted breeding (MAB) of the crop. For a non-model species, without a high-quality reference genome, transcriptome sequencing produces functional genetic information representing protein-coding genes and aids in the genome annotation and mapping. Identification and quantification of these transcripts play an integral role in the understanding of the molecular biology of a cell as it is the intermediate between genome and proteome. Analyzing the expressed transcriptomic information from both cultivated and weedy species would assist in developing molecular tools to improve the agronomic traits of cultivated species and to control their weedy relatives. In this section, we have documented amaranth transcriptome generated from different tissue types expressed in response to various treatment conditions from both cultivated and weedy species.

1.3.1 Amaranth Transcriptome on Herbicide Resistance

The genus *Amaranthus* comprises 70 species, distributed around the globe, predominantly found in the warm regions of Central and South America (Mosyakin and Robertson 2003). Although many of these annual herbs are used either as cultivated pseudocereals or leafy vegetables and ornamentals, many of them are agronomically important weeds (e.g. *A. retroflexus, A. hybridus, A. palmeri, A. tuberoculatus* etc.) (Holm et al. 1997; Steckel 2007). These aggressive and weedy summer annuals are highly resilient to environmental stresses, possess high fecundity and seed dispersal rate with an ability to form hybrids and, most concerning, these are resistant to herbicides (Basu et al. 2004; Costea et al. 2004; Steckel 2007) and frequently found in cultivated fields outcompeting many vegetable crops and cash crops, decreasing yield and quality of the desired crop. Understanding their underlying genetic traits is vital to develop a molecular approach to control the species. To understand the transcriptome of these species, Riggins et al. (2010) produced a*de novo* transcriptome for *Amaranthus tuberculatus* (weedy species) profiling preliminary transcripts on herbicide resistance genes. To achieve a higher coverage of transcriptome, the study pooled RNA samples from different species (*A. albus, A. hybridus* L., *A. palmeri* S., *A. powellii* S., *A, retroflexus* L., *A. spinosus* L., *A. tuberoculatus* S.), tissue types, developmental stages, sexes, along with herbicide and abiotic stress treatments. Using the 454 Pyrosequencing approach, the study generated 483,225 raw reads (length 231 bp) and the reads were assembled into 44,469 unique sequences. The study highlighted four herbicide target-site genes, 4-hydroxyphenylpyruvate dioxygenase (HPPD), glutamine synthetase (GS), 5-enolpyruvylshikimate-3-phosphate synthase (EPSPS*)*, acetolactate synthase (ALS), and their corresponding nucleotide sequences. Although the study generated large sequence data and was groundwork in the field of herbicide omics, their experimental design and the choice of sequencing platform (454 pyrosequencing) made it difficult to further investigate and confirm the genes and gene networks that were involved in herbicide tolerance and susceptibility. Nevertheless, since the publication of the work, several other studies were deciphering the transcriptome information from the weedy species especially focusing on herbicide resistance (Salas-Perez et al. 2018; Kohlhase et al. 2019; Guo et al. 2020), thus the data serve as supporting information for the ongoing genomics research in the field.

Among several weedy species, *A. palmeri* is one of the most noxious, herbicide-resistant weeds. Molecular information about genetic resistance was limited. Salas-Perez et al. (2018) conducted transcriptome analysis of *A. palmeri* to profile the underlying genetic imprints of glufosinate tolerance genes. Tissue samples were taken from seedlings that were untreated and treated with 0.55 kg/ha glufosinate for 24 h. Although there was a bit of uncertainty on the phenotype of the tolerant seedlings, global gene expression conducted using the Illumina platform revealed 72,780 assembled contigs. Of the 567 differentially expressed genes, there were a greater number of ATP-binding cassette (ABC) transporters, glutathione S-transferase (GST), NAC [NAM (no apical meristem), ATAF ½ (Arabidopsis transcription activator factor ½) and CUC2 (cup-shaped cotyledon) (Souer et al. 1996)] transcription factors, nitronate monooxygenase (NMO), etc. in a majority of the presumed glufosinate-tolerant seedlings. Many of the upregulated genes were involved in stress response, detoxification, stress signaling, and transport; and the study predicted that those genes were potentially bestowing some level of tolerance to glufosinate. Furthermore, the study showed that the relative copy number of the glufosinate target genes was not significantly different between the susceptible and resistant lines (Salas-Perez et al. 2018). However, the susceptible lines accumulated more ammonia than the resistant plants. While it is likely that the herbicide tolerance is controlled by multiple genes, the availability of the transcriptome information is an asset for the development of genetic markers to identify and eliminate the herbicide-tolerant plants from a population. There were also apparent abiotic stress-tolerant genes in the pool of differentially expressed transcripts, which makes this weed species a potential candidate that can donate resistant traits to other grain species.

Similarly, Kohlhase et al. (2019) conducted RNA-seq analysis on two waterhemp (*A. tuberculatus*) populations that were resistant or susceptible to an herbicide that inhibits 4-hydroxyphenylpyruvate-dioxygenase (HPPD). HPPD is an enzyme in tyrosine catabolism and is important in the biosynthesis of compounds such as tocopherols and carotenoids (Almsick 2009). There are a handful of broad-spectrum herbicides from different manufacturers that can inhibit HPPD (Witschel 2012). Kohlhase et al. (2019) conducted RNAseq 48 multiplexed libraries of resistant and susceptible genotypes, over a treatment period of 24 h, and generated 2.36 billion clean reads, which were assembled into 113,893 contigs, which was unusually high, compared with *A. thaliana* (~30,000 genes), or closely related species (*A. hypochondriacus* (~66,370 contigs), the number of predicted contigs was unusually high (Weigel and Mott 2009; Clouse et al. 2016). The study predicted 970 putative cytochrome P450s, thereof 79 and 34 transcripts were differentially expressed in the resistant and susceptible genotype, respectively. There was one transcript (TRI02135) that was induced significantly within three hours of treatment (Kohlhase et al. 2019). Thus, the study revealed that *A. tuberculatus* responded to the herbicide as quickly as 3 h post-treatment. The study identified 189 candidates for single nucleotide polymorphisms (SNPs) for mapping herbicide-resistant traits. Although a bit preliminary, the study showed an inducible nature of herbicide response and identified transcripts and gene networks that were regulated in the early stage of treatment in both resistant and susceptible genotypes. Concrete genetic inferences may not be derived from the presented de novo transcriptome information, but it provides general information on the putative herbicide resistance gene candidates and further validation is needed.

1.3.2 Amaranth Transcriptome on Stress Response

Expression analyses of plant responses to abiotic stress have been conducted in *Amaranthus* spp. (Clouse et al. 2016; Guo et al. 2020). Abiotic factors such as soil salinization, drought, and heavy metal contamination are global issues, deteriorating cultivable land, and leading to crop loss. Guo et al. (2020) investigated transcriptome changes triggered by salinity treatment of an edible amaranth variety (*A. mangostanus* L.). This study was conducted in two phenotypically different varieties (Quanhong and Liuye), using leaf and root tissue samples taken at 1-, 4-, 10-, 20-, and 45-day post-treatment, and documented biological mechanisms of heavy metal (cadmium) accumulation in plant tissues in response to high salinity. Using the Illumina platform, the study generated 148.65 million single-end raw reads from two pooled libraries representing the distinct cultivars. Their de novo assembly generated 104,483 contigs, which is ~1.6-fold higher than the predicted contigs in Plainsman (Clouse et al. 2016) and ~3.5-fold higher than the predicted genes in *A. thaliana* (Weigel and Mott 2009). About 3,000 differentially expressed genes were discovered in the salt-sensitive variety, which was approximately 4 times higher than in the salt-tolerant variety (Guo et al. 2020). The induced salinity stress repressed the expression of enzymes for cell wall biosynthesis in roots, reducing the barrier for cadmium uptake. Thus, the salt-sensitive variety was expected to accumulate more cadmium than the tolerant variety demonstrating a higher health risk (Guo et al. 2020). The study generated over 70 million raw reads per sample and is good preliminary transcriptome information on the stress response.

Similarly, Délano-Frier et al. (2011) performed a large-scale transcriptome analysis of *A. hypochondriacus*, grain amaranth, generating transcripts that are expressed in response to biotic stress. Using 454 pyrosequencing platform, the study generated 2,700,168 raw reads from six different cDNA libraries of vegetative tissues (leaf and colored stem), at four different treatment conditions (water stress, salt stress, insect herbivory, and bacterial infection), the study generated 21,207 assembled contigs, of which 1,971 transcripts were differentially expressed in response to the tested stress conditions. Some of these stress-inducible contigs were related to AREB-like protein, Dof-type zinc finger protein, MYB-like transcription factors, SCOF-1 zinc finger proteins, etc. (Délano-Frier et al. 2011).

A. hypochondriacus (grain species) and *A. tuberculatus* (weedy species) transcriptomes were compared using data from two different researchers (Riggins et al. 2010; Délano-Frier et al. 2011). The comparison found only low similarity in the transcriptomes of these two species, although these differences could be due to multiple factors. For example, the treatment conditions differed for *A. tuberculatus* and *A. hypochondriacus*: Riggins et al. (2010) developed their *A. tuberculatus* transcriptome by treating the seedlings with herbicide (lactofen, glyphosate, and atrazine) and cold stress, whereas the *A. hypochindriacus* (Délano-Frier et al. 2011) transcriptome was induced in response to insect herbivory, bacterial infection, and abiotic (water and salt) stresses. Other technical factors, including growth conditions, depth of sequencing, and use of different assemblers (Egassembler vs. Newbler), could have also contributed to variation. Thus, strong inferences cannot be made about this comparison.

1.3.3 Amaranth Transcriptome on Genome Rearrangement Assessment

Deciphering genome duplication events (Vanneste et al. 2014; Dohm et al. 2014), phylogenies (Zulawski et al. 2014; Walsh et al. 2015), and genome organization (Yoon et al. 2011; Ying et al. 2015) have been accelerated by advances in

NGS in both model and non-model species. Genome reorganization can affect gene regulation and could potentially bring about phenotypic changes (Nomura et al. 2003; Chen and Ni 2006). This is true also of amaranths. Sunil et al. (2017) conducted deep transcriptome sequencing of *A. hypochondriacus* and documented differential genome organization that is specific to betalains traits in the Caryophyllales order. The study produced 84.8 billion bases of which only 88.99 million bases were used to assemble 125,581 transcripts, which is ~4 fold higher than the predicted genes (~30,000) in *Arabidopsis* and ~2 fold higher than the predicted transcripts (66,370) in Plainsman (Weigel and Mott 2009; Clouse et al. 2016). The study predicted 13,529 putative transcripts that are unique to *A. hypochondriacus*; 581 of them were predicted as real chimeric, of which 406 and 351 were found in *A. hypochondriacus* and *Beta vulgaris* genome, respectively. Gene ontology analysis of the chimeric transcripts revealed that predominantly the transcripts were nuclear and were involved in biological processes, including stress-response, transport, signal transduction, protein metabolism, etc. (Sunil et al. 2017). Among the predicted chimeric transcripts, the study found a gene pair of 4-hydroxy-tetrahydrodipicolinate synthase (DHDPS1) (EC 4.3.3.7) and alpha-glucosidase (EC 3.2.1.20). In which the enzymes, DHDPS1 and alpha-glucosidase, were involved in the biosynthesis of lysine and sucrose catabolism, respectively (Craciun et al. 2000; Sunil et al. 2017). Corresponding genes for the enzymes were 612 bp apart in *A. hypochondriacus* and were potentially linked (Sunil et al. 2017). As the grain amaranths are reputed for high lysine content, the molecular information on the chimeric transcripts is an important contribution.

1.3.4 Amaranth Transcriptome on Seed Development

Embryogenesis and seed maturation are two of the most important phases in seed development (Locascio et al. 2014; Bian et al. 2019). The phases involve complex molecular changes, alteration in seed size, color, nutrient content, embryo size, anatomy, etc., and understanding the changes in the transcriptome level is crucial in the enhancement of seed quality and yield. Using Illumina platform, Adhikary (2019) produced transcriptome information from Plainsman representing eight different tissue types, including developmental stages (leaf, stem, root); pooled drought-stressed tissues (leaf, stem, and root); pooled flowers from both male and female flowers; immature seeds containing globular, heart, and torpedo stage embryos and mature seeds containing linear cotyledon stage embryos; and green cotyledons. All of these tissues were well watered and grown under optimal environmental conditions; however, watering was withheld for 10 days for the drought-stressed plants, and the leaves, stems, and roots of these noticeably wilted plants were harvested and were included in the study. The study generated the first large-scale transcriptome resources (31.7 Gbps) for *A. hypochondriacus* cv Plainsman representing seed developmental stages, vegetative, stressed, and reproductive tissues. Over 352 million clean reads of 90 bp were assembled into 65,947 contigs. This transcriptome assembly is the first high-quality transcriptomic resource for the study of developing seeds, and other developmental stages in amaranths and these data will be valuable in annotating whole-genome assemblies of *A. hypochondriacus* and its relatives. The study also revealed 283, 978, and 26 transcripts that are specific to seed, non-seed, and drought stress, respectively. Although the study was the first of its kind and has predicted important genes involved in seed development and seed storage proteins, the study lacks experimental repeats. Thus, further validation of the predicted transcripts is required.

1.3.5 Amaranth Transcriptome on Leaf Pigmentation

Many members of the amaranth family (Amaranthaceae) produce red (betacyanin) or yellow (betaxanthin) betalain pigments, which are chemically distinct from the anthocyanins responsible for red pigmentation (Packard et al., in press). Although the biological and ecological role of these secondary metabolites is still to be elucidated, presumably these pigments are involved in attracting pollinators and responding against biotic and abiotic stresses (Adhikary et al. 2020). Some of the amaranth species are decorated with multiple colors (red, yellow, green) on their leaf and are known as a tricolor leaf. While the molecular mechanism behind the mosaic-like leaf pigmentation is yet to be described in amaranths, the recent RNA-seq studies on *A. tricolor* leaf sectors have highlighted some molecular information on the phenomenon (Liu et al. 2019). Under an assumption that betalains were present only on the red-pigmented region of leaves, the study collected red and green tissues and isolated RNA and generated two independent cDNA libraries for sequencing. The study reported 15.08 Gb clean reads with 7.53 Gb (29.914 million reads) and 7.53 Gb (29.913 million reads) to the red and green leaf sectors, respectively. Reads were assembled into 80,026 unigenes with an N50 of 1,202 bp. Although the number of gene models predicted was a bit higher as compared with the *Arabidopsis* ($\sim$30,000), the study documented 2,178 differentially expressed unigenes between red and green sections (Liu et al. 2019). There are about 75 different betalain types reported from only 17 of the 34 families in the Caryophyllales order (Khan and Giridhar 2015), it may not be true to assume that the pigments are exclusive to the pigmented region on a tricolor leaf. Although the samples were pooled (10 individuals) and produced sequence reads to an acceptable number for assessing the gene expression, the study still lacks experimental repeats. With the data, strong conclusions cannot be derived; however, the study certainly developed preliminary transcriptomic information on betalain metabolism (Liu et al. 2019).

1.3.6 Amaranth Transcriptome on Flowering Genes

A. tricolor, one of the summer annuals primarily used as a leafy vegetable and ornamental, has gained importance in the horticultural field. Under high temperature and short-day length, these plants trigger early flowering; however, this phenomenon may not be preferred in horticulture because early maturing and flowering may lead to poor vegetative growth leading to a meagre yield of leafy vegetable and seeds. Liu et al. (2014) conducted a transcriptome study comparing differential gene expression between in vitro plantlet development at four stages (young, adult seedling, floral buds, and fully open flowers) deciphering floral regulatory mechanisms in the species. The large-scale transcriptome study generated 319 million high-quality reads assembled into 99,312 unigenes with N50 of 1,460 bp. The study reported 735, 17,184, 274, and 206 unique genes expressed at the young, adult, floral bud, and flowering stages, respectively. Some of the genes (XTH9, XTH33, ELF3, PIF3, etc.) related to photoperiod and plant circadian rhythm were identified and their expression was validated through qPCR (Liu et al. 2014). Other important genes—ubiquitin-mediated proteolysis proteins [constitutive photomorphogenic 1 (COP1) and peter pan syndrome (PPS)] play an important role in flowering (Lau et al. 2011; Pineiro and Jarillo 2013), especially transitioning from juvenile phase to flowering stage (Tanaka et al. 2011), the transcriptome study revealed COP1 and PPS orthologs in amaranths were relatively expressed higher in the young stage than in the adult seedling stage and least in the flowering bud stage indicating that the transcripts might have been involved in the transition phase. Similarly, other potential genes (GLP1 and HSP70) related to the transition from vegetative to the reproductive stage were regulated differentially at different tested stages (Liu et al. 2014). Although there were some inconsistencies in data presentation, sampling method, and absence of biological replicates, the study has certainly produced preliminary transcriptome information from different plant developmental stages maintained in an in vitro growth system and has highlighted some of the key pathways and putative genes involved in the developmental stages. Due to the lack of biological replicates in transcriptome sequencing, differential expression of genes at different developmental stages may not be calculated with confidence; however, the sequence data can certainly be re-analyzed and used for genome annotation and finding putative functional genes.

1.4 Amaranth Sequence Information in Public Databases

Gene expression analyses in amaranths have been published with increasing speed in the last few years; however, the transcriptomic information available for the species is still less than that of other economically important crops. Moreover, there is a dearth of reliable sequence information covering biotic and abiotic stress responses. For instance, at the time of this publication, there were 288 and 8,546 mRNA and genomic DNA/RNA sequence submissions, respectively, on the NCBI nucleotide database under the genus name '*Amaranthus*' (https://www.ncbi.nlm.nih.gov/nuccore/?term=Amaranthus). Likewise, there were 179 RNA submissions under Sequence Reads Archive (SRA) database representing 116 (paired) and 63 (single) library layout for *Amaranthus*; whereas there were 107,205 mRNA sequences under nucleotide and 13,688 RNA submissions under SRA database representing 7,559 (paired) and 6,129 (single) library layout for *Oryza sativa* (https://www.ncbi.nlm.nih.gov/sra/?term=Oryza+sativa). Analogously, there are 23 and 1,694 protein information available for *Amaranthus* in Swiss-Prot and TrEMBL (https://www.uniprot.org/uniprot/?query=Amaranthus&sort=score), whereas there were 5,582 and 468,648 protein information available for *Oryza sativa* in Swiss-Prot and TrEMBL, respectively (https://www.uniprot.org/uniprot/?query=oryza+sativa&sort=score). The usefulness of these sequences information is dependent upon the experimental design, sequencing depth, choice of sequencing platform, selection of assembly software, quality of assembly, and the experimental design. Thus, relatively, there is a dearth of high-quality molecular information for amaranth as compared with a mainstream crop such as rice. We expect these issues to be resolved in the next coming years as the drive to expand the economic value of amaranth increases.

1.5 Experimental Design and Transcriptome Study

For a successful transcriptome study, experimental design is crucial; however, the step is often overlooked. With advances in next-generation sequencing (NGS) technology, RNAseq has become a mainstream methodology, as it can provide holistic quantification of all RNA species present in a living cell at a developmental stage or in response to environmental perturbation. The technology is not only extremely sensitive but also produces highly reproducible data between technical and biological replicates, provides a precise quantification of transcripts, and isoforms than other available methodologies (microarray, qRT-PCR, etc.) (Wang et al. 2009; Nagalakshmi et al. 2008; Cloonan et al. 2008). A well-designed experiment considers library types (single and paired-end), sequencing depth (read depth), read length along with biological and technical replicates per treatment. If a study design aims to discover novel genes and their isoforms, paired-end with >200 million reads per sample was found to be highly effective (Tarazona et al. 2011). According to the Encyclopedia of DNA elements (ENCODE) (ENCODE Project Consortium), highly expressed gene (FPKM values >10) transcripts were quantified accurately using $\sim$36 million mapped reads (Trapnell et al. 2009; ENCODE Project Consortium); however, lowly expressed genes (FPKM values <10) were accurately quantified with $\sim$80 million mapped reads (Sims et al. 2014). Thus, if the research goal is to quantify across a broad range of genes including long non-coding RNAs (lncRNAs), sequencing depth >80 million reads per sample may be considered (Sims et al. 2014). Furthermore, good RNAseq data provide a higher number of reads that are mapped back to the reference genome. Paired-end read lengths, that is, >76 bp are expected to map by >80% along with >70% of the reads match with no mismatches and <1% matches to rRNA and ~15% map to splice junctions (Djebali et al. 2012; Sims et al. 2014). The majority of the amaranth transcriptome studies that were carried out were conducted using Illumina platform and generated paired-end read length between 90 and 125 bp; however, sequencing depth was variable (Table 1.1).

Despite multiple advantages and reliability of RNA-seq technology, it still has some limitations. During RNA-seq sample preparation, RNA and cDNA are fragmented; there are a series of steps until a sequenced fragment is generated, thereby increasing the possibility of artifacts and technical errors (Wang et al. 2009). One way to discriminate the potential error is by assessing multiple biological replicates of a sample. For a biological question to identify a majority of significantly differentially expressed (SDE) genes, at least 12 replicates per treatment were required to identify >90% truly SDE genes (Schurch et al. 2016). In the amaranth study, 8 out of 10 studies conducted differential gene expression assessment, but a majority of the studies have not reported clear information on their biological replicates and experimental repeats (Table 1.1). Despite statistical strength and biological relevance, multiple biological replicates become a compromise concerning cost and precision. Let us take a scenario where a study was carried out with a good number of biological replicates ($n > 5$), but the sequencing depth was fairly low (<36 million reads); although, the study will result in decreased technical precision in the RNAs level (for lowly expressed genes, FPKM <10), it will still provide accurate variability in the biological samples producing robust information on DEG (for highly expressed genes, FPKM >10) (Trapnell et al. 2009; ENCODE Project Consortium). Many of the reported amaranth transcriptome studies were carried out by pooling the tissue samples before or after RNA extraction (Table 1.1); however, this still does not minimize the possibility of introducing errors and artifacts that may arise during the sample preparation (for more details, Wang et al. 2009). Depending on the complexity of the transcriptome of an organism and how well a chosen differential gene expression (DGE) bioinformatics tool can address the presence of transcript isoforms, non-coding small RNA transcription, etc. higher number of replicates is extremely crucial to assess the statistical power and biological significance of the data. Thus, a suitable number of biological replicates developed from independent experimental repeats are strongly recommended.

1.6 Final Remarks

Amaranth was a staple of the Aztec diet and is described now as a 'superfood', in part because of its high protein content and balanced amino acid profile. However, this crop is still not in the mainstream cultivation practices in North America and in many parts of the world. As we are heading towards a massive increase in human population in the next few decades, global climate change, we strongly believe this crop has a huge potential to support food security. It is important that we prepare this crop for our future generations.

After decades of neglect, the crop was re-domesticated and brought to the attention of farmers and researchers. In recent years, *A. hypochondriacus* and *A. tricolor* along with other grain, vegetable, and weedy species have gained commercial and agricultural importance. For such non-model species, without a reference genome, improved knowledge of their transcriptome will facilitate further crop improvement and weed management. In the last 10 years, omics work has developed some molecular information; however, it is not enough. Especially, the dynamics of RNA species have not been fully explored. Research on long non-coding RNAs (lncRNAs) has not been initiated yet; although there is one preliminary study on micro-RNA (miRNA) (Liu et al. 2020), there is more to be done. NGS work on amaranth transcriptome has generated a fair amount of sequence information; however, there is a dearth of data curation, and that is well reflected in the

Table 1.1 Publicly available transcriptome sequence information

NCBI Accession	Species	Experimental design			Methods			Results			References
		Tissue type	Biological replicate	Sample	Sequencing platform	Sequence type	Assembly software	Assembled contigs	Assembled reads/bases	N50 or average seq. length	
SRA012013	*A. tuberculatus*	Seedling and flower	No replicate	*Pooled*	GS-FLX 454 pyrosequencing	–	CAP319 and Egassembler (de novo)	44,469	442,772 reads	434 bp	Riggins et al. (2010)
PRJNA79625, PRJNA65409	*A. hypochondriacus* and *A. tuberculatus*	Leaf and stem	No replicate	–	454 Pyrosequencing	–	Newbler	21,207	270,0168 reads	1064 bp	Délano-Frier et al. (2011)
PRJNA65409	*A. hypochondriacus*	Leaf, stem, and seed	No clear information	*Pooled* (*n = 10*)	Illumina	Paired end (PE 72)	Trinity (de novo)	57,658	119,875,998 reads	1884 bp	Sunil et al. (2014)
PRJNA202395	*A. tricolor*	Young, adult, floral buds, and fully open flower	No replicate	*Pooled*	Illumina Hiseq 2000	Paired end (PE 100)	Trinity (de novo)	99,312	319,199,714 reads	1460 bp	Liu et.al (2014)
PRJNA263128	*A. hypochondriacus*	Leaf, stem, root, and flowers	No replicate	–	Illumina Hiseq	Paired end (PE 90)	ABySS(de novo)	66,370	352,557,994 reads	1491 bp	Clouse et al. (2016)
SRP111547	*A. hypochondriacus*	Leaf, stem, root, buds, and flowers, and seeds	Likely 3 bioreplicates (no clear information)	*Pooled* (*n = 10, 15*)	Illumina Hiseq 2500	Paired end (PE 100)	Oasis, SOAPdenovo-trans, Trinity	125,581	88, 999,254 bases	834 bp	Sunil et al. (2017)
PRJNA390774	*A. palmeri*	Leaf	3 bioreplicates	–	Illumina Hiseq 2500	Paired end (PE 125)	Trinity (de novo)	72,780	1,667,277,670 reads	675 bp	Salas-Perez et al. (2018)
PRJNA432348	*A. tuberculatus*	Leaf	Likely 3 bioreplicates (no clear information)	–	Illumina Hiseq 2500	Paired end (PE 100)	Trinity (de novo)	113,893	2.36 billion reads	1709 bp	Kohlhase et al. (2019)

(continued)

Table 1.1 (continued)

NCBI Accession	Species	Experimental design			Methods			Results			References
		Tissue type	Biological replicate	Sample	Sequencing platform	Sequence type	Assembly software	Assembled contigs	Assembled reads/bases	N50 or average seq. length	
SRR5930345	*A. tricolor*	Leaf	No replicate	*Pooled (n = 10)*	Illumina Hiseq 2500	Paired end (PE 125)	Trinity (de novo)	80.026	59,827,883 reads	1202 bp	Liu et al. (2019)
SRP151898	*A. mangostanus*	Leaf and root	No replicate	*Pooled* (n = 2)	Illumina HiSeq 2000 or 4000	Single end	Tgicl (de novo)	104,483	74.58 and 74.07 million reads	1661 bp	Guo et al. (2020)

SwissProt and other public databases with their limited protein information. Besides fundamental science research, the molecular data have a higher impact when breeders access the omics information and leverage it into the molecular breeding program; however, amaranth research and extension programs are not there yet. Amaranth genome mapping is still far from completion, 100% of the transcriptome studies conducted until this day performed de novo assembly because the research area lacks a high-quality reference genome for the species. A limited tissue source and treatment conditions have been used to generate transcriptome information (Table 1.1). Considering the dynamic nature of transcriptomes, diverse tissue types, and environmental variables and stresses, biotic and abiotic need to be included in future transcriptome studies. As noted above, there were flaws in the experimental designs thereof, further studies with improved methods, number of biological replicates, experimental repeats, and read depth need consideration. These amendments will greatly enhance the quality and confidence in the molecular data for future research and crop improvement.

References

Adhikary D (2019) Development of genetic resources and functional analysis of betalains in *Amaranthus* species. Dissertation. UBC, p 156. https://hdl.handle.net/2429/69760

Adhikary D, Forsyth JA, Murch SJ, Deyholos MK (2020) Impact of betacyanins on responses to ultraviolet radiation in *Amaranthus tricolor* L. J Plant Interact 15(1):117–126. https://doi.org/10.1080/17429145.2020.1766584

Almsick VA (2009) New HPPD-Inhibitors-a proven mode of action as a new hope to solve current weed problems. Outlooks Pest Manag 20:27–30. https://doi.org/10.1564/20feb09

Alvarez-Jubete L, Holse M, Hansen Å, Arendt EK, Gallagher E (2009) Impact of baking on vitamin e content of pseudocereals amaranth, quinoa, and buckwheat. Cereal Chem 86(5):511–515. https://doi.org/10.1094/CCHEM-86-5-0511

Basu C, Halfhill MD, Mueller TC, Stewart CN (2004) Weed genomics: new tools to understand weed biology. Trends Plant Sci 9:391–398. https://doi.org/10.1016/j.tplants.2004.06.003

Bennett MD, Smith JB (1976) Nuclear dna amounts in angiosperms. Philos Trans R Soc Lond B Biol Sci 274(933):227–274. https://doi.org/10.1098/rstb.1976.0044

Berger A, Gremaud G, Baumgartner M, Rein D, Monnard I, Kratky E, Geiger W, Burri J, Dionisi F, Allan M (2003) Cholesterol-lowering properties of amaranth grain and oil in hamsters. Int J Vitam Nutr Res 73:39–47. https://doi.org/10.1016/S0308-8146(02)00387-4

Betschart AA, Irving DW, Shepherd AD, Saunders RM (1981) *Amaranthus cruentus*: milling characteristics, distribution of nutrients within seed components, and the effects of temperature on nutritional quality. J Food Sci 46:1181–1187. https://doi.org/10.1111/j.1365-2621.1981.tb03019.x

Bian J, Deng P, Zhan H, Wu X, Nishantha MDLC, Yan Z, Du X, Nie X, Song W (2019) Transcriptional dynamics of grain development in barley (*Hordeum vulgare* L.). Int J Mol Sci 20(4):962. https://doi.org/10.3390/ijms20040962

Chen ZJ, Ni Z (2006) Mechanisms of genomic rearrangements and gene expression changes in plant polyploids. Bio Essays 28(3):240–252. https://doi.org/10.1002/bies.20374

Cloonan N, Forrest AR, Kolle G, Gardiner BB, Faulkner GJ, Brown MK, Taylor DF, Steptoe AL, Wani S, Bethel G, Robertson AJ, Perkins AC, Bruce SJ, Lee CC, Ranade SS, Peckham HE, Manning JM, McKernan KJ, Grimmond SM (2008) Stem cell transcriptome profiling via massive-scale mRNA sequencing. Nat Methods 5(7):613–619. https://www.nature.com/articles/nmeth.1223

Clouse JW, Adhikary D, Page JT, Ramaraj T, Deyholos MK, Udall JA, Fairbanks DJ, Jellen EN, Maughan PJ (2016) The amaranth genome: genome, transcriptome, and physical map assembly. Plant Genome 9(1). https://doi.org/10.3835/plantgenome2015.07.0062

Conesa A, Madrigal P, Tarazona S, Gomez-Cabrero D, Cervera A, McPherson A, Szcześniak MW, Gaffney DJ, Elo LL, Zhang X, Mortazavi A (2016) A survey of best practices for RNA-seq data analysis. Genome Biol 17:13. https://doi.org/10.1186/s13059-016-0881-8

Costea M, Weaver SE, Tardif FJ (2004) The biology of Canadian weeds. 130. *Amaranthus retroflexus* L., *A. powellii* S. Watson and *A. hybridus* L. Can J Plant Sci 84:631–668. https://doi.org/10.4141/P02-183

Craciun A, Jacobs M, Vauterin M (2000) *Arabidopsis* loss-of-function mutant in the lysine pathway points out complex regulation mechanisms. FEBS Letters 487(2):234–238. https://doi.org/10.1016/S0014-5793(00)02303-6

Délano-Frier JP, Avilés-Arnaut H, Casarrubias-Castillo K, Casique-Arroyo G, Castrillón-Arbeláez PA, Herrera-Estrella L, Massange-Sánchez J, Martínez-Gallardo NA, Parra-Cota FI, Vargas-Ortiz E, Estrada-Hernández MG (2011) Transcriptomic analysis of grain amaranth (*Amaranthus hypochondriacus*) using

454 pyrosequencing: comparison with *A. tuberculatus*, expression profiling in stems and in response to biotic and abiotic stress. BMC Genom 12:363. https://doi.org/10.1186/1471-2164-12-363

Djebali S, Davis CA, Merkel A, Dobin A, Lassmann T, Mortazavi A, Tanzer A, Lagarde J, Lin W, Schlesinger F, Xue C, MarinovGK KJ, Williams BA, Zaleski C, Rozowsky J, Röder M, Kokocinski F, Abdelhamid RF, Alioto T, Antoshechkin I, Baer MT, Bar NS, Batut P, Bell K, Bell I, Chakrabortty S, Chen X, Chrast J, Curado J, Derrien T, Drenkow J, Dumais E, Dumais J, Duttagupta R, Falconnet E, Fastuca M, Fejes-Toth K, Ferreira P, Foissac S, Fullwood MJ, Gao H, Gonzalez D, Gordon A, Gunawardena H, Howald C, Jha S, Johnson R, Kapranov P, King B, Kingswood C, Luo OJ, Park E, Persaud K, Preall JB, Ribeca P, Risk B, Robyr D, Sammeth M, Schaffer L, See LH, Shahab A, Skancke J, Suzuki AM, Takahashi H, Tilgner H, Trout D, Walters N, Wang H, Wrobel J, Yu Y, Ruan X, Hayashizaki Y, Harrow J, Gerstein M, Hubbard T, Reymond A, Antonarakis SE, Hannon G, Giddings MC, Ruan Y, Wold B, Carninci P, Guigó R, Gingeras TR (2012) Landscape of transcription in human cells. Nature 489(7414):101–108. https://doi.org/10.1038/nature11233

Dohm JC, Minoche AE, Holtgräwe D, Capella-Gutiérrez S, Zakrzewski F, Tafer H, Rupp O, Sörensen TR, Stracke R, Reinhardt R, Goesmann A, Kraft T, Schulz B, Stadler PF, Schmidt T, Gabaldón T, Lehrach H, Weisshaar B, Himmelbauer H (2014) The genome of the recently domesticated crop plant sugar beet (*Beta vulgaris*). Nature 505(7484):546–549. https://doi.org/10.1038/nature12817

ENCODE Project Consortium (2004) The ENCODE (ENCyclopedia Of DNA Elements) project. Science 306(5696):636–640. https://doi.org/10.1126/science.1105136

Guo SH, Jiang LY, Xu ZM, Li QS, Wang JF, Ye HJ, Wang LL, He BY, Zhou C, Zeng EY (2020) Biological mechanisms of cadmium accumulation in edible Amaranth (*Amaranthus mangostanus* L.) cultivars promoted by salinity: a transcriptome analysis. Environ Pollut 262:114304. https://doi.org/10.1016/j.envpol.2020.114304

He HP, Corke H (2003) Oil and squalene in amaranthus grain and leaf. J Agric Food Chem 51(27):7913–7920. https://doi.org/10.1021/jf030489q

Holm L, Doll J, Holm E, Pancho J, Herberger J (1997) World weeds: natural histories and distribution. Wiley, Inc, p 1152

Huerta-Ocampo JA, Barrera-Pacheco A, Mendoza-Hernández CS, Espitia-Rangel E, Mock HP, Barba de la Rosa AP (2014) Salt stress-induced alterations in the root proteome of *Amaranthus cruentus* L. J Proteome Res 13(8):3607–3627. https://doi.org/10.1021/pr500153m

Khan MI, Giridhar P (2015) Plant betalains: chemistry and biochemistry. Phytochemistry 117:267–295. https://doi.org/10.1016/j.phytochem.2015.06.008

Kietlinski KD, Jimenez F, Jellen EN, Maughan PJ, Smith SM, Pratt DB (2014) Relationships between the weedy *Amaranthus hybridus* (Amaranthaceae) and the grain amaranths. Crop Sci 54:220. https://doi.org/10.2135/cropsci2013.03.0173

Kohlhase DR, O'Rourke JA, Owen MDK, Graham MA (2019) Using RNA-seq to characterize responses to 4-hydroxyphenylpyruvate dioxygenase (HPPD) inhibitor herbicide resistance in waterhemp (*Amaranthus tuberculatus*). BMC Plant Biol 19(1):182. https://doi.org/10.1186/s12870-019-1795-x

Kong X, Bao J, Corke H (2009) Physical properties of Amaranths starch. Food Chem 113:371–376. https://doi.org/10.1016/j.foodchem.2008.06.028

Lau OS, Huang X, Charron JB, Lee JH, Li G, Deng XW (2011) Interaction of *Arabidopsis* DET1 with CCA1 and LHY in mediating transcriptional repression in the plant circadian clock. Mol Cell 43(5):703–712. https://doi.org/10.1016/j.molcel.2011.07.013

Lightfoot DJ, Jarvis DE, Ramaraj T, Lee R, JellenEN MPJ (2017) Single-molecule sequencing and Hi-C-based proximity-guided assembly of amaranth (*Amaranthus hypochondriacus*) chromosomes provide insights into genome evolution. BMC Biol 15(1):74. https://doi.org/10.1186/s12915-017-0412-4

Liu S, Kuang H, Lai Z (2014) Transcriptome analysis by Illumina high-throughput paired-end sequencing reveals the complexity of differential gene expression during In Vitro plantlet growth and flowering in *Amaranthus tricolor* L. PLoS ONE 9(6):e100919. https://doi.org/10.1371/journal.pone.0100919

Liu S, Zheng X, Pan J, Peng L, Cheng C, Wang X, Zhao C, Zhang Z, Lin Y, XuHan X, Lai Z (2019) RNA-sequencing analysis reveals betalains metabolism in the leaf of *Amaranthus tricolor* L. PLoS ONE 14(4):e0216001. https://doi.org/10.1371/journal.pone.0216001

Liu S, Peng L, Pan J, Wang X, Zhao C, Cheng C, Zhang Z, Lin Y, XuHan X, Lai Z (2020) Identification of microRNAs involved in betalain metabolism in the green and red sectors of amaranth leaves based on Illumina sequencing data. J Hortic Sci Biotechnol 95:454–459. https://doi.org/10.1080/14620316.2019.1702481

Locascio A, Roig-Villanova I, Bernardi J, Varotto S (2014) Current perspectives on the hormonal control of seed development in *Arabidopsis* and maize: a focus on auxin. Front Plant Sci 5:412. https://doi.org/10.3389/fpls.2014.00412

Mallory MA, Hall RV, McNabb AR, Pratt DB, JellenEN MPJ (2008) Development and characterization of microsatellite markers for the grain amaranths. Crop Sci 48:1098. https://doi.org/10.2135/cropsci2007.08.0457

Martirosyan DM, Miroshnichenko LA, Kulakova SN, Pogojeva AV, Zoloedov VI (2007) Amaranth oil application for coronary heart disease and hypertension. Lipids Health Dis 6:1. https://doi.org/10.1186/1476-511X-6-1

Maughan PJ, Sisneros N, Luo M, Kudrna D, Ammiraju JSS, Wing RA (2008) Construction of an *Amaranthus hypochondriacus* bacterial artificial chromosome library and genomic sequencing of herbicide target genes. Crop Sci 48:85–94. https://doi.org/10.2135/cropsci2007.08.0434tpg

Montgomery JS, Giacomini D, Waithaka B, Lanz C, Murphy BP, Campe R, Lerchl J, Landes A, Gatzmann F, Janssen A, Antonise R, Patterson E, Weigel D, Tranel PJ (2020) Draft Genomes of *Amaranthus tuberculatus*, *Amaranthus hybridus*, and *Amaranthus palmeri*. Genome Biol Evol 12 (11):1988–1993. https://doi.org/10.1093/gbe/evaa177

Mosyakin SL, Robertson KR (2003) Amaranthus in flora of North America north of Mexico. OUP, pp 410–435

Nagalakshmi U, Wang Z, Waern K, Shou C, Raha D, Gerstein M, Snyder M (2008) The transcriptional landscape of the yeast genome defined by RNA sequencing. Science 320(5881):1344–1349. https://doi.org/10.1126/science.1158441

Nomura T, Ishihara A, Imaishi H, Ohkawa H, Endo TR, Iwamura H (2003) Rearrangement of the genes for the biosynthesis of benzoxazinones in the evolution of Triticeae species. Planta 217(5):776–782. https://doi.org/10.1007/s00425-003-1040-5

Okada M, Yoshida K, Nishijima R, Michikawa A, Motoi Y, Sato K, Takumi S (2018) RNA-seq analysis reveals considerable genetic diversity and provides genetic markers saturating all chromosomes in the diploid wild wheat relative *Aegilops umbellulata*. BMC Plant Biol 18(1):271. https://doi.org/10.1186/s12870-018-1498-8

Packard EE, da Costa Ribeiro Quintans ILA, Adhikary D (nd in press) Genetics of betalain pigments in amaranth species. In: Adhikary D, Deyholos MK, Delano-Frier JP (eds) The Amaranth genome. Springer Nature

Pedersen B, Hallgren L, Hansen I, Eggum BO (1987) The nutritive value of amaranth grain (*Amaranthus caudatus*). Plant Food Hum Nutr 36:325–334. https://doi.org/10.1007/BF01892353

Pineiro M, Jarillo JA (2013) Ubiquitination in the control of photoperiodic flowering. Plant Sci 198:98–109. https://doi.org/10.1016/j.plantsci.2012.10.005

Qureshi AA, Lehmann JW, Peterson DM (1996) Amaranth and its oil inhibit cholesterol biosynthesis in 6-week-old female chickens. J Nutr 126(8):1972–1978. https://doi.org/10.1093/jn/126.8.1972

Riggins CW, Peng Y, Stewart CN Jr, Tranel PJ (2010) Characterization of de novo transcriptome for waterhemp (*Amaranthus tuberculatus*) using GS-FLX 454 pyrosequencing and its application for studies of herbicide target-site genes. Pest Manag Sci 66 (10):1042–1052. https://doi.org/10.1002/ps.2006

Salas-Perez RA, Saski CA, Noorai RE, Srivastava SK, Lawton-Rauh AL, Nichols RL, Roma-Burgos N (2018) RNA-Seq transcriptome analysis of *Amaranthus palmeri* with differential tolerance to glufosinate herbicide. PLoS ONE 13(4):e0195488. https://doi.org/10.1371/journal.pone.0195488

Sauer JD (1949) The grain Amaranths: a survey of their history and classification. Ann Mo Bot Gard 37:561–632. https://doi.org/10.2307/2394403

Schurch NJ, Schofield P, Gierliński M, Cole C, Sherstnev A, Singh V, Wrobel N, Gharbi K, Simpson GG, Owen-Hughes T, Blaxter M, Barton GJ (2016) How many biological replicates are needed in an RNA-seq experiment and which differential expression tool should you use? RNA 22(6):839–851. https://doi.org/10.1261/rna.053959.115

Steckel LE (2007) The dioecious *Amaranthus* spp.: here to stay. Weed Technol 21:567–570. https://doi.org/10.1614/WT-06-045.1

Silva-Sanchez C, Barba de la Rosa AP, Leon-Galván MF, de Umen BO, de Leon-Rodríguez A, Gonzalez de Mejía E (2008) Bioactive peptides in amaranth (*Amaranthus hypochondriacus*) seed. J Agric Food Chem 56:1233–1240. https://doi.org/10.1021/jf072911z

Sims D, Sudbery I, Ilott NE, Heger A, Ponting CP (2014) Sequencing depth and coverage: key considerations in genomic analyses. Nat Rev Genet 15(2):121–132. https://doi.org/10.1038/nrg3642

Souer E, van Houwelingen A, Kloos D, Mol J, Koes R (1996) The no apical meristem gene of Petunia is required for pattern formation in embryos and flowers and is expressed at meristem and primordia boundaries. Cell 85(2):159–170. https://doi.org/10.1016/S0092-8674(00)81093-4

Sunil M, Hariharan N, Dixit S, Choudhary B, Srinivasan S (2017) Differential genomic arrangements in Caryophyllales through deep transcriptome sequencing of *A. hypochondriacus*. PLoS One 12(8): e0180528. https://doi.org/10.1371/journal.pone.0180528

Sunil M, Hariharan AK, Nayak S, Gupta S, Nambisan SR, Gupta RP, Panda B, Choudhary B, Srinivasan S (2014) The draft genome and transcriptome of *Amaranthus hypochondriacus*: a C4 dicot producing high-lysine edible pseudo-cereal. DNA Res 21 (6):585–602. https://doi.org/10.1093/dnares/dsu021

Tarazona S, García-Alcalde F, Dopazo J, Ferrer A, Conesa A (2011) Differential expression in RNA-seq: a matter of depth. Genome Res 21(12):2213–2223. https://doi.org/10.1101/gr.124321.111

Tanaka N, Itoh H, Sentoku N, Kojima M, Sakakibara H, Izawa T, Itoh J, Nagato Y (2011) The COP1 ortholog PPS regulates the juvenile-adult and vegetative-reproductive phase changes in rice. Plant Cell 23 (6):2143–2154. https://doi.org/10.1105/tpc.111.083436

Tandang-Silvas MR, Cabanos CS, Carrazco Peña LD, De la Rosa AP, Osuna-Castro JA, Utsumi S, Mikami B, Maruyama N (2012) Crystal structure of a major seed storage protein, 11S proglobulin, from *Amaranthus hypochondriacus*: insight into its physico-chemical properties. Food Chem 135(2):819–826. https://doi.org/10.1016/j.foodchem.2012.04.135

Tosi EA, Ré E, Lucero H, Masciarelli R (2001) Dietary fiber obtained from amaranth (*Amaranthus cruentus*) grain by differential milling. Food Chem 73:441–443. https://doi.org/10.1016/S0308-8146(00)00326-5

Trapnell C, Pachter L, Salzberg SL (2009) TopHat: discovering splice junctions with RNA-Seq. BIX 25 (9):1105–1111. https://doi.org/10.1093/bioinformatics/btp120

Vanneste K, Baele G, Maere S, Van de Peer Y (2014) Analysis of 41 plant genomes supports a wave of successful genome duplications in association with the Cretaceous-Paleogene boundary. Genome Res 24 (8):1334–1347. https://doi.org/10.1101/gr.168997.113

Vecchi B, Añón MC (2009) ACE inhibitory tetrapeptides from *Amaranthus hypochondriacus* 11S globulin. Phytochemistry 70(7):864–870. https://doi.org/10.1016/j.phytochem.2009.04.006

Venskutonis PR, Kraujalis P (2013) Nutritional components of Amaranth seeds and vegetables: a review on composition, properties, and uses. Compr Rev Food Sci Food Saf 12(4):381–412. https://doi.org/10.1111/1541-4337.12021

Walsh BM, Adhikary D, Maughan PJ, Emshwiller E, Jellen EN (2015) *Chenopodium* polyploidy inferences from Salt Overly Sensitive 1 (SOS1) data. Am J Bot 102(4):533–543. https://doi.org/10.3732/ajb.1400344

Wang Z, Gerstein M, Snyder M (2009) RNA-Seq: a revolutionary tool for transcriptomics. Nat Rev Genet 10(1):57–63. https://doi.org/10.1038/nrg2484

Weigel D, Mott R (2009) The 1001 genomes project for *Arabidopsis thaliana*. Genome Biol 10(5):107. https://doi.org/10.1186/gb-2009-10-5-107

Witschel M (2012) Bioactive heterocyclic compound classes, pp 61–68. https://doi.org/10.1002/9783527664412.ch5

Ying BW, Matsumoto Y, Kitahara K, Suzuki S, Ono N, Furusawa C, Kishimoto T, Yomo T (2015) Bacterial transcriptome reorganization in thermal adaptive evolution. BMC Genom 16:802. https://doi.org/10.1186/s12864-015-1999-x

Yoon SH, Reiss DJ, Bare JC, Tenenbaum D, Pan M, Slagel J, Moritz RL, Lim S, Hackett M, Menon AL, Adams MW, Barnebey A, Yannone SM, Leigh JA, Baliga NS (2011) Parallel evolution of transcriptome architecture during genome reorganization. Genome Res 21(11):1892–1904. https://doi.org/10.1101/gr.122218.111

Zulawski M, Schulze G, Braginets R, Hartmann S, Schulze WX (2014) The Arabidopsis Kinome: phylogeny and evolutionary insights into functional diversification. BMC Genom 15(1):548. https://doi.org/10.1186/1471-2164-15-548

Organelle Genomics: The Chloroplast Genome of Amaranth

2

Thiruvarangan Ramaraj
and Sathish K. Ponniah

Abstract

Amaranth is an emerging pseudocereal native to the New World that has gained increased attention in recent years because of its nutritional quality, particularly its seed protein, with high levels of essential amino acid lysine. This chapter will concentrate on the research advancements made on organelle genomes, particularly the chloroplast genome. The chloroplast genome is an integral part of the plant genome. They are circular and conserved in size, genome structure, and gene content. Albeit chloroplast genomes have less than one percent of the genes in the nucleus, they encode essential genes responsible for energy production, respiration, photosynthesis, and genes that control agronomically crucial traits. Here, we discuss the characteristic features of the amaranth chloroplast genome, such as length, structure, repeats, gene and protein contents, and, finally, its function and importance in phylogenetic and genetic diversity studies in the *Amaranthaceae* family.

T. Ramaraj (✉)
School of Computing, College of Computing & Digital Media, DePaul University, Chicago, USA
e-mail: tramaraj@depaul.edu

S. K. Ponniah
Department of Agriculture, University of Arkansas at Pine Bluff, Pine Bluff, USA

2.1 Introduction

The grain amaranths, *A. hypochondriacus* L., *A. caudatus L.*, and *A. cruentus* L., belongs to the *Amaranthaceae* family (Caryophyllales: Amaranthaceae). The putative progenitor species, *A. hybridus* L., A. quitensis H.B.K., and *A. powellii* S. Wats., are classified as part of the *A. hybridus* complex. Amaranths are considered paleoallotetraploids (2n = 4x = 32) (Pal et al. 1982; Greizerstein and Poggio 1994, 1995; Clouse et al. 2016) that possibly ascended from multiple domestication events (Sauer 1950; Mallory et al. 2008; Rastogi and Shukla 2013). Before the Spanish Conquest in the later part of the fifteenth century and early part of the sixteenth century, the grain amaranths were an important staple food crop that also played an important cultural role in pre-Columbian civilizations in their regions of domestication (Sauer 1950, 1967, 1993). The Aztec emperor Moctezuma II required nearly equal annual tributes of amaranth, maize, and beans (Sauer 1950, 1967, 1993). The Europeans' arrival after the Spanish conquest of the New World suppressed the cultivation of the grain amaranths because of their deeply rooted use in indigenous religious ceremonies (Sauer 1976, 1993; Iturbide et al. 1994). Their use continued to decline until the end of the nineteenth century and not listed on a survey of commercial grain crops in Latin America (Sauer 1950). Their cultivation and use

D. Adhikary et al. (eds.), *The Amaranth Genome*, Compendium of Plant Genomes,
https://doi.org/10.1007/978-3-030-72365-1_2

Table 2.1 Details of origin and accession number of amaranth germplasms

No.	Accession number	Species	Collection place
1	PI558499	*A. hypochondriacus* L. cv Plainsman	USA
2	PI477913	*A. cruentus L.*	Mexico
3	PI481125	*A. hypochondriacus*	India
4	PI511731	*A. hypochondriacus*	Mexico
5	PI540446	*A. hypochondriacus*	Pakistan
6	PI605351	*A. hybridus L.*	Greece
7	PI619259	*A. hypochondriacus*	Nepal
8	PI642741	*A. caudatus L.*	Bolivia

persisted locally in intermediate-elevation inter-Andean valleys (*A. caudatus*) and the Mesoamerican Highlands. Now, amaranth is an emerging pseudocereal native to the New World that has garnered increased attention in recent years because of its nutritional quality with an estimated genome size of 466 Mb (Bennett and Smith 1976). We concentrate on the amaranth chloroplast genome in this chapter.

Chloroplasts are green plastid organelle found in land plants, eukaryotic algae, and some protists. Chloroplasts contain the required machinery for the process of photosynthesis. Photosynthesis, "synthesis of light," is a biological process in which the sun's radiation energy is captured and altered into chemical energy that can then be used by organisms. Studies have recently shown that chloroplasts have evolved from endosymbiotic, cyanobacterial ancestors (Ziehe et al. 2017). During evolution, the chloroplast genome lost most of its genes (Kress et al. 2005; Chase et al. 2007). Most of the chloroplast genes are encoded in the nucleus and post-translationally imported into the organelle (Blankenship 2014; Sabater 2018). The chloroplast genome in most land plants has a quadripartite structure consisting of a small single copy (SSC), a large single copy (LSC), and two copies of inverted repeat (IR) regions (Chen et al. 2019). Besides, the chloroplast genome has a highly conserved gene content and a nearly collinear gene order (Jansen et al. 2005). It has a compact size, with no recombination, and maternal inheritance (Birky 2001), has served as an essential DNA sequence for plant identification and the reconstruction of plant phylogenies (Parks et al. 2009; Moore et al. 2010). In this chapter, we will discuss the chloroplast genome assembly of an economically important grain amaranth cultivar (*A. hypochondriacus* cv "Plainsman") and also chloroplast genomes of four additional *A. hypochondriacus* accessions as well as one accession each of *A. caudatus*, *A. cruentus*, and *A. hybridus*. We will further discuss the genome size, structure, and gene content of the chloroplast genome for each of these accessions (Table 2.1) and report simple sequence repeats (SSRs), repetitive sequences, number of IRs, single-nucleotide polymorphisms (SNPs), and insertion/deletion polymorphisms (Indels). The published chloroplast genomes in the genus *Amaranthus* serve as an important reference resource for phylogenetic studies in the *Amaranthaceae* family (Schmitz-Linneweber et al. 2001; Li et al. 2014).

2.2 Chloroplast Genome Sequencing

Seeds of amaranth germplasm were obtained from the publicly available U.S. Department of Agriculture National Plant Germplasm System (USDA-NPGS) *Amaranthus* germplasm collection (http://www.ars-grin.gov/) (Table 2.1). The sequencing libraries were generated and sequenced using second—(HiSeq platform, Illumina, San Diego, California, USA) and third—

(PacBio RS II, Pacific Biosciences, Menlo Park, California, USA) generation sequencing technologies. De novo genome assembly programs were used to assemble the chloroplast genome. All eight chloroplast genomes reported were composed of a single circular double-stranded DNA molecule with the typical quadripartite structure as represented in Fig. 2.1. The genomes were all similar in length ranging between 150,499 and 150,559 base pairs and GC content of 37.0%. Also, the assembled genomes were identical in length to the other chloroplast genome sequences in the *Amaranthaceae* family (*S. oleracea*: 150,725 and *B. vulgaris* subsp. vulgaris: 149,637) (Schmitz-Linneweber et al. 2001; Li et al. 2014).

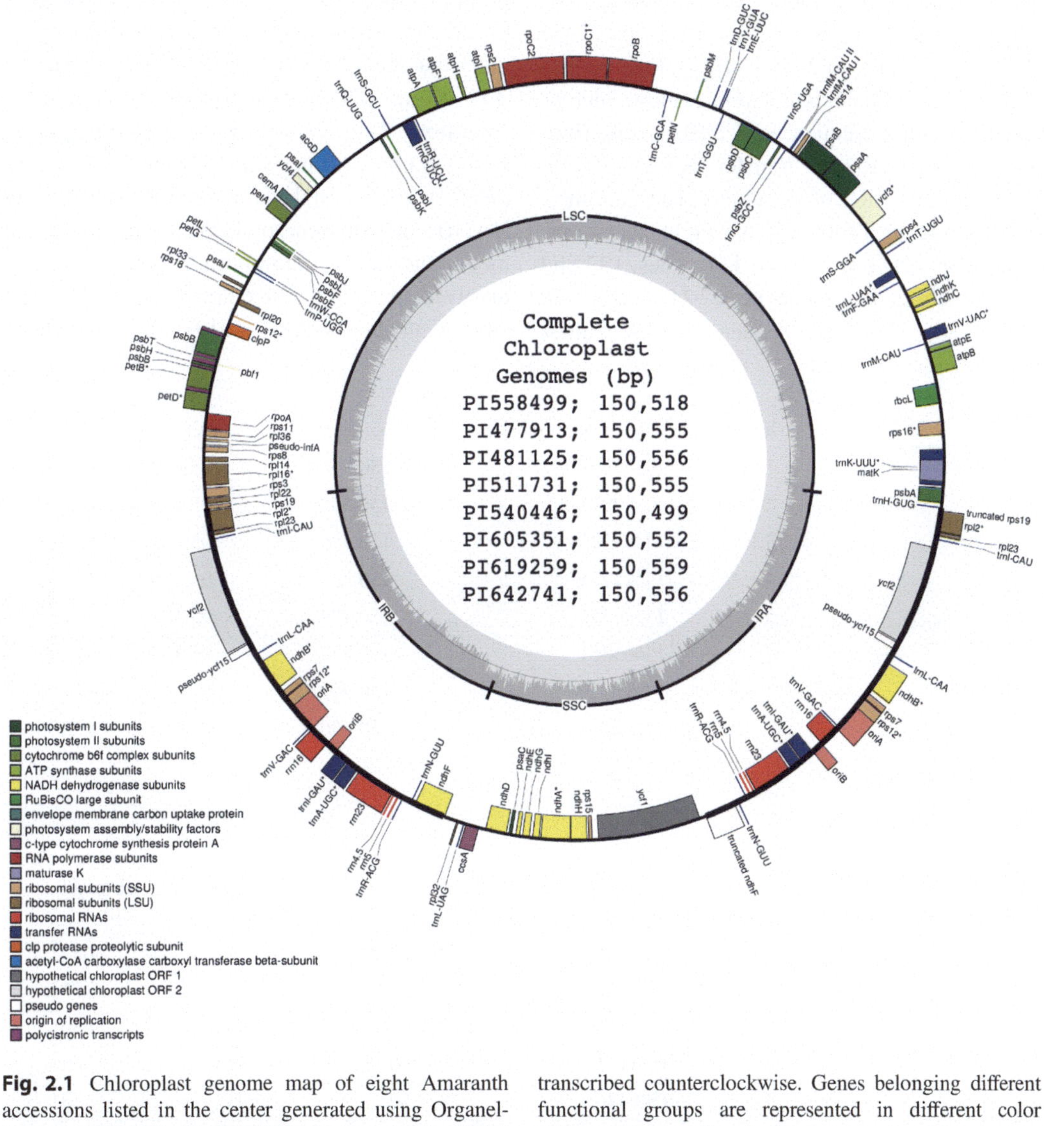

Fig. 2.1 Chloroplast genome map of eight Amaranth accessions listed in the center generated using OrganellarGenomeDRAW (OGDRAW) (Lohse et al. 2007) software package. Genes shown inner circle are transcribed clockwise and those on the outer circles are transcribed counterclockwise. Genes belonging different functional groups are represented in different color schemes. Gray spikes in the innermost circle represent GC content

2.3 Chloroplast Genome Assembly Details

All sequenced genomes were assembled to one circular sequence with length ranging between 150,499 and 150,559 base pairs and GC and AT contents of 37.0% and 63.0%, respectively (Table 2.2). GeSeq (Tillich et al. 2017) was run online using version 2.03 with default parameter settings to generate annotations. The length of the LSC and the SSC and the two IR ranged from 83,850 to 83,919 (~56%), 17,941 to 17957 (12%), and 24,340 to 24,391 (16%), respectively (Table 2.3). Different annotation features such as Coding Domain Sequences (CDS), genes, ribosomal RNA (rRNA), transfer RNA (tRNA), exons, introns, IRs, of the chloroplast genomes, are reported in Table 2.4. All eight accessions have genes between 130 and 131 genes, 122 and 123 protein-coding sequences, 24 exons, 78 tRNA, and 8 rRNA sequences. All values reported in Tables 2.3 and 2.4 were generated from GeSeq. Table 2.5 also represents genes contained in the sequenced amaranth chloroplast genomes. The chloroplast genome contains three major gene categories that encode (i) photosynthesis, (ii) self-replication, (iii) other genes, and (iv) unknown function (Table 2.5) across the eight accessions. It is very similar to the other published chloroplast genomes in the Amaranthaceae family. The ATP synthases are multi-protein complexes that are found in the energy-transducing membranes of the chloroplast genomes that are mainly responsible for synthesizing ATP from ADP and phosphate under aerobic conditions using the proton-motive force generated by photosynthesis as a source of energy (Walker 2013). NDH is a non-proton pumping NADH dehydrogenase located in the inner membrane of the chloroplast genome primarily involved in carbon utilization (Vamshi Krishna and Venkata Mohan 2019). Photosynthetic

Table 2.2 Genome sequence metrics

S. No.	Accession number	Genome length (bp)	AT	GC
1	PI558499	150,518	95,408	55,110
2	PI477913	150,555	95,458	55,097
3	PI481125	150,556	95,453	55,103
4	PI511731	150,555	95,450	55,105
5	PI540446	150,499	95,405	55,094
6	PI605351	150,552	95,454	55,098
7	PI619259	150,559	95,460	55,099
8	PI642741	150,556	95,450	55,106

Table 2.3 Key genome features

S. No.	Accession number	LSC length (bp)	SSC length (bp)	IR length (bp)
1	PI558499	83,873	17,941	24,352
2	PI477913	83,873	17,945	24,351
3	PI481125	83,908	17,946	24,361
4	PI511731	83,906	17,947	24,351
5	PI540446	83,850	17,947	24,351
6	PI605351	83,850	17,947	24,351
7	PI619259	83,910	17,947	24,391
8	PI642741	83,919	17,957	24,340

Table 2.4 Annotation features

Features	PI558499	PI477913	PI481125	PI511731	PI540446	PI605351	PI619259	PI642741
CDS	123	123	122	122	122	123	122	122
Gene	131	131	130	130	130	131	130	130
rRNA	8	8	8	8	8	8	8	8
Exon	24	24	24	24	24	24	24	24
Intron	12	12	12	12	12	12	12	12
IRs	2	2	2	2	2	2	2	2
tRNA	78	78	78	78	78	78	78	78

Table 2.5 Annotation details (Gene names, Gene Group, and Gene Category)

Gene names	Gene group	Gene category
atpA,atpF,atpH,atpI,atpE,atpB	Subunits of ATP synthase	Photosynthesis
ndhJ,ndhK,ndhC,ndhB,ndhF,ndhD,ndhE,ndhG,ndhI,ndhA,ndhH,ndhB	Subunits of NADH-dehydrogenase	
petN,petA,petL,petG,petB,petD	Subunits of cytochrome b/f complex	
psaB,psaA,psaI,psaJ,psaC	Subunits of photosystem I	
psbA,psbK,psbI,psbM,psbD,psbC,psbZ,psbJ,psbL,psbF,psbE,psbB,psbT,psbN,psbH	Subunits of photosystem II	
rbcL	Subunit of RuBisCO	
rpl33,rpl20,rpl36,rpl14,rpl16,rpl22,rpl2,rpl32,rpl2	Large subunit of ribosome	Self-replication
rpoC2,rpoC1,rpoB,rpoA	DNA-dependent RNA polymerase	
rps16,rps2,rps14,rps4,rps18,rps12,rps12,rps11,rps8,rps3,rps19,rps7,rps12,rps12,rps15, rps7	Small subunit of ribosome	
rrn16,rrn23,rrn4.5,rrn5,rrn5,rrn4.5,rrn23,rrn16	rRNA genes	
trnH-GUG,trnK-UUU,trnQ-UUG,trnS-GCU,trnG-UCC,trnR-UCU,trnC-GCA,trnD-GUC,trnY-GUA,trnE-UUC,trnT-GGU,trnS-UGA,trnG-GCC,trnfM-CAU,trnS-GGA, trnT-UGU,trnL-UAA,trnF-GAA,trnV-UAC,trnM-CAU,trnW-CCA,trnP-UGG,trnI-CAU,trnL-CAA,trnV-GAC,trnI-GAU,trnA-UGC,trnR-ACG,trnN-GUU,trnL-UAG, trnN-GUU,trnR-ACG,trnA-UGC,trnI-GAU,trnV-GAC,trnL-CAA,trnI-CAU	tRNA genes	
accD	Subunit of acetyl-CoA-carboxylase	Other genes
ccsA	C-type cytochrome synthesis gene	
cemA	C-type cytochrome synthesis gene	
clpP	Protease	
infA	Translational initiation factor	
matK	Maturase	
ycf3,ycf4,ycf2,ycf1,ycf1,ycf2	Conserved open reading frames	Unknown function

Electron Transport (PET) genes are part of the subunit composition of cytochrome b_6f complex resides in the thylakoid and cytoplasmic membranes of the chloroplast genomes and are primarily responsible for the transportation of electrons between photosystems (UniProt Consortium 2008). PSA (Photosystem I) genes encode a polypeptide of photosystems I and II reaction centers, which utilizes light energy to transfer electrons from plastocyanin to ferradoxin (Fromme et al. 2001). This gene is also used in the transformation of chloroplasts where researchers have used *Chlamydomonas reinhardtii* system to develop a site-directed mutagenesis to test predictions of structure–function relationships in the photosystem I reaction center (Bingham et al. 1991).

2.3.1 Codon Usage Pattern

This section focuses on codon usage, which is an essential feature of all genomes, especially chloroplasts (Zhou et al. 2016). Based on the standard codon usage analysis, there are overall 64 (3 out of the 64 represents translation stop signals) codons were present across the eight accessions focused on this chapter. Eight accessions contained 122 or 123 number of CDS constituting a total of 90,987–91,070 bps and were composed of 29,834–30,335 codons (Table 2.6). Total codons number reported in Table 2.6 was generated using the "Codon Usage," program from the sequence manipulation suite (Stothard 2000).

2.3.2 Simple Sequence Repeat Content

Simple sequence repeats (SSRs), referred to as microsatellites, play a very crucial role in the reorganization and stabilization of chloroplast genomes (Milligan et al. 1989; Maréchal et al. 2009; Chaney et al. 2016). SSRs have high mutation rates and can be easily genotyped using PCR and more frequently used as markers for phylogenetic analysis and in marker-assisted breeding (Chaney et al. 2016). There are 29–37 SSRs in the eight sequenced chloroplast genomes. SSRs were identified using the following standard threshold specifically for chloroplast genomes, minimum stretch of 12 for mono-, 6 for di-, 4 for tri-, and 3 for tetra-, penta-, and hexa nucleotide repeats and a minimum distance of 100 bp between compound SSRs (Table 2.7) (Sablok et al. 2015; Chaney et al. 2016). Table 2.7 values were generated using MISA-Web, a web server for SSR prediction (Thiel 2016; Beier et al. 2017) Around 60.0–63.0% of the identified SSRs are found in the LSC region of the chloroplast genomes (Chaney et al. 2016).

2.3.3 Repeated Sequences

Repeated elements in the *Amaranthus* chloroplast genomes were identified using REPuter (Kurtz et al. 2001). A minimum repeat size of 30 bp and a sequence identity of $\geq 90\%$, hamming distance of 3, and maximum computed repeats of 50 were used to identify forward and

Table 2.6 Codon usage analysis

S. No.	Accession number	Total CDS	Total bps in CDS	Total codons
1	PI558499	123	91,062	30,019
2	PI477913	123	91,061	30,335
3	PI481125	122	90,987	29,952
4	PI511731	122	90,996	30,313
5	PI540446	122	91,051	29,834
6	PI605351	123	91,070	30,277
7	PI619259	122	90,981	30,309
8	PI642741	122	90,989	30,310

Table 2.7 Simple sequence repeats (SSRs)

Accessions	Mono	Di	Tri	Tetra	Penta	Hexa	Total
PI558499	15	2	5	11	0	2	37
PI477913	11	2	3	11	1	1	29
PI481125	11	2	3	11	1	1	29
PI511731	11	2	2	11	1	1	28
PI540446	11	2	2	12	1	1	29
PI605351	11	2	3	11	1	1	29
PI619259	11	2	5	11	1	1	32
PI642741	13	1	4	11	1	2	33

Table 2.8 Number of repeated sequences by repeat type in the eight amaranth chloroplast accessions

Accession	Forward	Palindrome	Total
PI558499	14	20	34
PI477913	16	21	37
PI481125	15	21	36
PI511731	15	21	36
PI540446	16	21	37
PI605351	15	21	36
PI619259	15	21	36
PI642741	15	21	36

palindromic repeats. Based on the parameters used, no reverse or complement repeats were present. Forward repeats were between 14 and 16 bp and palindromic repeats were between 20 and 21 bp long (Table 2.8).

2.3.4 Variation Content

There are several different types of variations that can be inferred using nucleotide sequence data. They mainly include (1) Single-Nucleotide Polymorphisms (SNPs), (2) small-scale and large-scale insertion and deletions (Indels), and (3) SSRs. There are several standard computational workflows such as SAMToolsmpileup (Li et al. 2009), GATK (McKenna et al. 2010) to identify these aforementioned polymorphisms. In Chaney et al. 2016, PI558499, *A. hypochondriacus* L. "Plainsman" cultivar was used as the reference and the remaining seven accessions were mapped back to the reference to infer SNPs, indels, and SSR polymorphisms. On an average, 210 SNPs and 122 indels were identified across the seven accessions. Indels ranged in size from 2 bp to 75 bp. Of all the SNPs reported, 18.6% represents T/G base substitutions that accounted for the highest percentage of all SNPs, while G/C base substitutions were the least frequent at only 2.9% (Chaney et al. 2016).

2.4 Phylogenetic Analysis

Phylogenetic relationships among the eight accessions were examined using chloroplast whole-genome sequences. Along with eight amaranth accessions collected from different locations, two other angiosperms (Spinach and Beet), chloroplast genomes, from the same *Amaranthaceae* family were grouped. The resulting phylogenetic tree was with high bootstrap supports and provided the evolutionary relationship of the grain amaranth (*A. hypochondriacus* cv Plainsman) species (Fig. 2.2).

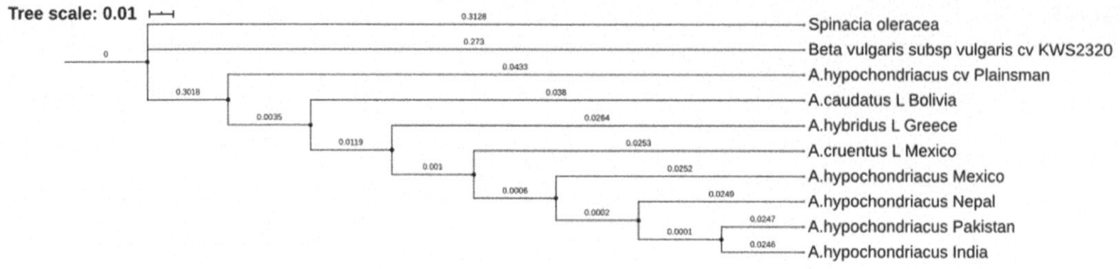

Fig. 2.2 Phylogeny analysis

The spinach and beet species genomes are grouped separately from the eight amaranth genomes with high bootstrap values. Within the amaranth species, a close relationship was observed with the origin of the species. All the Asian amaranth species, such as *A. hypochondriacus* (India, Pakistan, and Nepal), showed close relationship with low bootstrap values. Furthermore, there was a divergence observed within two Mexican species *A. hypochondriacus* and *A. cruentus* species. Also, chloroplast genome helped to group *A. hybridus* L. Greece and *A. caudatus* L. Bolivia with the *A. hypochondriacus* cv. Plainsman species (grain amaranth species). The dendrogram showed clear evidence that the grain amaranth collected from the USA is closely related to the *A. caudatus* L. Bolivia and *A. hybridus* L. Greece species. Both the Mexican and Asian *A. hypochondriacus* species were evolved from different ancestor(s) in comparison with grain amaranth species (*A. hypochondriacus* cv Plainsman).

2.5 Genetic Diversity in Amaranths

It is evident that chloroplasts play a crucial role in sustaining life on earth. As more and more chloroplast genomes become available for genus *Amaranthus*, it will enhance our understanding of chloroplast biology, diversity, gene transfer, conservation, and the genetic basis by which chloroplast transgenes can be engineered to enhance plant agronomic traits like salt and drought tolerance (Kress et al. 2005; Chase et al. 2007; Parks et al. 2009; Yang et al. 2013) to produce high-value agricultural products. The available chloroplast genomes, along with the reported SSRs, SNPs, and Indels, will be an invaluable genetic resource for studying the phylogeny and genetic diversity within the amaranth genus.

2.6 Perspectives on Future Genetic Improvements

In this rapidly changing climate and environmental conditions, there are several projects underway to engineer improvements in photosynthesis to meet the challenges of increasing demands for food and fuel (Hanson et al. 2013). Several transgenes have been introduced into chloroplast genomes in attempts to increase photosynthetic efficiency. The first transformation of the chloroplast genome was shown in *Chlamydomonas* and tobacco (Boynton et al. 1988; Svab and Maliga 1993). With the availability of complete high-quality chloroplast genomes, genetic characterizations should facilitate improved transformation efficiency and foreign gene expression in land plants through the utilization of endogenous flanking sequences and regulatory elements (Ruhlman et al. 2006). Chloroplast transformation has several advantages over nuclear transformation. Some of them include but are not limited to (i) absence of gene silencing, (ii) multigene transformation, (iii) site-directed gene integration, and (iv) high level of transgene expression. Chloroplast genome contains several copies of DNA and, hence, gene expression can be several times greater in chloroplast genomes than nuclear genomes. Because of this special characteristic, plants have the ability to produce large amounts of foreign proteins. Some of these foreign proteins can be pharmaceutical proteins (e.g., vaccine antigens and antibiotics) and industrial enzymes (e.g., hydrolases, redox enzymes, and transferases). The main point here is to emphasize the fact that chloroplast transformation has been used to control agronomic traits, including resistance to insects, diseases, herbicides, and abiotic stress tolerance of plants (Jin et al. 2015; Jin and Daniell 2015). It has been shown that double-stranded (ds) RNA expressed in chloroplasts can induce RNA interference (RNAi) of target genes in insect hosts (Zhang et al. 2015; Jin et al. 2015), demonstrating a new approach to protecting crops without chemicals. With all the remarkable progress in chloroplast transformation technologies and all the chloroplast genome resources becoming available for amaranth will be indispensable for chloroplast transformation in amaranth.

More chloroplast genomes are becoming available for the genus *Amaranthus*. In this chapter, we discussed chloroplast genomes of *Amaranthus hypochondriacus* L. cv Plainsman, representatives of the grain amaranths, *A. hypochondriacus, A. cruentus, and A. caudatus*—and their putative wild ancestor, *A. hybridus*. The amaranth chloroplast genome maintains the quadripartite structure that is highly conserved among most land plants and is highly conserved at the nucleotide level among the grain amaranths. Even given this conservation, there are also several SSRs, SNPs, and indels that will serve as invaluable genetic markers for the future genetic improvements of the emerging crop species like grain amaranths. Furthermore, the available complete chloroplast genomes can be an invaluable resource for broader comparative analysis and also open the door to biotechnology applications within the genus.

Conflict of Interest None declared.

References

Beier S, Thiel T, Münch T et al (2017) MISA-web: a web server for microsatellite prediction. Bioinformatics 33:2583–2585

Bennett MD, Smith JB (1976) Nuclear dna amounts in angiosperms. Philos Trans R Soc Lond B Biol Sci 274 (933):227–274

Bingham SE, Xu R, Webber AN (1991) Transformation of chloroplasts with thepsaB gene encoding a polypeptide of the photosystem I reaction center. FEBS Lett 292:137–140

Birky CW Jr (2001) The inheritance of genes in mitochondria and chloroplasts: laws, mechanisms, and models. Annu Rev Genet 35:125–148

Blankenship RE (2014) Molecular Mechanisms of Photosynthesis. John Wiley & Sons

Boynton JE, Gillham NW, Harris EH et al (1988) Chloroplast transformation in Chlamydomonas with high velocity microprojectiles. Science 240:1534–1538

Chaney L, Mangelson R, Ramaraj T et al (2016) The complete chloroplast genome sequences for four Amaranthus species (Amaranthaceae). Appl Plant Sci 4. https://doi.org/10.3732/apps.1600063

Chase MW, Cowan RS, Hollingsworth PM et al (2007) A proposal for a standardised protocol to barcode all land plants. Taxon 56:295–299

Chen Y, Hu N, Wu H (2019) Analyzing and characterizing the chloroplast genome of Salix wilsonii. BioMed research international

Clouse JW, Adhikary D, Page JT et al (2016) The Amaranth genome: genome, transcriptome, and physical map assembly. Plant Genome 9

Fromme P, Jordan P, Krauß N (2001) Structure of photosystem I. Biochimica et Biophysica Acta (BBA)-Bioenergetics 1507(1–3):5–31

Greizerstein EJ, Poggio L (1995) Meiotic studies of spontaneous hybrids of Amaranthus: genome analysis. Plant Breed 114:448–450

Greizerstein EJ, Poggio L (1994) Karyological studies in grain Amaranths. CYTOLOGIA 59:25–30

Hanson MR, Gray BN, Ahner BA (2013) Chloroplast transformation for engineering of photosynthesis. J Exp Bot 64:731–742

Iturbide GA, Gispert M, Hernandez-Bermejo JE, Leon J (1994) Grain amaranths (Amaranthus spp.). Neglected Crops 1492:93–101

Jansen RK, Raubeson LA, BooreJL et al (2005) Methods for obtaining and analyzing whole chloroplast genome sequences. Methods Enzymol 395:348–384

Jin S, Daniell H (2015) The engineered chloroplast genome just got smarter. Trends Plant Sci 20:622–640

Jin S, Singh ND, Li L et al (2015) Engineered chloroplast dsRNA silences cytochrome p450 monooxygenase, V-ATPase and chitin synthase genes in the insect gut and disrupts Helicoverpazea larval development and pupation. Plant Biotechnol J 13:435–446

Kress WJ, Wurdack KJ, Zimmer EA et al (2005) Use of DNA barcodes to identify flowering plants. Proc Natl Acad Sci U S A 102:8369–8374

Kurtz S, Choudhuri JV, Ohlebusch E, Schleiermacher C, Stoye J, Giegerich R (2001) REPuter: the manifold applications of repeat analysis on a genomic scale. Nucleic Acids Res 29:4633–4642. https://doi.org/10.1093/nar/29.22.4633

Li H, Cao H, Cai Y-F et al (2014) The complete chloroplast genome sequence of sugar beet (Beta

vulgaris ssp. vulgaris). Mitochondrial DNA 25:209–211
Li H, Handsaker B, Wysoker A et al (2009) The sequence alignment/Map format and SAMtools. Bioinformatics 25:2078–2079
Lohse M, Drechsel O, Bock R (2007) OrganellarGenomeDRAW (OGDRAW): a tool for the easy generation of high-quality custom graphical maps of plastid and mitochondrial genomes. Curr Genet 52(5–6):267–274
Mallory MA, Hall RV, McNabb AR et al (2008) Development and characterization of microsatellite markers for the grain Amaranths. Crop Sci 48:1098
Maréchal A, Parent J-S, Véronneau-Lafortune F et al (2009) Whirly proteins maintain plastid genome stability in Arabidopsis. Proc Natl Acad Sci USA 106:14693–14698
McKenna A, Hanna M, Banks E et al (2010) The genome analysis toolkit: a MapReduce framework for analyzing next-generation DNA sequencing data. Genome Res 20:1297–1303
Milligan BG, Hampton JN, Palmer JD (1989) Dispersed repeats and structural reorganization in subclover chloroplast DNA. Mol Biol Evol 6:355–368
Moore MJ, Soltis PS, Bell CD et al (2010) Phylogenetic analysis of 83 plastid genes further resolves the early diversification of eudicots. Proc Natl Acad Sci USA 107:4623–4628
Pal M, Pandey RM, Khoshoo TN (1982) Evolution and improvement of cultivated amaranths. J Hered 73:353–356
Parks M, Cronn R, Liston A (2009) Increasing phylogenetic resolution at low taxonomic levels using massively parallel sequencing of chloroplast genomes. BMC Biol 7:84
Rastogi A, Shukla S (2013) Amaranth: a new millennium crop of nutraceutical values. Crit Rev Food Sci Nutr 53:109–125
Ruhlman T, Lee S-B, Jansen RK et al (2006) Complete plastid genome sequence of Daucus carota: implications for biotechnology and phylogeny of angiosperms. BMC Genom 7:222
Sabater B (2018) Evolution and function of the chloroplast. Current investigations and perspectives. Int J Mol Sci 19. https://doi.org/10.3390/ijms19103095
Sablok G, Padma Raju GV, Mudunuri SB et al (2015) ChloroMitoSSRDB 2.00: more genomes, more repeats, unifying SSRs search patterns and on-the-fly repeat detection. Database 2015. https://doi.org/10.1093/database/bav084
Sauer JD (1976) Grain amaranth. Ewolution of crop plants. Simmonds NW, London
Sauer JD (1967) The grain Amaranths and Their Relatives: A Revised Taxonomic And Geographic Survey. Ann Mo Bot Gard 54:103–137
Sauer JD (1993) Historical geography of crop plants: a select roster. CRC Press
Sauer JD (1950) The grain Amaranths: a survey of their history and classification. Ann Mo Bot Gard 37:561–632
Schmitz-Linneweber C, Maier RM, Alcaraz JP et al (2001) The plastid chromosome of spinach (Spinacia oleracea): complete nucleotide sequence and gene organization. Plant Mol Biol 45:307–315
Stothard P (2000) The sequence manipulation suite: JavaScript programs for analyzing and formatting protein and DNA sequences. Biotechniques 28 (1102):1104
Svab Z, Maliga P (1993) High-frequency plastid transformation in tobacco by selection for a chimeric aadA gene. Proc Natl Acad Sci USA 90:913–917
Thiel T (2016) MISA—microsatellite identification tool. 2003. http://pgrc.ipk-gatersleben.de/misa/
Tillich M, Lehwark P, Pellizzer T et al (2017) GeSeq–versatile and accurate annotation of organelle genomes. Nucl Acids Res 45:W6–W11
UniProt Consortium (2008) The universal protein resource (UniProt). Nucl Acids Res 36:D190–5
Vamshi Krishna K, Venkata Mohan S (2019) Purification and characterization of NDH-2 protein and elucidating its role in extracellular electron transport and bioelectrogenic activity. Front Microbiol 10:880
Walker JE (2013) The ATP synthase: the understood, the uncertain and the unknown. Biochem Soc Trans 41:1–16
Yang J-B, Yang S-X, Li H-T et al (2013) Comparative chloroplast genomes of camellia species. PLoS ONE 8:
Zhang J, Khan SA, Hasse C et al (2015) Full crop protection from an insect pest by expression of long double-stranded RNAs in plastids. Science 347:991–994
Zhou T, Chen C, Wei Y et al (2016) Comparative transcriptome and chloroplast genome analyses of two related dipteronia species. Front Plant Sci 7:1512
Ziehe D, Dünschede B, Schünemann D (2017) From bacteria to chloroplasts: evolution of the chloroplast SRP system. Biol Chem 398:653–661

3 Genetic and Phenotypic Odyssey: Voyage of the Grain Amaranths from the Americas to the Old World

Daniel J. Fairbanks

Abstract

The grain amaranths consist of three species, *Amaranthus hypochondriacus* L., *Amaranthus cruentus* L., and *Amaranthus caudatus* L., which were domesticated in the Americas in pre-historic times. European botanists from the eighteenth through twentieth centuries described grain amaranths observed over a vast region of Asia where they were cultivated for grain and as a potherb, a smaller region in Africa, where they were mostly cultivated as an ornamental and as a potherb, and also in Europe, where they were cultivated as an ornamental and a commercial source of dyes. These botanists recorded that the grain amaranths were under intensive cultivation for grain in two regions of the Indian subcontinent: (1) in the north in the Himalayas and adjoining foothills and plains and (2) in the extreme south surrounding the cities of Coimbatore, Salem, and Madura. So entrenched were the grain amaranths in the culture of the people who cultivated them, botanists at the time presumed that these amaranth species were native to the region. Comparison of the Asian grain amaranths with their American counterparts led twentieth century botanists to conclude that the Asian grain amaranths were, in fact, a subset of their American counterparts and had been introduced into Asia through human contact. Jonathan D. Sauer conducted a fully revised taxonomy of the grain amaranths and concluded that the Asian amaranths were probably introduced from the Americas in the sixteenth century. Others have speculated, on the basis of limited evidence, that the Asian amaranths are among a number of plant species domesticated in the Americas that were introduced to Asia through pre-Columbian contact.

3.1 Early Evidence of Grain Amaranths in the Old World

Towards either Throne they bow, and to the ground
With solemn adoration down they cast
Their Crowns inwove with Amarant and Gold,
Immortal Amarant, a Flour which once
In Paradise, fast by the Tree of Life
Began to bloom, but soon for mans offence
To Heav'n remov'd where first it grew, there grows,
And flours aloft shading the Fount of Life
(Milton 1674, pp. 71–72)

The grain amaranths (*A. hypochondriacus* L., *A. cruentus* L., and *A. caudatus* L.) constitute a group of domesticated species of undisputed

D. J. Fairbanks (✉)
Department of Biology, Utah Valley University, 800 W. University Parkway, Orem, UT 84058, USA
e-mail: daniel.fairbanks@uvu.edu

D. Adhikary et al. (eds.), *The Amaranth Genome*, Compendium of Plant Genomes,
https://doi.org/10.1007/978-3-030-72365-1_3

American origin, although they have been cultivated for centuries throughout much of the Old World. In Milton's famous lines above, from his epic poem *Paradise Lost*, the poet praises the "Immortal Amarant" as he imagines it blooming "In Paradise, fast by the Tree of Life." Such poetry evidences the universal admiration of amaranth for its immense beauty, which is especially true of the tall grain amaranths, blooming in fields awash with color. In Milton's seventeenth century England, amaranth was raised as a popular ornamental and utilized commercially as a source of red betalain dyes. One of the cultivated grain species, *A. hypochondriacus*, bears the English common name "Prince-of-Wales feather," a name dating to the early nineteenth century (Sauer 1967). The presence of amaranths in the Old World, however, clearly predates Milton's time with evidence of its introduction to Europe in the sixteenth century (Sauer 1995).

Jonathan Deininger Sauer (1918–2008) was a pioneering American botanist with specialties in plant domestication and geography whose 1950 Ph. D. dissertation at the University of Wisconsin explored in detail the history of the grain amaranths. From that time forward, he became the foremost expert on the history of the amaranths, the grain amaranths in particular, publishing several reviews on the topic (Sauer 1950, 1967, 1976, 1995). As he explained in the first of these reviews, the amaranths "are firmly entrenched in the religious and economic lives of various Old and New World peoples as wild or cultivated grains, pot-herbs, dye-plants, fetishes, and ornamentals" (Sauer 1950, p. 561). By the nineteenth century, grain amaranth production had declined in its native regions of the Americas, whereas in India, it had dramatically increased. In Sauer's (1967) words, in reference to *A. hypochondriacus*, "for at least a hundred years, the species has been a far more widespread and important crop in Asia than in its homeland" (p. 115), and that "India alone undoubtedly harvests more grain from this one species than is produced by all the relics of grain amaranth cultivation in the New World" (p. 132). A principal reason for this decline in the Americas is the historical observation that "in Christian eyes the Mexican grain amaranth was the very symbol of heathen idolatry and its cultivation declined drastically during the colonial period" (Sauer 1967, p. 115).

After addressing the historical evidence of domestication and ceremonial uses of the grain amaranths in the pre-colonial Americas, Sauer (1950) turned his attention to Asia, remarking that "the crop is scattered so widely through Asia and is so firmly entrenched among remote peoples that it give a powerful impression of great antiquity in the area" (p. 588). He further pointed out that a number of European botanists had made excursions to Asia in the eighteenth century and collected specimens, which were cataloged in European herbaria and also collected seeds for planting in European botanical gardens and glasshouses. Carolus Linnaeus (1707–1778) named the three grain species: *A. caudatus* L. in 1753, *A. hypochondriacus* L. also in 1753, and *A. cruentus* L. in 1759 (Sauer 1967). These species designations, with modern revisions, currently remain in effect.

The grain amaranths differ from all other *Amaranthus* species notably by their pale seeds, a Mendelian character due to recessive alleles of the same orthologous gene in all three grain species (Kulakow et al. 1985). As a consequence, interspecific grain-amaranth hybrids retain this unique pale-seeded character. By contrast, other species tend to have small glossy black seeds. Due to the high degree of morphological variation, species may be easily misclassified, especially when only seeds are examined. DNA analysis, however, provides definitive distinctions, in particular, given that complete nuclear and chloroplast genome sequences are now available (Clouse et al. 2016; Chaney et al. 2016; Transue et al. 1994).

The nineteenth century botanists Francis Buchanan-Hamilton (1762–1829), Alphonse Pyramus De Candolle (1806–1893), and Joseph Dalton Hooker (1817–1911), as well as early twentieth century botanists Oakes Ames (1874–1950), Nicolai Vavilov (1887–1943), and Elmer Drew Merrill (1876–1956), made extensive expeditions throughout the Indian subcontinent and the Himalayas where they observed and researched the widespread cultivation of grain amaranths. These renowned botanists concluded

—mistakenly as it turns out—that "the crop has certainly been cultivated in southern Asia from time immemorial and probably originated there" (Sauer 1950, p. 588).

The grain amaranths were cultivated in Africa as a potherb, an ornamental, and for medicinal purposes, and to a lesser extent as a grain crop, interestingly at one time for export of grain from Africa to India in the 1940s (Sauer 1950, 1967). Thellung (1914) scrupulously cataloged and characterized the European amaranths and arrived at the conclusion that all amaranth species in Europe were non-native, introduced no earlier than the sixteenth century.

3.2 The Asian Grain Amaranths

The Asian grain amaranths were cultivated across a wide triangular range extending as far northeast as Heilongjang Province, north of the Korean Peninsula, in China, from there southwestward to the southern parts of the Indian Subcontinent, thence northwest into what is now eastern Iran, and in an eastward swath from there across the Himalayas. European botanists, who explored the range of these cultivated crops, noted that there were two regions of intense Asian cultivation where the grain amaranths had become principal staple foods steeped in tradition. The northern of these two regions lies in modern-day Nepal, Bhutan, northern Bangladesh, and northeast India, on the foothills and plains and into the southern Himalayan slopes reaching elevations as high as 3,500 m. Highland cultivation was observed in environmentally extreme parts of the Himalayas on land described by Sauer (1950) as "tiny steep plots studded with glacial boulders" where "the boulders aid growth of the plants by storing the sun's heat" (p. 592). In the mountainous regions, the grain amaranths were cultivated as a dryland crop with relatively meager seed production, whereas in the plains, they were often irrigated, cultivated in larger fields, with correspondingly higher seed yield. In this northern region, landraces were what Sauer defined as "rich and variable in color," bearing "bewilderingly variable" common names, among them *chuyá*, *anardana*, *amardana*, *ramdana*, *báthu*, and *sag*, some of which applied not only to amaranths but also to cultivated chenopods (*Chenopodium spp.*) (Sauer 1950, 1967).

The other region of intense cultivation is in the southern part of India surrounding the cities of Coimbatore, Salem, and Madura. Nineteenth century botanists recorded local common names of the specimens they collected there, among them *kiery*, *keerai*, or*kira*, *rajgeerah* or *rajgira*, *ramdana*, and *landesi* (Sauer 1950; National Academy of Sciences 1984).

Beyond these two regions of major cultivation in Asia, grain amaranths were documented as minor field crops, potherbs, and ornamentals. Aitchison (1891) recorded that in the region of modern-day Afghanistan and western Iran, plants bearing the common name *taj-kharus* were "thinly scattered" and "sparsely spread," and that the "seed is eaten and cooked" (p. 10). Hu (1948, 2005) wrote of grain amaranths in western China, which she characterized as a traditional minor crop used as a potherb, an infusion for teas, and popped seeds mixed with sugary syrups and hardened as a confection (as also widely practiced in India and the Americas). She gave the common names as *shien*, *xian-cai*, *ren-xian*, and *jen-hsien*, and Sauer (1950) noted the common name *tien-shu-tze*.

3.3 Taxonomic Revision of the Amaranths

As aforementioned, early botanists considered the Asian cultivated grain amaranths to be native to the region, and as such classified them with a wide and contradictory range of unique Latin binomial names, as was also the case in the Americas. The resulting taxonomic confusion led Sauer (1950, 1967) to mostly discard previous classifications and construct an entirely new taxonomic revision of the genus based largely on floral morphology. This revision led him at first to conclude that there were four grain amaranth species (1950), which he later reduced to three (1967), using the original Linnaean names: *A. hypochondricus* L., *A. cruentus* L., and *A.*

caudatus L. Two species, *A. hypochondricus* and *A. cruentus*, were domesticated in Middle America, whereas *A. caudatus* was domesticated in Andean South America. With respect to the oldest Asian types, Sauer's (1950) conclusion is quite striking:

> Taxonomically, all of the grain amaranths cultivated in the Old World are indistinguishable from certain of those cultivated in the New World. Not only in terms of species, but also in terms of subspecific entities, the available Old World specimens represent nothing but a small sample of the diversity present in the American grain amaranths (p. 613).

He identified *A. hypochondriacus* and *A. caudatus* as the two predominant species cultivated in Asia, while noting that *A. cruentus* had been introduced more recently. Notably, he also recognized that "no evidence was found, either in the literature or in the herbarium, of any non-cultivated entity closely related to the grain amaranths which is peculiar to the Old World" (p. 613), implying that there is no evidence that the cultivated types could have been independently domesticated there. This initial conclusion, that all cultivated Asian grain types were, in fact, all derived from American introductions, has been amply confirmed through subsequent botanical and genetic analysis (see Singh (2017) and references therein).

Sauer (1950) also recognized powerful similarities in Asia and the Americas in cultivation methods, such as highland farming in locations unsuitable for most other crop plants, lowland production in relatively large monocultural fields, and mixed intercropping especially with maize. The preparation of the harvested seed was also remarkably similar in Asia and the Americas, where harvested seeds were often ground into flour for a number of different food preparations, or commonly popped and mixed with syrup or honey as a binder then allowed to congeal in variously shaped molds through partial desiccation. This latter confectionary preparation is known as *laddoos* in India (National Academy of Sciences 1984), *Xiam-zidian-xin* in China (Hu 2005), and *alegría* in Middle America (Aguilar and Alatorre 1978; Sauer 1967).

3.4 Pre-Columbian or Post-Columbian Introduction of the Grain Amaranths

The abundant and indisputable evidence of American origins for widespread indigenous grain amaranths in Asia raises the inevitable question of whether the cultivated amaranths were introduced to that continent in pre-Columbian times. Sauer (1950) was initially reluctant to voice an opinion, stating that "the question as to whether the crop reached Asia before the European expansion cannot be answered with certainty," although he did point out that "if the crop was introduced into Asia after Columbus, it must be credited with a remarkable achievement in making itself very much at home among strangers within a few generations" (pp. 614–615). With respect to the timing of their introduction to Europe, he noted that specimens of *A. hypochondriacus* "do not appear in the European herbals until about 1700, more than a century after *A. caudatus* had been brought to Europe as an ornamental" (p. 615).

In Sauer's second review (1967), he again addressed the perplexing question of pre-Columbian as opposed to post-Columbian introduction of the grain amaranths to Asia in the closing passages of his text:

> As for the intriguing problem of when and how the species were introduced to the Old World, the evidence remains inconclusive. There is no proof that any of them were in the Old World before Columbus. On the other hand, their spread in Asiatic and African cultivation certainly outran European contacts and could scarcely have been initiated by any direct transfer between Old and New World colonies of the same European power. Introduction through Europe itself seems most likely in the case of the Andean *A. caudatus*. Pale-seeded forms evidently were taken there directly from South America in the sixteenth century, long before they were found in Asia or Africa. However, dark seeded forms that are minor components of the Andean-crop somehow became established in remote parts of Asia and Africa before they were known in Europe (p. 131).

An ad hoc panel, convened by the National Research Council of the National Academy of Sciences, USA (1984) to address modern development of grain amaranth production and use, claimed that "although grain amaranth is a crop of the Americas, some time ago (probably since Columbus) *A. hypochondriacus* underwent a remarkable migration to Asia" (p. 10). Sauer was a member of that panel, which, in 1984, remained reluctant to attribute the well-entrenched cultivation of grain amaranths in India to possible pre-Columbian introduction.

Merrill (1954) in his documentation of botanical aspects of the voyages of Captain Cook raised the possibility that Portuguese traders introduced the grain amaranths from Brazil to India. His claim was speculative, based only on the indirect evidence of his observations of grain amaranths growing "in the mountains back of the old Portuguese-controlled areas on the Malabar coast, and [that] the Abyssinian localities are not too far inland from ports on the west African coast often visited by the early Portuguese sailing to the Orient by way of Brazil" (p. 302). From these observations, he surmised: "I may state that I have the impression that the Asiatic forms [of the grain amaranths] were certainly brought from Brazil to Goa by the Portuguese over that apparently forgotten trade route from 1500 to 1665; but I know nothing about *Amaranthus* in Brazil" (p. 302). He then made a plea for others to conduct additional research on the topic: "In further studies on Amaranthus the possible role played by Portuguese colonizers (post-1500) should receive more attention" (p. 302). Singh (1917), dismissed the possibility of Portuguese traders introducing the grain amaranths arriving in India, stating that "this visualization is not at all satisfying, because there is no evidence that amaranths were ever cultivated for grain in Brazil or any other Portuguese areas of the New World" (p. 322).

The most extensive argument favoring pre-Columbian introduction of the grain amaranths to Asia is a lengthy review by Sorenson and Johannessen (2004) consisting of a broad collection of evidence to support their proposal that a number of cultivated plant species domesticated in the Americas were likewise cultivated in Asia in pre-Columbian times. They addressed the amaranths as one of several "especially cogent cases" (p. 6). Their summary of evidence began with quotation of Sauer's (1950) comment that Bretschneider's (1896) *Botanicon Sinicum* includes "what seems to be a clear reference to a grain amaranth" (p. 589 in Sauer 1950; p. 7 in Sorenson and Johannessen 2004). Bretschneider's *Botanicon Sinicum* consists of a lengthy compilation of numerous references to various plants in ancient Chinese writings. The passages in question are from pages 404 to 407 of Part III Bretschneider's book, and refer to *hien*, a Chinese common name for amaranths, which Bretschneider (1896) determined was "*Amarantus*," naming as species *A. Blitum*, *A. tricolor* L., *A. polygamous* L., *A. spinosus*, and *A. melancholicus*. On page 405, Bretschneider cited texts from the third, seventh, tenth, and eleventh centuries with references to various types of *hien* and their vegetable and medicinal uses. Expanding on the statement by Sauer (1950), Sorenson and Johannessen (2004) pointed to the tenth century Chinese text in Bretschneider (1896), which cited six types of *hien* with medicinal properties. They then surmised that "the pre-Columbian presence of grain amaranths in Asia that was indicated by their wide distribution is thus confirmed by this tenth century mention" (Sorenson and Johannessen 2004, p. 7). Bretschneider's (1896) actual text, however, does not refer to seeds consumed as food, rather vegetable consumption of leaves and the use of seeds for medicinal purposes, a fact also noted by Sauer (1950). The seeds are characterized in the texts as "small and black," and "small, black, and shining" (Bretschneider 1896, p. 406), which is not typical of the grain amaranths. Merrill (1954) had raised justifiable skepticism that Bretschneider (1896) was referring to grain amaranths 50 years before Sorenson and Johannessen (2004) revisited the reference:

> I am not impressed by the young Sauer's reference to Amaranthus in China in 950 A.D. as "a clear reference to a grain amaranth." My view is based on the reason that the early Chinese record appears in a work on *materia medica*, and at that time there were several species of Amaranthus, other than the grain-producing ones, widely cultivated in China, some of which may have been used for medicinal purposes. As far as I know, the tender parts of all

> species of Amaranthus are edible as pot herbs, and their seeds are also edible; but they are not noted for their medicinal qualities. I suspect that the suggestion that the figure to which Sauer refers was clearly a grain amaranth does not appear in Bretschneider's work (p. 302).

It, therefore, is unlikely that the apparent *Amaranthus* species referred to in Bretschneider (1896) were, in fact, domesticated American grain amaranths. As mentioned by Sauer (1967), a number of wild *Amaranthus* species are native to Asia, and the ancient Chinese references may have been to some of these.

Sorenson and Johannessen (2004) then turned their attention to archeological evidence with the following claim:

> Archaeology now has confirmed a dramatically earlier date for the grain amaranths in Asia. Seeds of *A. caudatus*, along with *A. spinosus*, a thorny weed that also grew in pre-Columbian Mesoamerica ... have been excavated in India. They date to the 1000–800 BC period at the site of Narhan (Gorakhpur Dist., Uttar Pradesh)... (p. 7).

Sorenson and Johannssen (2004) based this claim on Sarasawat et al. (1994, pp. 282, 284, 331) as the source of evidence, unfortunately misinterpreting this source. Sarasawat et al. (1994) provided the following under the heading "Amaranthus L.":

> Five circular to lenticular seeds in carbonised state, with their edges narrowed into a thin rim, measure 1.00–1.50 mm across the utricles. Their surface is smooth and a notch on the margin is rather obscure. On a comparative basis, the seeds have been found comparable with those of *Amaranthus* species (p. 282).

Sarasawat et al. (1994) then named five *Amaranthus* species growing in the region in modern times. The first two listed are *A. caudatus* and *A. spinosus*, which is probably why Sorenson and Johannssen (2004) mistook the archeological seeds as belonging to these two species. By contrast, Sarasawat et al. (1994) tentatively attributed the archaeological seeds to *Amaranthus viridis* L., a wild and weedy species: "*A. viridis* is most common all over the region, particularly in the cultivated fields. The ancient seeds recovered in association of the crop-remains, may be of this species" (p. 282). The origin of *A. viridis* is uncertain, although it is considered to be "possibly of Asian origin" (Jansen 2004). Sorenson and Johannssen (2004) also cite pages 284 and 331 of Sarasawat et al. (1994). There is no mention of amaranth on page 284, and a note on page 331 briefly addresses modern amaranths cultivated in the region with no reference to ancient amaranths.

The remaining argument in support of pre-Columbian introduction of grain amaranths to Asia by Sorenson and Johannessen (2004) is based on the parallel evidence of pre-Columbian introductions of maize (*Zea mays*, L.), which indirectly implies that, if maize had been introduced, grain amaranths may have concurrently been introduced with maize given that they were cultivated together in the Americas.

Sauer (1995), after reviewing the available evidence, summed up what he perceived as the most likely scenario for the introduction and spread of grain amaranths in the Old World:

> During the early colonial period, while amaranth production was declining in the New World with disruption of native cultures, the three domesticates began spreading through the Old World. All three were probably introduced to Europe as curiosities during the sixteenth century. Pale-seeded forms of both *A. hypochondriacus* and *A. caudatus* were recorded there about 1600; the Dutch probably introduced both to Ceylon early in the eighteenth century. By the early nineteenth century, *A. hypochondriacus*, with *A. caudatus* as an occasional, minor associate, had been adopted as a staple grain in the Nilgiri hills of south India and in the Himalayas; they have since spread over increasingly wide regions of China to Manchuria and eastern Siberia (p. 10).

More recently, Singh (2017) has reviewed the available literature and determined that "whether the crop reached Asia before the European expansion has not been answered with certainty" (p. 322). DNA analysis is also unlikely to provide a definitive answer. The few studies comparing Asian and American accessions with DNA markers' result, as expected, with Asian accessions clustering with their American species counterparts with no distinct geographical cladistic separation (Park et al. 2014), confirming previous conclusions based on morphological classification.

3.5 Modern Dissemination and Use of Amaranths in the Old World

The worldwide popularity of the grain amaranths has exploded in the past 40 years, in part as a concerted scientific effort to promote their nutritional value and adaptability for cultivation in a wide range of extreme climates, especially relevant in an era of climate change (National Academy of Sciences 1984; Jacobsen et al. 2006; Weerasekara and Waisundara 2020; Adhikary et al. 2020). Grain amaranths are now available worldwide on grocery shelves as whole and popped seeds, and in a large number of processed food products. Moreover, their appeal as an ornamental, as so poetically articulated by Milton in the seventeenth century, continues to expand as plant breeders develop an astonishingly diverse number of ornamental varieties for an increasing market demand.

3.6 Conclusions

Two of the three grain amaranth species, *A. hypochondriacus* and *A. caudatus*, have been under intensive cultivation for grain in two regions of the Indian subcontinent, probably since the sixteenth century where their production during the twentieth century exceeded that of the Americas, where the crop was domesticated. The observation by European botanists of the degree to which grain amaranths were a staple food and an integral part of the culture in the Indian subcontinent led them to conclude that the crop had been domesticated there. However, modern taxonomical revision has undisputedly revealed that the Asian grain amaranths are descended from a subset of the American types, having been introduced to Asia. The timing of the first introduction probably dates to the sixteenth or seventeenth century, although some authors have speculated, on the basis of limited evidence, that they may have been introduced in pre-Columbian times. Due to concerted efforts at promotion and dissemination, popularity of the grain amaranths has dramatically increased in the late twentieth and early twenty-first centuries.

References

Adhikary D, Khatri-Chetri U, Slaski J (2020) Amaranth: an ancient and high-quality wholesome crop. In: Waisundara VY (ed) Nutritional value of amaranth. IntertechOpen. https://doi.org/10.5772/intertechopen.82910

AguilarJ AG, Alatorre G (1978) Monografia de la planta de la alegría. Memoria 1978. Grupo De Estudios Ambient 1:156–203

Aitchison JET (1891) Notes to assist in a further knowledge of the products of western Afghanistan and of northeastern Persia. Trans & Proc Bot Soc Edinburgh 18:1–228

Bretschneider E (1896) Botanicon sinicum: notes on Chinese botany from native and western sources. part III: botanical investigations into the materia medica of the ancient Chinese. Kelly & Walsh, Shanghai

Chaney L Mangelson R, Ramaraj T, Jellen EN, Maughan PJ (2016) The complete chloroplast genome sequences for four *Amaranthus* species (Amaranthaceae). Appl Plant Sci. 4(9):1600063. https://doi.org/10.3732/apps.1600063

Clouse JW, Adhikary D, Page JT, Ramaraj T, Deyholos MK, Udall JA, Fairbanks DJ, Jellen EN, Maughan PJ (2016) The amaranth genome: genome, transcriptome, and physical map assembly. Plant Genome 9(1). https://doi.org/10.3835/plantgenome2015.07.0062

Hu S-Y (1948) Shien—some noteworthy edible herbs of China. The Herbalist 14:30–36

Hu S-Y (2005) Food plants of China. The Chinese University Press, Hong Kong

Jacobsen SE, Mujica A, Ortiz R (2006) The global potential for quinoa and other Andean crops. Food Rev Int 19:139–148. https://doi.org/10.1081/FRI-120018880

Jansen PCM (2004) *Amaranthus viridis* L. record from PROTA4U. In: Grubben GJH, Denton, OA (eds) Plant resources of tropical Africa, Wageningen, Netherlands. https://uses.plantnet-project.org/en/Amaranthus_viridis_(PROTA). Accessed 30 October 2020

Kulakow PA, Hauptli H, Jain SK (1985) Genetics of grain amaranths. I. Mendelian analysis of six color characters. J Hered 76:27–30. https://doi.org/10.1093/oxfordjournals.jhered.a110012

Merrill ED (1954) The botany of Cook's voyages. Chron Bot 14(5/6):161–384

Milton J (1674) Paradise lost: a poem in twelve books, 2nd edn revised and augmented by the same author. S. Simmons, London

National Academy of Sciences (1984) Amaranth: modern prospects for an ancient crop. National Academy Press, Washington DC. https://doi.org/10.17226/19381

Park Y-J, Nemoto K, Tomooka N (2014) Molecular characterization and genetic diversity of the enzyme (SBE) gene from *Amaranthus*: the evolutionary origin of grain amaranths. Mol Breed 34:1975–1985. https://doi.org/10.1007/s11032-014-0156-6

Saraswat KS, Sharma NK, Saini DC (1994) Plant economy at ancient Narhan (ca. 1,300 BC – 300/400 AD). In: Singh P (ed) Excavations at Narhan (1984–1989). Banaras Hindu University, Varanasi

Sorenson JL, Johannessen CL (2004) Scientific evidence for pre-Columbian transoceanic voyages. Sino-Platonic Papers, No. 133. University of Pennsylvania, Philadelphia

Sauer JD (1950) The grain amaranths: a survey of their history and classification. Ann Mo Bot Gard 37:561–632. https://doi.org/10.2307/2394403

Sauer JD (1967) The grain amaranths and their relatives: a revised taxonomic and geographic survey. Ann Mo Bot Gard 54:103–137. https://doi.org/10.2307/2394998

Sauer JD (1976) Grain amaranths. In: Simmonds NW (ed) Evolution of crop plants. Longman, London

Sauer JD (1995) Grain amaranths. In: Smartt J, Simmonds NW (eds) Evolution of crop plants, 2nd ed. Longman, London

Singh AK (2017) Early history of crop introductions into India: II. *Amaranthus* (L.) spp. Asian Agrihist 21:319–324

Thellung A (1914) Amarantus. In: Ascherson P, Graebner KORPP (eds) Synopsis der mitteleuropäischen flora. 5. band, 1, abtheilung. Gebrüder Bornträger Leipzig

Transue DK Fairbanks DJ, Robison LR, Andersen WR (1994) Species identification by RAPD analysis of grain amaranth genetic resources. Crop Sci 34:1385–1389. https://doi.org/10.2135/cropsci1994.0011183X003400050044x

Weerasekara AC, Waisundara VY (2020) Amaranth as a pseudocereal in modern times: nutrients, taxonomy, morphology and cultivation. In: Waisundara VY (ed) Nutritional value of amaranth. IntertechOpen. https://doi.org/10.5772/intertechopen.82910

Genetics of Betalain Pigments in Amaranth Species

4

Erica E. Packard,
Isadora Louise Alves da Costa Ribeiro Quintans,
and Dinesh Adhikary

Abstract

Amaranth is an ancient pseudo-cereal that has been gaining research attention in recent years due to its tolerance to harsh climates, a high seed protein content and its secondary metabolites, such as betalain pigments. Betalains have been reported to have a broad spectrum of applications, ranging from human nutrition, pharmaceutical potential, crop enhancement, and antimicrobial properties. Several members of the Amaranthaceae family produce red (betacyanin) and/or yellow (betaxanthin) betalain pigments, which are chemically distinct from anthocyanins, the pigment responsible for red and blue coloration in most other plants. Although the specific roles of betalains in plant physiology are not fully understood, they are known to be involved in attracting pollinators and further hypothesized to be involved in responses against biotic and abiotic stresses. With a recent focus on sequencing and genomics in amaranth, reverse genetic tools are being developed and applied for functional gene annotation and characterization of betalain biosynthesis pathway genes via transient transformation methods. To better understand betalain genetics, this chapter presents the current status of research in *Amaranthus* species concentrating on the types of betalains found, their biosynthesis pathway, the evolution of pathway genes, and their ecological and physiological significance in the plant. Together, this information will create a spotlight for future research on betalain biosynthesis and will lead to the development of improved strategies to fully uncover the biological role of these pigments.

4.1 Introduction

The ancient grain amaranth crop is gaining renewed research interest not only as a nutrient-rich "super-food" that can grow in a variety of climates but also as a source of bioactive compounds such as betalains (Adhikary et al. 2020a; Joshi et al. 2018). Betalains are derivatives of the chromophore betalamic acid [4-(2-oxoethylidene)-1,2,3,4- tetrahydropyridine-2, 6-dicarboxylic acid] (Fig. 4.1c) (Khan and Giridhar 2015; SlimenIB and Abderrabba 2017; Miguel 2018).There are two main betalain groups: betaxanthins, which are yellow pigments having

E. E. Packard
Department of Biology, The University of British Columbia, Kelowna, BC, Canada

I. L. Alves da Costa Ribeiro Quintans
Universidade Federal Rural do Semi-árido, Mossoró, Brazil

D. Adhikary (✉)
Department of Agricultural, Food & Nutritional Sciences, University of Alberta, Edmonton, AB, Canada

D. Adhikary et al. (eds.), *The Amaranth Genome*, Compendium of Plant Genomes,
https://doi.org/10.1007/978-3-030-72365-1_4

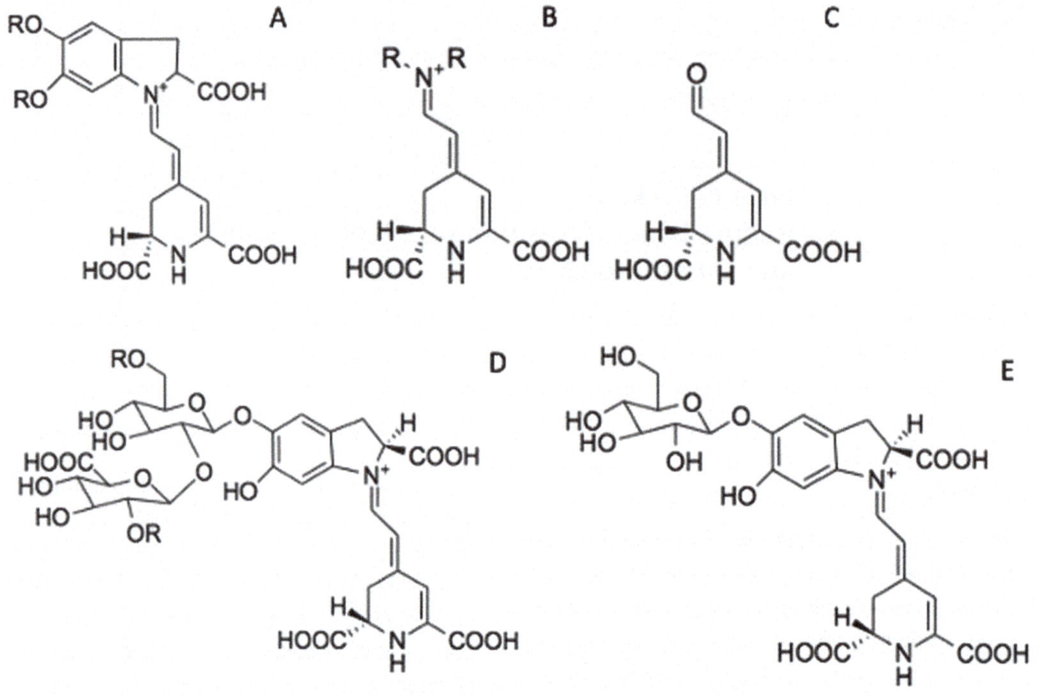

Fig. 4.1 Chemical structures of **a** betacyanin, **b** betaxanthins, **c** betalamic acid, **d** amaranthin and **e** betanin. Drawn with ChemDraw v19.0 (https://chemdrawdirect.perkinelmer.cloud/js/sample/index.html)

maximal absorbance at 480 nm, and betacyanins, which are responsible for red–violet pigmentation that have maximal absorbance at 536 nm (Figs. 4.1a, b and 4.2) (Gandía-Herrero and García-Carmona 2013; SlimenIB and Abderrabba 2017). Condensation of betalamic acid with an amino group or their derivatives results in betaxanthins, while the condensation with *cyclo-*

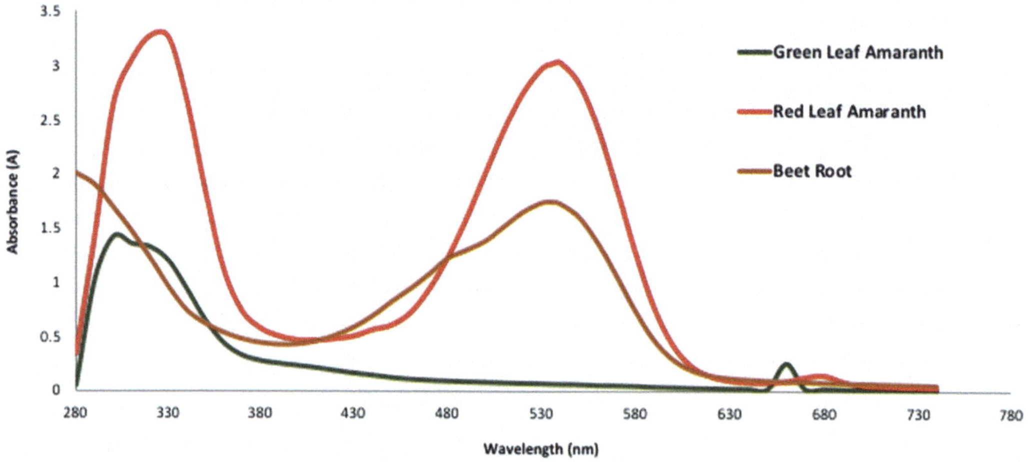

Fig. 4.2 Comparison of absorption spectra produced from betalain extracts produced from the leaves of the betalainic species Beet (*Beta vulgaris*) and "green" and "red" vegetable amaranths using UV–Vis spectrophotometry. The UV–Vis absorbance spectra of the leaf extracts ranged between 280 and 730 nm

DOPA [cyclo-L-(3,4-dihydroxyphenylalanine)] or its glycosyl derivatives results in betacyanins (SlimenIB and Abderrabba 2017).To date, 75 different betalains have been reported from 17 of the 34 families in the Caryophyllales order, some species of Basidiomycetes, such as *Amanita muscaria* and *Hygrocybeconica*, and one species of bacteria, *Gluconacetobacterdiazotrophicus* (Gill 1996; Khan and Giridhar 2015; Miguel 2018; Contreras-Llano et al. 2019). Interestingly, no plant species outside of the Caryophyllales has been found to produce betalains, a property that makes it very scarce in nature (Strack et al. 2003). Betalains are found to accumulate in the vacuoles of reproductive and vegetative tissues of plants, creating a wide spectrum of vibrant colors in organs including roots, stems, leaves, and flowers (Wink 1997; Strack et al. 2003).

The ability of betalains to scavenge radical oxygen species (ROS) makes them very important for assisting in many different types of environmental stresses (Cai et al. 2003; SlimenIB and Abderrabba 2017; Sarker and Oba 2018a, 2019a, 2019b). In-depth knowledge of the ecological and biological role of betalains is important for developing crops resilient to changing environmental conditions (Joshi et al. 2018; Adhikary et al. 2019). In addition, their bioactivity makes them valuable to the human health and food industry, as there is substantial research validating the antioxidant and free radical scavenging abilities of betalains both in vitro and in vivo (Kanner et al. 2001; Cai et al. 2003; Gliszczyńska-Świgło et al. 2006; Sarker and Oba 2019a). Betalains are considered a class of dietary antioxidants (Kanner et al. 2001) and although currently they are primarily sourced from beetroot (*Beta vulgaris* subsp. *vulgaris*), *Amaranthus* species may be considered a source for commercial use (Cai et al. 2003, 2005; Gengatharan et al. 2015). Cai et al. (2003) suggested that some forms of betalains from the Amaranthaceae family are even more potent than other common antioxidants such as ascorbic acid, rutin, and catechin. Furthermore, there is evidence to suggest that betalains could be beneficial in preventing different types of cancer (Esatbeyoglu et al. 2014; Gandía-Herrero et al. 2016 and references therein). Sreekanth et al. (2007) showed that betalain induces dose- and time-dependent cell death of the human chronic myeloid leukemia cell line (K562). Moreover, betalain extract from *A. spinosus* has shown significant antimalarial activity in mice (Hilou et al. 2006). Pigment extract from beetroot pomace also inhibited the growth of Gram-negative bacteria (Čanadanović-Brunet et al. 2011) and inhibited growth in *Salmonella typhymurium* and *Bacilulus cereus* (Velićanski et al. 2011).

As plant pigments, betalains are often presumed to have similar physiological and ecological functions as anthocyanins. Betalains and anthocyanins have similar optical properties (Jain and Gould 2015) and are induced by similar environmental factors, including both biotic and abiotic stresses (Hughes and Smith 2007). Both of these biological pigments can be found in the root and shoot systems of many land plants; however, they have never been found to naturally co-occur within an organism (Jain and Gould 2015). The puzzle of the evolution of betalains and their divergence from anthocyanins has long been an unanswered question to researchers, but recent studies are providing insights that could help resolve this long-standing enigma (Brockington et al. 2011; Timoneda et al. 2019; Sheehan et al. 2020). Similar to the role of anthocyanins, flowers colored by betalain pigments are attractive to pollinators and animal dispersers (Stintzing and Carle 2004). Both pigments have a clear role in protecting plants against environmental stresses (Jain and Gould 2015), but it is still not fully understood what kind of advantages betalanic plants may have over anthocyanic plants. There is yet to be a unified answer on the evolutionary relationship of betalains and anthocyanins, which incorporates the origin of betalains, the homoplastic distributions of the two pigments within the core Caryophyllales, and the apparent mutual exclusivity (Brockington et al. 2011). The betalain biosynthesis pathway has been under extensive research and multiple reviews have investigated what is currently known (Gandía-Herrero and García-Carmona 2013; Khan and Giridhar 2015;

Miguel 2018; Adhikary 2019; Timoneda et al. 2019 and references therein). Yet, many details regarding the evolution of the involved proteins and enzymes are still to be elucidated. Overall, *Amaranthus* species present a viable option for crop improvement and commercial production of betalains. Therefore, this chapter will discuss the specific gene isoforms present in *Amaranthus* species, their evolution, and divergence from anthocyanins and the latest developments in the understanding of their biosynthesis pathway.

4.2 Betalains in *Amaranthus* Species

4.2.1 Type of Betalains in Amaranthus Species

In *Amaranthus* species, there are two predominant betacyanins: amaranthin and isoamaranthin (Cai et al. 2001) (Fig. 4.1d). These two pigments were first isolated from the leaves of *A. tricolor* (Piattelli et al. 1964). Amaranthin is also the name of one of the four structural types of betacyanins (SlimenIB and Abderrabba 2017). Amaranthin has two glucosyl groups bonded at the C5 carbon of the cyclo-DOPA moiety (SlimenIB and Abderrabba 2017). The remaining three structural types are betanin, gomphrenin and bougainvillein (Strack et al. 2003; SlimenIB and Abderrabba 2017). Gomphrenin and betanin are structural isomers; gomphrenin has a hydroxyl group attached to the C5 carbon and a glucosyl attached to the C6 carbon, while in betanin, the two groups are on opposite carbons (Strack et al. 2003; SlimenIB and Abderrabba 2017). Aclation and glycosylation of betacyanins result in the immense diversity of natural betacyanins found. Yet, these final decoration steps and the genes and enzymes that control them are not known (Timoneda et al. 2019; Polturak and Aharoni 2019). Within betalain isoforms, there are variations in the bioactivity caused by protonation and deprotonation at varying pH levels (Gliszczyńska-Świgło et al. 2006). The concentration of betalains fluctuates greatly among species, genotypes, plants of different ages, and even between apical and basal leaves of a single plant (Cai et al. 1998, 2001; Khandaker et al. 2009; Khanam and Oba 2013; Sarker and Oba 2019a). In addition, some studies found that the cultivated species have a greater content of betalains than wild types (Cai et al. 1998, 2001). Within *Amaranthus* species, betacyanic pigmentation can be present in the leaves (e.g. *A. tricolor, A. cruentus L.*) (Fig. 4.3), inflorescences (e.g. *A. hybridus*, *A. hypochondriacus*, *A. caudatus*), stems (e.g. *A. palmeri*), and/or seedlings (e.g. *A. lividis*, *A. powellii*, *A. spinosus*) (Cai et al. 1998, 2001). Betanin is also marginally present in many *Amaranthus* species, as was found by Cai et al. (2001), which utilized high-performance liquid chromatography (HPLC), spectral analyses, and mass spectrometry (MS) to isolate and identify betacyanins from a large group of species in the Amaranthaceae family (Fig. 4.1e). Cai et al. (2001) also found isobetanin in *A. cruentus*, *A. hypochondriacus*, *A. tricolor,* and *A. powellii*, but never in amounts greater than 1% of the total betacyanins isolated. Betalain pigments have been extracted and identified from plants since the early 1900s, but with advances in technology, more pigment forms are consistently being identified (Khan and Giridhar 2015). Molecular techniques could also be used to simplify purification and isolation by generating cell cultures able to secrete the desired pigments or accumulate them in cellular vacuoles but, so far, the mechanisms responsible for transporting betalains within and to the extracellular space of the cell are unknown (Polturak and Aharoni 2019). It is important to consider the variation in betalain content and type across species and cultivars as this aspect affects the potential bioactivities of the plant and the ability for those pigments to be utilized.

In *Amaranthus* species, the presence of betacyanins and their functional properties have been reviewed more extensively than those of betaxanthins (Jain and Gould 2015). The most commonly studied betaxanthins are indicaxanthin and vulgaxanthin I, which have not been identified in *Amaranthus* species (SlimenIB and Abderrabba 2017). However, the presence and content of other betaxanthins appear to vary

Fig. 4.3 Leaf extracts: *Amaranthustricolor* produces a spectrum of betalain pigments at different absorption spectra. Red pigments are betacyanins and yellow pigments are betaxanthins

greatly across species. Non-naturally occurring methylated arginine-betaxanthin was identified in *A. tricolor* leaves using reversed phase HPLC and liquid chromatography–mass spectrometry (LC–MS) (Biswas et al. 2013). Despite the presence of betaxanthins, the content of amaranthin and betalamic acid was both found to be over three times greater than methylated arginine-betaxanthin (Biswas et al. 2013). Another study investigated the bioactivity of four cultivars of both *A. tricolor* and *A. hypochondriacus* and found that total betalain and betacyanin content was generally higher in *A. tricolor* than in *A. hypochondriacus* (Khanam and Oba 2013). Surprisingly, betaxanthins were found in greater amounts than betacyanins in nearly all cultivars analyzed except for two, which were both *A. tricolor* (Khanam and Oba 2013). Overall, the known diversity of betaxanthins is limited in comparison to betacyanins and unlike betacyanins, there are no known glycosylated betaxanthins. A study generated novel betaxanthins by feeding different amines to transgenic strains of *Saccharomyces cerevisiae*, which created a new spectrum of pigments ranging from orange to blue–violet (Grewal et al. 2018). Furthermore, Polturak and Aharoni (2019) suggested in their recent review that the engineering of heterologous plant hosts could result in a novel production of betalains, such as glycosylated betaxanthins, through catalysis with endogenous enzymes. A new strategy to increase the relative amounts of betaxanthins over betacyanins could be via the controlled expression of either the α-

and β-clades of cytochrome P450 enzymes (Polturak and Aharoni 2019). Thus, fine-tuning the expression of the CYP76AD1-like subfamily of cytochrome P450 enzymes in the pathway could allow for precise control over the yellow to red spectrum, for use in the design of ornamentals or related biotechnological objectives (Polturak and Aharoni 2019). Betaxanthins are known to be less stable than betacyanins (SlimenIB and Abderrabba 2017); however, they show promising bioactivity (Gandía-Herrero et al. 2016). Therefore, further investigation into betaxanthins in *Amaranthus* species would be worthwhile.

There are many avenues to control and modify betalain pigment accumulation in *Amaranthus* species and other organisms through genetic engineering. Increasing the yield of desired betalain isoforms or betalain intermediates could be done by modifying the regulation of either the shikimate pathway or other downstream steps (Polturak and Aharoni 2019). The production of tyrosine and subsequent betalain pigments was increased sevenfold in heterologous *Nicotiana benthamiana* by deregulating the key-enzyme arogenate dehydrogenase, which is involved in a negative feedback loop with tyrosine (Timoneda et al. 2019; Polturak and Aharoni 2019). Silencing of *CYP76AD1* resulted in the loss of red pigmentation and in the accumulation of L-DOPA precursor in *A. tricolor* leaves (Adhikary et al. 2019). Similarly, the heterologous expression of tyrosinase gene and a DOPA 4,5-dioxygenase gene in non-betalainic plants, tobacco and *Arabidopsis*, resulted in the accumulation of betaxanthin in the cultured cells (Nakatsuka et al. 2013). Modifying the transcriptional regulation of the betalain biosynthesis pathway could be an avenue to increase pigment yields but, so far, only one MYB-type transcription factor, i.e. *Bv*MYB1 from beetroot (*B. vulgaris*), is known, and no research has been done specifically in amaranth plants (Hatlestad et al. 2012, 2015; Polturak and Aharoni 2018). Techniques designed to alter the yield of betalains are yet to be utilized to their full potential in *Amaranthus* species.

4.2.2 Betalains in Other Plant Species

Other genera in the Amaranthaceae family contain a diverse range of betalain isoforms (Table 4.1). Betacyanins were first discovered in beetroot, and they are one of the primary sources of betanin as they produce this isoform in large amounts in their taproot and surrounding peel tissue (Kanner et al. 2001; Strack et al. 2003; SlimenIB and Abderrabba 2017; Miguel 2018). Betalains from beetroot have been studied extensively and have been found, not only to have in vivo/in vitro antioxidant capacity but also anti-microbial properties (Kanner et al. 2001; Gliszczyńska-Świgło et al. 2006; Gengatharan et al. 2015; Čanadanović-Brunet et al. 2011; Velićanski et al. 2011). The species *Gomphrena globosa* contains betalains in its purple and red inflorescences with histidine-betaxanthin (histidine-bx) and gomphrenin III being the predominant betaxanthin and betacyanin pigments, respectively (Cai et al. 2001; Kugler et al. 2007; Miguel 2018). Gomphernin-type betacyanins are rare and have only been found in *G. globosa*, *Iresineherbstii,* and *Bougainvillea glabra* (from the Nyctaginaceae family). As such, the genes that control their production are still unknown (Heuer et al. 1994; Strack et al. 2003; Polturak and Aharoni 2019). *Iresineherbstii* Hook. f. ex Lindl. is the main source of the iresine and isoiresinebetacyanins (Cai et al. 2001, 2005). Recently, a study by Spórna-Kucab et al. (2020) identified 22 different betacyanins in *I. herbstii* but still, no research is documented on betaxanthins in this species. The genus *Althernathera* is known to contain betalains; however, most of the reported bioactivity (e.g. anti-microbial, anti-inflammatory, anti-hyperalgesic, and anti-parasitic) and, consequently, the medical applications of *Alternanthera* are associated with other secondary metabolites such as phenols, terpenes, and sterols (Deladino et al. 2017; Miguel 2018). The extracts of *Celosia* are found to have diverse health properties (Miguel 2018 and references therein), but similar to the genus *Althernathera*, most properties are related to the presence of other interacting antioxidants and secondary metabolites. Quinoa (*Chenopodium quinoa*) was already considered a nutritional grain

Table 4.1 Betalains in some genera of the Amaranthaceae

Species	Organ	Types of betacyanins	Types of betaxanthins	Reference
Beta vulgaris subsp. *vulgaris*	rt, lv	Prebetanin Betanin Isoprebetanin Betanidin Isobetanin 17-decarboxy-neobetanin 17-decarboxy-betanidin/isobetanidin Neobetanin 15-decarboxy-betanidin/isobetanidin 17-decarboxy-betanin 17-decarboxy-isobetanin 2-decarboxy-neobetanin 2,17-bidecarboxy-betanin/isobetanin2,17-bidecarboxy-neobetanin 6′-O- feruloylbetanin 6′-O-feruloyl-isobetanin 2-decarboxy-isobetanin 2-decarboxy-betanin	Glutamine-bx (vulgaxanthin I) valine-bx Leucine-bx (vulgaxanthinIV) asparagine-bx (vulgaxanthinIII) phenylalanine-isobxtryptophan-bx Tyramine-bx (miraxanthin III) 3-methoxytyramine-bx phenylalanine-bx Threonine-bx aspartic acid-bx - (miraxanthin II) γ-aminobutyric acid-bx	Miguel 2018
Gomphrena globosa	in	Amaranthin Isoamaranthin Betanin Isobetanin Gomphrenin I/II/III Isogomphrenin I/II/III Betanidin Isobetanidin Celosianin II Isocelosianin II	Histidine-bx (muscaarin VII) arginine-bx Glutamine-bx (vulgaxanthin I) Lysine-bx Isoleucine-bx Tryptophan-bx	Kugler et al. 2007; Miguel 2018
Iresineherbstii Hook. f. ex Lindl	lv	Amaranthin Isoamaranthin Betanin Isobetanin Iresine I** Isoiresine Hylocerenin Isohylocerenin Celosianin II** Isocelosianin lindenin Isolindenin 2′''-O-E-feruloyl-iresinin 2′''-O-E-sinapoyl-iresinin 2′''-O-E-feruloyl-isoiresinin 2′''-O-E-sinapoyl-isoiresinin Sinapoyl-gomphrenin I** Sinapoyl-isogomphrenin Gomphrenin II/III** Isogomphrenin II/III	None identified	Cai et al. 2001, 2005; Spórna-Kucab et al. 2020
*Althernathera*spp.	lv, st	Amaranthin Isoamaranthin Betanin Isobetanin	Dopamine-betaxanthine* 3-methoxytyramine-betaxanthin*	Deladino et al. 2017; Miguel 2018

(continued)

Table 4.1 (continued)

Species	Organ	Types of betacyanins	Types of betaxanthins	Reference
Celosia spp.	in	Amaranthin Isoamaranthin Betalamic acid Betanin Celosianin I/II Isocelosianin I/II	Dopamine-bx 3-methoxytyramine-bx S-tryptophan-bx	Schliemann et al. 2001; Cai et al. 2005
*Chenopodium quinoa*Willd	sd	Betanin Isobetanin Amaranthin	Dopaxanthin Dopamine-bx (miraxanthin V) proline-bx (indicaxanthin)	Escribano et al. 2017; Miguel 2018

*Tentative identifications
**Names have been revised in recent papers
Abbreviations: rt, root; in, inflorescences; lv, leaves; st, stems; sd, seeds

with wide usage before betalains were identified in the colored seeds, in 2015 (Tang et al. 2015; Miguel 2018). More unique betalains have been identified in other plant families in the Caryophyllales but as of 2015, only 20 of 472 total genera have known pigment profiles (reviewed in Khan and Giridhar 2015). Other notable plant families include: Cactaceae, which contains *Opuntia* species known for betaxanthin accumulation in their edible yellow and red fruits; Nyctaginaceae, which includes *Bougainvillea,* one of the three gomphrenin-type containing plants, and Portulacaceae, which contains *Portulaca grandiflora,* a flowering succulent in which the enzyme 4,5-DOPA-extradiol-dioxygenase, a key component of the betalain pathway, was first identified (Gandía-Herrero et al. 2016).

4.3 Divergence of Betalains from Anthocyanins

4.3.1 Betalains Versus Anthocyanins

Betalains were originally mistaken for flavonoids and "nitrogenous anthocyanins" (Jain and Gould 2015; SlimenIB and Abderrabba 2017), but it is now understood that anthocyanins are derived from the amino acid phenylalanine, while betalains are derived from tyrosine (Li et al. 2019; Timoneda et al. 2019). Anthocyanin is the dominant form of pigmentation in flowering plants (occurring in all members of the Angiospermae), whereas only 6 out of the 25 core Caryophyllales families produce anthocyanins: Molluginaceae, Caryophyllaceae, Kewaceae, Macarthuriaceae, Limeaceae, and Simmondsiaceae (Clement and Mabry 1996; Brockington et al. 2011; Thulin et al. 2016; Timoneda et al. 2019). Betalains and anthocyanins are considered mutually exclusive, since there is no evidence of plants producing these pigments simultaneously in nature (Miguel 2018). The mechanisms for this phenomenon are still unclear, but evidence suggests that this phenomenon is not due to inherent biochemical incompatibility, as natural anthocyanic species can be made to co-produce betalain by combining plant genetic engineering and L-DOPA substrate feeding (Harris et al. 2012).

Both betalain and anthocyanin play similar roles in plant physiology and are found to have benefits for plants and humans (Stintzing and Carle 2004; Li et al. 2019). A review by Jain and Gould (2015) confirmed that both betalains and anthocyanins are largely capable of filling the same protective role in plants under abiotic stresses (UV-radiation, drought, oxidative, and salinity). It is less understood whether betalains have complementary roles to anthocyanins for protection against heavy metal stress and deterrence of herbivory (Jain and Gould 2015). Anthocyanins are known to be excellent antioxidants yet, in some cases, betalains are found to be even more effective than anthocyanins (Gliszczyńska-Świgło et al. 2006; Jain and

Gould 2015). A study by Gliszczyńska-Świgło et al. (2006) found that betanin was between 1.5 and 2.0 times more effective than anthocyaninsat scavenging free radicals in solutions with a pH higher than four. A recent commentary by Gandía-Herrero and García-Carmona (2020) suggested that the differences in redox potential when acting as antioxidants in cell processes may explain the prevalence of betalains over anthocyanins in the core Caryophyllales. The vibrant colors of betalains are more stable than anthocyanins at a greater range of pH (Strack et al. 2003; Celli and Brooks 2017), which could be beneficial for betalainic plants that switch to Crassulacean acid metabolism (CAM) under drought conditions and, therefore, would be able to maintain photoprotection when vacuolar pH fluctuates (Jain and Gould 2015). The broader stability of betalains is also important for human use in the food industry, as it strengthens their potential for use as a colorant in foods and natural supplements (Stintzing and Carle 2004; Celli and Brooks 2017). More research has been done on anthocyanins compared with betalains (Polturak and Aharoni 2018); therefore, further studies focusing on genetics and evolutionary divergence are still required to clarify the differences in functional roles of these two pigments.

4.3.2 Evolutionary Relationship of Betalains and Anthocyanins

The evolutionary relationship of betalains and anthocyanins in Caryophyllales is very intriguing and remains uncertain. It is still an open question whether the ancestor of the order Caryophyllales bore both pigments or if only one type predominated. It has been proposed that some enzymes in betalain biosynthesis are homologous to genes in the anthocyanin biosynthetic pathway (Christinet et al. 2004). As noted by Brockington et al. (2011), there are three key aspects of the betalain and anthocyanin evolutionary relationship in Caryophyllales that need to be addressed by evolutionary theories: (1) the singular origin of betalain pigmentation, (2) the presence of both pigments among the plant lineages belonging to the core Caryophyllales, although (3) these pigments have never been found simultaneously in a plant in natural conditions. For instance, Brockington et al. (2011) extnsively reviewed the hypotheses that support each of the three unique aspects that define the pigment biosynthesis evolution in the Caryophyllales. The first, related to the singular origin of the derived betalain character, is usually justified by the selection of these pigments to suit pollinators (Brockington et al. 2011). However, the arguments supporting this idea are scant. The second, related to the presence of both pigments in the core Caryophyllales (Brockington et al. 2011), was explained by Clement and Mabry (1996) as a result of the coevolution of both betalain and anthocyanin pigments in Caryophyllales ancestors, with subsequent loss of one of the pigments (Brockington et al. 2011). The main issue with this hypothesis is that anthocyanin and betalain pigments have never been found to co-occur in nature, even though these two pigments have complementary roles (Brockington et al. 2011). Furthermore, the distribution of the anthocyanin families across the Caryophyllales is scattered, indicating the homoplastic character of these two pigments (Timoneda et al. 2019). There is still not a strong hypothesis to support the third aspect, the mutually exclusive nature of the two pigments (Brockington et al. 2011). Clement and Mabry (1996) attempted to justify the pigment's exclusiveness based on metabolic cost, but their study was not conclusive. Betaxanthin production was induced in transgenic tobacco and *Arabidopsis* (both anthocyanin-producing plants) using heterologous expression, which was the first report of the accumulation of betalain in species outside of the core Caryophyllales (Nakatsuka et al. 2013). This study demonstrated that betalain and anthocyanin pigments are not chemically incompatible, which added increased controversy to the prevailing mutual exclusion in plants.

Brockington et al. (2015) claimed that the origin of the betalain-specific isoforms (i.e., DODA -α and CYP 76AD 1-α) in the early evolutionary history of the Caryophyllales came from gene duplication and neo-functionalization that occurred before the evolution of the betalain biosynthesis pathway; therefore, creating alpha and beta lineages of the genes (Timoneda et al. 2019). The key enzymatic step, catalyzed by L-DOPA 4,5- dioxygenase activity, involving the conversion of L-DOPA to betalamic acid, was determined to have originated from one of the duplicated genes, DODAα (Brockington et al. 2015; Timoneda et al. 2019). These genes were later lost twice through the evolution of Caryophyllales in anthocyanic lineages (Brockington et al. 2015).

The recent research progress from studies on the betalain biosynthesis pathway and phylogenetic analysis has contributed to characterizing the betalain phenomenon in the Caryophyllales. At the time of the study reported by Brockington et al. (2011), there was insufficient evidence to fully address the unanswered questions. However, a new study by Sheehan et al. (2020) has made significant developments on how the enzymeL-DOPA 4,5- dioxygenase, or DODA, evolved by characterizing the activity of 23 DODA proteins in four different betalainic species and five different anthocyanic species. Using updated phylogenetic data, they found that marginal L-DOPA 4,5-dioxygenase activity is present across the entire DODAα lineage, including anthocyanic plants of the Caryophyllales, and that high activity is parallel to the phylogenetic patterns of betalains (Sheehan et al. 2020). Interestingly, high L-DOPA 4,5-dioxygenase activity appears to be polyphyletic since marginal and high activity clades are sister to one another, which suggests further gene duplication. There are at least two hypotheses, but still not a clear answer, to explain the loss of the DODAα loci in anthocyanic lineages (Sheehan et al. 2020).

The theories regarding the evolution of betalains are not mutually exclusive and may be complementary. Because betalains and anthocyanins do not appear to occur simultaneously in plant tissues, it has been suggested that these pigments might have other biological roles beyond pollination and dispersal of seeds (Clement and Mabry 1996).

4.4 Functional Analysis of Betalain Pigments in Biotic and Abiotic Stresses

Plants respond to environmental challenges with molecular and physicochemical mechanisms (Isah and Isah 2019). Plants are able to minimize the effects of stresses on their growth and development by producing a complex network of chemicals, such as secondary metabolites (e.g. terpenoids, sesquiterpenoids, indoles, isoflavonoids, flavonoids, quinones, and polyphenols) (Koche 2014; Isah and Isah 2019). The role of pigments such as betalains in plant defense has been currently addressed due to its strong antioxidant effects (Li et al. 2019). However, there is not enough evidence to support this hypothesis yet (Adhikary 2019). Information on the role of betalains in biotic and abiotic stresses is still insufficient. Exploring the bioactivity of betalains in response to different types of stresses will improve the use of these pigments in research and in the pharmaceutical industry (Polturak and Aharoni 2018; Isah and Isah 2019). Moreover, a better understanding of the ecological role of betalains will potentially contribute to the development of stress-tolerant crops through breeding techniques.

4.4.1 Abiotic Stress

Plants accumulating betalains are often acclimatized to grow in arid and semi-arid regions, and in saline and alkaline soils, indicating adaptive benefits to unfavorable environmental conditions (Jain and Gould 2015; Polturak and Aharoni 2018; Li et al. 2019). Betalains actively scavenge ROS, and, therefore, alleviate the oxidative stress in plants caused by adverse environmental effects (Jain et al. 2015). Although betalain accumulation in plants responds to drought, salinity, exposure to light or UV radiation (Polturak and Aharoni 2018), the responses to salinity stress are particularly remarkable (Davies et al. 2018). However, few studies have shown this effect in *Amaranthus* species. It has been shown that betalain accumulates in quinoa in response to salinity, suggesting a potential role of this pigment in salt stress tolerance (Imamura et al. 2018). Likewise, betalain and other pigments were augmented in *A. tricolor* under salt stress (Sarker and Oba 2019b). A report by Wang et al. (2008) found that there was higher tonoplast H+-ATPase and superoxide dismutase activity in high betacyanin-pigmented seedlings of *Suaeda salsa*. These results indicated a greater tolerance to salt stress in the betacyaninic plants, even though the betacyanin content did not increase with the salt treatment (Wang et al. 2008). The coastal succulent plant *Disphymaaustrale*, native to New Zealand, also responds to salinity by increasing betalain content, apparently to photoprotect the chloroplasts affected by the salt stress (Jain and Gould 2015). Betacyanin pigments were also found to contribute to the photoprotection of *A. cruentus* by regulating the excitation pressure on the photosystem II (PSII) due to an attenuation of the higher incidence of light during water stress (Nakashima et al. 2011). A recent study on the role of betacyanin protection against UV-B radiation in *A. tricolor* showed an increased production of secondary metabolites with known photo-protective effects, such as melatonin. However, there was no evidence of the UV-B protective effect of betalains (Adhikary et al. 2020b). Conversely, earlier studies with *Mesembryanthemum crystallinum* showed an accumulation of betalain pigments in response to ultraviolet radiation (Ibdah et al. 2002). Further studies could assess the long-term effects of betalains in the protection from abiotic stresses across several plant species.

4.4.2 Biotic Stress

Knowledge of the role of betalains in plant resistance to biotic stresses is even smaller than the fragmentary information regarding their contribution to abiotic stress tolerance (Polturak and Aharoni 2018). Betalains have an apparent role in plant defense against insect herbivores (Niveyro et al. 2013) and pathogenic fungi (Polturak et al. 2017). However, there is not enough evidence to support these defensive functions, yet. A transcriptomic study on *A. hypochondriacus* documented 1,971 genes that were differentially expressed in response to biotic and abiotic stresses (Délano-Frier et al. 2011). Of those, one coded for a subunit of a putative dopa dioxygenase, an enzyme involved in the biosynthesis of betacyanin and betalain-related glycosyltransferases (Délano-Frier et al. 2011). These results were consistent with the highly pigmented stem tissue used for this transcriptomic study. Nonetheless, this finding could not be attributed to a biotic stress response (Délano-Frier et al. 2011). Additional research from the same group found that betacyanin biosynthetic genes were induced by insect-herbivory in *A. hypochondriacus*, even in non-pigmented plants, suggesting that these genes might have a role beyond betacyanin biosynthesis (Casique-Arroyo et al. 2014). In another case, transgenic tobacco plants accumulating high levels of betalain presented increased resistance to *Botrytis cinerea*, a necrotrophic fungus that causes gray mold in plants, leading to major crop losses (Polturak et al. 2017). More importantly, Polturak and colleagues (2017) concluded from their results that betalain roles in plants cannot be attributed only to pollination attraction and may contribute to other purposes in plants, such as fungal resistance. This work also showed that transgenic plants are suitable for the production of high levels of betalains through heterologous expression, which holds great promise for increased commercial betalain production (Polturak et al. 2017).

4.5 Betalain Biosynthesis Pathway Genes

4.5.1 Betalain Biosynthesis Pathway

The biosynthesis of betacyanins and betaxanthins has largely been resolved with the recent characterization of the core enzymes of the biosynthetic pathway (Timoneda et al. 2019; Polturak and Aharoni 2018). L-tyrosine, which originates from the Shikimate pathway, is the main precursor of the betalain pathway (Timoneda et al.

2019). The first step of the betalain pathway is the hydroxylation of tyrosine to L-DOPA, catalyzed by tyrosine hydroxylase (CYP76AD1/ CYP76AD5/ CYP76AD6/ CYP76AD15) (Fig. 4.4, step a) (Adhikary et al. 2019; Timoneda et al. 2019). The conversion of tyrosine to L-DOPA was initially associated with a putative polyphenol oxidase (PPO), as reviewed by Timoneda et al. (2019). This hypothesis was later challenged by molecular and reverse genetics, which indicated that the cytochrome P450 enzyme family was responsible for the catalyzation of the first step and other subsequent steps in the pathway: the conversion of tyrosine to L-DOPA (Sunnadeniya et al. 2016; Polturak et al. 2016) and, subsequently, of, L-DOPA into the *cyclo*-DOPA intermediate (Hatlestad et al. 2012; Timoneda et al. 2019). L-DOPA is converted to *cyclo*-DOPA by L-DOPA oxidase (CYP76AD1 and CYP76AD3) (Fig. 4.4, step b) (Timoneda et al. 2019). L-DOPA is alternatively oxidized by the enzyme 4,5-DOPA dioxygenase extradiol (DODA) to 4,5-seco-DOPA, which ultimately produces betalamic acid (Fig. 4.4, step c) (Timoneda et al. 2019). Betalamic acid may conjugate with *cyclo*-DOPA leading to the

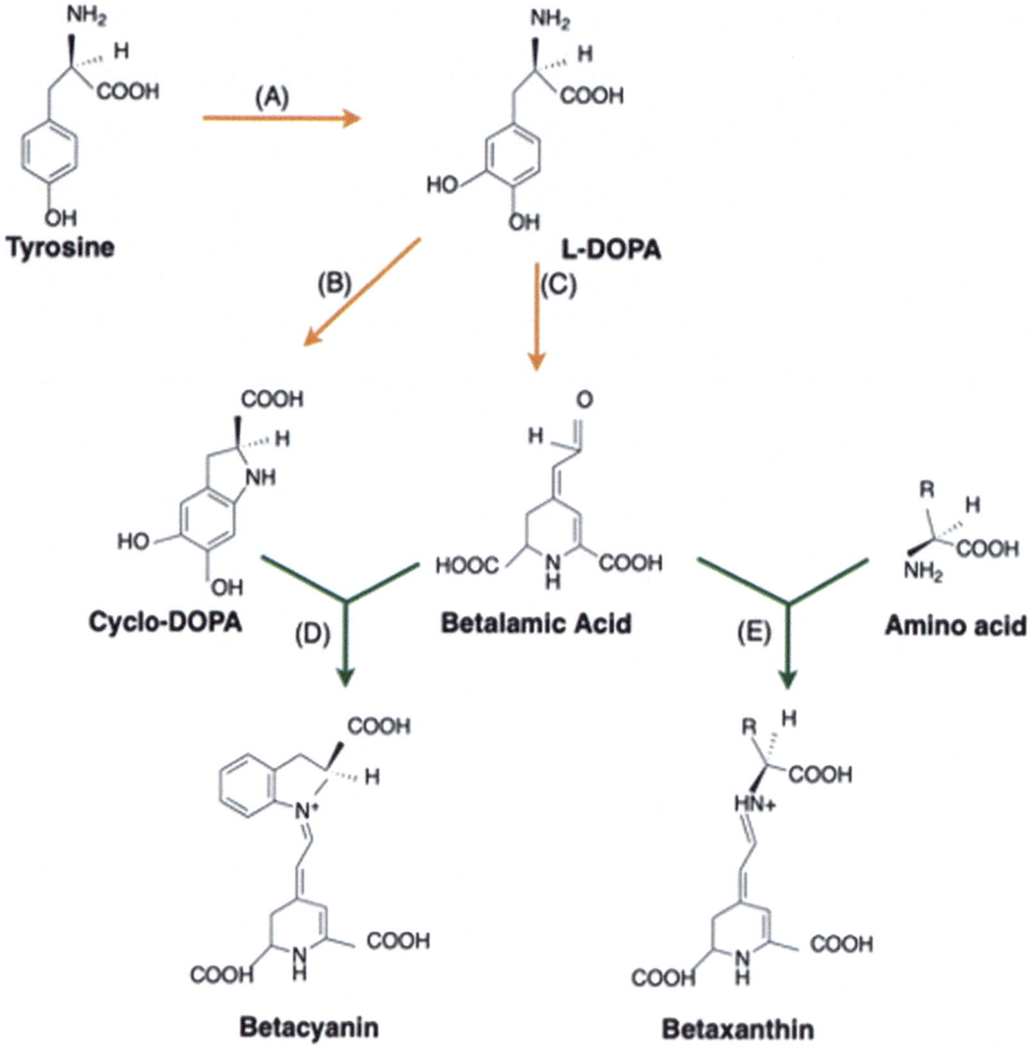

Fig. 4.4 Betalain biosynthetic pathway. Steps a–c are controlled by multiple enzymes, whereas steps d and e are spontaneous. Pathway was redrawn from Hatlestad et al. (2012)

formation of the red–violet betacyanin pigments (Fig. 4.4, step d), or condense with amines and amino acids, to form yellow betaxanthins (Fig. 4.4, step e). The presence of betalamic acid as the central chromophore of betacyanin and betaxanthins pigments is what defines the betalain molecules biochemically (Timoneda et al. 2019). The steps that lead to the formation of L-DOPA, *cyclo*-DOPA and betalamic acid are enzymatically catalyzed, whereas the processes that lead to the formation of betanidin, betacyanin, and betaxanthins are spontaneous (Nakatsuka et al. 2013; Timoneda et al. 2019). Furthermore, betalain structural modifications, such as glycosylation, can occur in some steps of the biosynthesis pathway forming a diversity of molecules (Timoneda et al. 2019). Also, betacyanins can undergo acylation reactions that increase the number of unique betalain moieties, but the enzymes that catalyze these additional reactions are not well studied (Timoneda et al. 2019).

Several genes involved in the betalain biosynthesis pathway have been revealed using the new generation sequencing technologies (Hatlestad et al. 2012; Sunil et al. 2014; Clouse et al. 2016; Liu et al. 2019; Adhikary et al. 2019). Key genes in *Amaranthus* have been identified through the sequencing of the *Amaranthus* genome and transcriptome (Sunil et al. 2014; Clouse et al. 2016), and reverse genetics studies (Adhikary 2019; Adhikary et al. 2019), that will be further detailed. A keyword search for betalain-related proteins that have been annotated in the public database Uniprot (https://www.uniprot.org) (The UniProt Consortium 2018) reveals a total of 531 proteins. Of these, 32 proteins belong to the *Amaranthus* genus, 13 of which belong to *A. tricolor*, 11 to *A. cruentus*, 7 to *A. viridis*, and 1 to *A. retroflexus* (Table 4.2) (Accessed on July 15, 2020). These genetic data are essential to the advancement of betalain research.

Table 4.2 Betalain-related genes from *Amaranthus* retrieved from the Uniprot database

Entry	Protein names	Gene names	Organism
A0A0D5CP38	Tyrosine/DOPA decarboxylase		*Amaranthus tricolor*
A0A5P9NG88	Putative cytochrome P450 76AD21 (Fragment)	CYP76AD21	*Amaranthus viridis*
A0A5P9NE19	Putative cytochrome P450 76AD21 (Fragment)	CYP76AD21	*Amaranthus tricolor*
A0A5P9NEZ5	Putative cytochrome P450 76AD20 (Fragment)	CYP76AD20	*Amaranthus viridis*
A0A5P9NE42	Putative cytochrome P450 76AD20 (Fragment)	CYP76AD20	*Amaranthus tricolor*
A0A5P9NE18	Putative 4,5-DOPA dioxygenase (Fragment)	DODA	*Amaranthus tricolor*
A0A5P9NEE1	Polyphenol oxidase (Fragment)	PPO	*Amaranthus viridis*
A0A5P9NE28	Polyphenol oxidase (Fragment)	PPO	*Amaranthus tricolor*
A0A0D5MEW8	Glycosyltransferase (EC 2.4.1.-)	DOPA5-GT	*Amaranthus tricolor*
A0A0D5CNB7	Glycosyltransferase (EC 2.4.1.-)		*Amaranthus tricolor*
A0A0D5MEP1	Glycosyltransferase (EC 2.4.1.-)	B5-GT	*Amaranthus tricolor*
Entry	Protein names	Gene names	Organism
A0A0D5MFF2	Glycosyltransferase (EC 2.4.1.-)	B6-GT	*Amaranthus tricolor*
A0A0U1YJV0	DODA2 (Fragment)		*Amaranthus cruentus*
A0A0U1YJM3	DODA1		*Amaranthus cruentus*
A0A0G2USW9	Cytochrome P450 76AD1-like protein (Fragment)		*Amaranthus cruentus*
A0A0G2UWS8	Cytochrome P450 76AD1-like protein (Fragment)		*Amaranthus cruentus*
A0A0G2URI4	Cytochrome P450 76AD1-like protein (Fragment)		*Amaranthus cruentus*
A0A0G2URE1	Cytochrome P450 76AD1-like protein (Fragment)		*Amaranthus cruentus*
I3PFJ7	CYP76AD2		*Amaranthus cruentus*

(continued)

Table 4.2 (continued)

Entry	Protein names	Gene names	Organism
A0A5P9NEE0	Cyclo-DOPA-5-O-glucosyltransferase (Fragment)	cD5GT	*Amaranthus viridis*
A0A5P9NFD8	Cyclo-DOPA-5-O-glucosyltransferase (Fragment)	cD5GT	*Amaranthus tricolor*
A0A5P9NEE9	Cyclo-DOPA 5-O-glucosyltransferase (Fragment)	cD5GT	*Amaranthus viridis*
A0A2Z0PTH5	Catalase		*Amaranthus cruentus*
A0A5P9NEF0	Betanidin-6-O-glucosyltransferase (Fragment)	cD5GT	*Amaranthus viridis*
A0A5P9NEU5	Betanidin-6-O-glucosyltransferase (Fragment)	B6GT	*Amaranthus viridis*
Entry	Protein names	Gene names	Organism
A0A5P9NEF7	Betanidin-6-O-glucosyltransferase (Fragment)	AtrB6GT	*Amaranthus tricolor*
A0A0D5ME21	Anthocyanidin 3-O-glucosyltransferase-like protein	A3-GT	*Amaranthus tricolor*
A0A0D5CP49	4,5-DOPA dioxygenase extradiol		*Amaranthus tricolor*
A0A0G2UR02	4,5-dioxygenase-like protein (Fragment)		*Amaranthus retroflexus*
A0A0G2USB7	4,5-dioxygenase-like protein (Fragment)		*Amaranthus cruentus*
A0A0G2UR37	4,5-dioxygenase-like protein (Fragment)		*Amaranthus cruentus*
A0A0G2UYU4	4,5-dioxygenase-like protein (Fragment)		*Amaranthus cruentus*

**Source* Uniprot (https://www.uniprot.org) (The UniProt Consortium 2018)

4.5.2 Reverse Genetics Work in Amaranth

With the advent of next-generation sequencing, the number of sequenced genes with unknown gene function has rapidly increased in both model and crop species including *Amaranthus.* Therefore, reverse genetics approaches have become increasingly valuable. It involves a known DNA or RNA sequence and disruption of its function in vivo followed by observation of the resulting phenotype (Griffiths et al. 2000). Multiple reverse genetics tools have been developed for plants including RNA-mediated interference (RNAi), insertional mutagenesis, chemical mutagenesis paired with high-throughput sequencing or genotyping (e.g. Targeting Induced Local Lesions in Genomes [TILLING]), Clustered Regularly Interspaced Short Palindromic Repeats CRISPR-associated (CRISPR/Cas), Zinc Finger Nucleases (ZFNs), Transcription-Like Effector Nucleases (TALENs)-mediated gene editing, etc. Their implementation has made a tremendous impact in reverse genetics studies in higher eukaryotes. Most of these powerful techniques rely, however, on the previous knowledge of the precise nucleotide sequences to be manipulated (Gurumurthy et al. 2016).

Several studies using reverse genetics have characterized genes of the betalain biosynthesis pathway in the Caryophyllales (Adhikary et al. 2019; Timoneda et al. 2019). The silencing of *CYP76AD1* in *B. vulgaris* led to the loss of red pigmentation and the production of the yellow pigments only, whereas the expression of the same gene in transgenic anthocyanic lines recovered the betacyanin pigmentation (Hatlestad et al. 2012). The tyrosine hydroxylase activity of the *CYP76AD1, CYP76AD5,* and *CYP76AD6* genes was also shown using reverse genetics, as extensively reviewed by Timoneda et al. (2019). For instance, the loss-of-function effects of *CYP76AD6* from *B. vulgaris* and recombinant expression in *Nicotiana benthamiana* revealed its specific tyrosine hydroxylase activity, participating only in the L-DOPA formation, whereas CYP76AD1 also catalyzes L-DOPA formation and its subsequent conversion to *cyclo*-DOPA (Polturak et al. 2016). A recent study established a virus-induced gene silencing (VIGS) method for functional studies of genes in *Amaranthus,* using genes of the betalain biosynthesis pathway as a proof of concept (Adhikary et al. 2019). In this study, the researchers successfully silenced the *AtriCYP76AD1* gene belonging to the cytochrome P450 family. It was observed that the expression of the transcripts of *AtriCYP76AD1* and related genes (*AtriCYP76AD1, CYP76AD5,* and *CYP76AD6*) diminished significantly, accompanied by a visible reduction in red pigmentation, and emergence of green leaves due to loss-of-function of the target gene (Adhikary et al. 2019). This study provided avenues to use VIGS technology within amaranth and related species as well as the capacity to functionally characterize other genes having agronomical importance. Furthermore, Zheng et al. (2016) cloned and expressed putative betalain pathway genes in *A. tricolor* and assessed the effect of phytohormones. Although the study demonstrated the expression of the *DOPA5-GT*, *AmaDODA*, *AmaB5-GT*, *AmaTyDC*, and *AmaB6-GT* genes and showed pigment accumulation in response to phytohormone treatments, it was unable to establish a correlation between gene expression and pigment accumulation. Thus, it could not be concluded with certainty that these genes were involved in the betalain biosynthetic pathway.

4.6 Perspectives on Future Research and Genetic Improvements

Betalain-related research in *Amaranthus* species is limited. A complete understanding of the mechanisms by means of which certain genes regulate the presence and concentration of betalain pigments can ultimately lead to increased yields, improved cultivated varieties, and the generation of other plants/systems able to produce and accumulate betalains. Until recently, transgenic plants could only produce betalains when fed the necessary substrates; now, characterization of the main enzymatic steps allows for the betalain biosynthesis pathway to be engineered into other desirable plants or microbes (Polturak et al. 2016; Polturak and Aharoni 2019). There are still only a few examples of heterologous expression of betalains in other organisms and an expansion of the species capable of producing betalains could create novel ornamentals as well as new dietary sources of betalains (Polturak and Aharoni 2019). Not only would heterologous expression of betalain biosynthesis genes in other plants be important for the many applications of betalains but would also permit the targeted synthesis of pathway intermediates, such as L-DOPA, which is a therapeutic compound in the treatment of Parkinson's disease (Schwinn 2016; Timoneda et al. 2019). Optimization of betalain accumulation in crops could be realized by inhibiting regulatory feedback loops within the shikimate pathway, which produces tyrosine, and by selecting preferential isoenzymes (Polturak and Aharoni 2019). Control of the feedback inhibitions that regulate the betalain biosynthesis pathway has yet to be attempted but could be another valuable method to increase the yield of betalains and other pathway intermediates (Polturak and Aharoni 2019). A positive regulator of the betalain biosynthesis pathway, *Bv*MYB1, has been identified in beetroot, but no heterologous genes have been found in other betalain-producing plants (Hatlestad et al. 2012; Lloyd et al. 2017; Polturak and Aharoni 2019). As technologies to isolate and identify unique moieties have improved, the diversity of betalains across species, cultivars, genotypes, and even within individual plants has become clear. Although research has been performed to allow the identification of the diverse betalain isoforms present in *Amaranthus*, full pigment profiles are still lacking in many species. Betalain forms are diverse; therefore, comparing the bioactivity of newly gleaned forms is worthwhile to optimize genetically engineered crops. In addition, new types of betalains could be created by feeding different amines into heterologous organisms or by expanding what is known about the decoration of betalains in the later steps of the biosynthesis pathway. Heterologous expression of betalain genes in plants still needs to be explored more; however, some of the basic aspects of the betalain biosynthesis pathway have not been fully elucidated yet. *Amaranthus* species have received much less research attention than *B. vulgaris* as a source of betalains, and now with the completed genome, some of the new techniques mentioned above can be considered.

Amaranthus genome and some transcriptome information are now available (Clouse et al. 2016); however, functional annotation of the agronomical important genes is yet to be characterized. Furthermore, reverse genetics technologies such as CRISPR/Cas system are yet to be developed within the species. Determining the origin of betalains and their divergence from anthocyanins is a key question that needs to be answered comprehensively. To date, what is understood about the evolutionary history of pigments in the Caryophyllales has largely been reviewed and determined by Brockington's research group (Brockington et al. 2011, 2015; Timoneda et al. 2019; Sheehan et al. 2020). Resolving the function of key proteins, such as DODAα, in some lineages that express only marginal L-DOPA 4,5-dioxygenase activity will be the focus of future investigations into the evolution of betalains (Sheehan et al. 2020). Still many of the questions regarding the complex evolutionary story of these pigments remain unanswered: e.g. is it evolutionary advantageous for plants to use betalain biosynthesis pathway instead of the anthocyaninic pathway. Moreover, could these two pigments have complementary roles in vivo?

An investigation into any adaptive advantages derived from the accumulation of betalains over anthocyanins has the potential of generating essential information for creating of more resilient crops, which may endure the negative effects of the intensified climate changes. Jain and Gould (2015) concluded, from their review centered on the betalains versus anthocyanins conundrum, that not only the antioxidant and possible photoprotective abilities of betalains but also their potential for NaCl sequestration could be key factors able to create stronger resilience against salinity stress. However, there is little to no research regarding the mechanisms linking betalains with the storage and Na^+ sequestration in the tonoplast (Wang et al. 2008; Jain and Gould 2015). Betalains in *Amaranthus* species show promising evidence that they are important in the protection against both abiotic and biotic stresses, but more data are needed. For example, betalains may have a role in defense against herbivory, but very few studies have focused on this topic (Délano-Frier et al. 2011; Casique-Arroyo et al. 2014). Furthermore, gene annotation within the betalain biosynthesis pathway could lead to transcriptomic studies better suited to investigate betalains contribution to the stress tolerance that characterizes most *Amaranthus* species.

4.7 Conclusions

Betalain pigments have great potential for commercial applications due to their nutritional value, antioxidant, and stress-protective properties. They can also be used as natural colorants and in the production of ornamental plants. However, the occurrence of these pigments in plants is limited to some lineages of the Caryophyllales order. Therefore, fully understanding the betalain metabolic pathway and its biological role in plants is necessary to expand the production of these pigments and the pathway intermediates beyond the order of the Caryophyllales. Research efforts have been made to reveal the core enzymes involved in the betalain metabolic pathway. These studies have enlightened the steps of pigment evolution within the core Caryophyllales. Nevertheless, the knowledge regarding the betalain biosynthesis pathway is still limited in comparison with other classes of pigments, such as anthocyanins and carotenoids. As the development of high-throughput sequencing technologies and computational analysis tools keep improving, at lower costs, we expect to advance in the production of higher amounts of betalains using in vitro systems (such as microorganisms and plants). Moreover, the development of new tools for gene editing, such as CRISPR/Cas, will allow us to better characterize these gene functions. More information regarding the genes related to the betalain biosynthesis pathway will enhance our understanding of its regulation, the biological role of betalains, and the evolutionary history of the pathway. Altogether, this research will ultimately aid in selecting target genes and strategies for genetic modification of plants, to provide support to basic research and to commercial applications.

Acknowledgements We would like to thank Dr. Michael Deyholos for reading and providing valuable feedback on the manuscript.

References

Adhikary D (2019) Development of genetic resources and functional analysis of betalains in *Amaranthus* species. Dissertation, University of British Columbia

Adhikary D, Khatri-Chhetri U, Slaski J (2020a) Amaranth: an ancient and high-quality wholesome crop. In: Tech Open. https://doi.org/10.5772/intechopen.88093

Adhikary D, Forsyth JA, Murch SJ, Deyholos MK (2020) Impact of betacyanins on responses to ultraviolet radiation in *Amaranthus tricolor* L. J Plant Interac 15:117–126. https://doi.org/10.1080/17429145.2020.1766584

Adhikary D, Khatri-Chhetri U, TymmFJM et al (2019) A virus-induced gene-silencing system for functional genetics in a betalainic species, *Amaranthus tricolor* (Amaranthaceae). Appl Plant Sci 7:e01221. https://doi.org/10.1002/aps3.1221

Biswas M, Dey S, Sen R (2013) Betalains from *Amaranthus tricolor* L. J PharmacognPhytochem 1:87–95

Brockington SF, Yang Y, Gandia-Herrero F et al (2015) Lineage-specific gene radiations underlie the evolution of novel betalain pigmentation in Caryophyllales. New Phytol 207:1170–1180. https://doi.org/10.1111/nph.13441

Brockington SF, Walker RH, Glover BJ et al (2011) Complex pigment evolution in the Caryophyllales. New Phytol 190:854–864. https://doi.org/10.1111/j.1469-8137.2011.03687.x

Cai YZ, Sun M, Corke H (2005) Characterization and application of betalain pigments from plants of the Amaranthaceae. Trends Food Sci Technol 16:370–376. https://doi.org/10.1016/j.tifs.2005.03.020

Cai Y, Sun M, Corke H (2003) Antioxidant activity of betalains from plants of the Amaranthaceae. J Agric Food Chem 51:2288–2294. https://doi.org/10.1021/jf030045u

Cai Y, Sun M, Corke H (2001) Identification and distribution of simple and acylated betacyanins in the Amaranthaceae. J Agric Food Chem 49:1971–1978. https://doi.org/10.1021/jf000963h

Cai Y, Sun M, Wu H et al (1998) Characterization and quantification of betacyanin pigments from diverse Amaranthus species. J Agric Food Chem 46:2063–2070. https://doi.org/10.1021/jf9709966

Čanadanović-Brunet JM, Savatović SS, Ćetković GS et al (2011) Antioxidant and antimicrobial activities of beet root pomace extracts. Czech J Food Sci 29:575–585. https://doi.org/10.17221/210/2010-cjfs

Casique-Arroyo G, Martínez-Gallardo N, González de la Vara L, Délano-Frier JP (2014) Betacyanin biosynthetic genes and enzymes are differentially induced by (a)biotic stress in *Amaranthus hypochondriacus*. PLoS ONE. https://doi.org/10.1371/journal.pone.0099012

Celli GB, Brooks MSL (2017) Impact of extraction and processing conditions on betalains and comparison of properties with anthocyanins—a current review. Food Res Int 100:501–509. https://doi.org/10.1016/j.foodres.2016.08.034

Christinet L, Burdet FX, Zaiko M et al (2004) Characterization and functional identification of a novel plant 4,5-extradiol dioxygenase involved in betalain pigment biosynthesis in *Portulaca grandiflora*. Plant Physiol 134:265–274. https://doi.org/10.1104/pp.103.031914

Clement JS, Mabry TJ (1996) Pigment evolution in the caryophyllales: asystematic overview. Botanica Acta 109:360–367. https://doi.org/10.1111/j.1438-8677.1996.tb00584.x

Clouse JW, Adhikary D, Page JT et al (2016) The amaranth genome: genome, transcriptome, and physical map assembly. Plant Genome. https://doi.org/10.3835/plantgenome2015.07.0062

Contreras-Llano LE, Guerrero-Rubio MA, Lozada-Ramírez JD et al (2019) First betalain-producing bacteria break the exclusive presence of the pigments in the plant kingdom. MBio. https://doi.org/10.1128/mBio.00345-19

Davies KM, Albert NW, Zhou Y, Schwinn KE (2018) Functions of flavonoid and betalain pigments in abiotic stress tolerance in plants. In: Annual plant reviews online. American Cancer Society, pp 21–62

Deladino L, Alvarez I, De Ancos B et al (2017) Betalains and phenolic compounds of leaves and stems of *Alternanthera brasiliana* and *Alternanthera tenella*. Food Res Int 97:240–249. https://doi.org/10.1016/j.foodres.2017.04.017

Délano-Frier JP, Avilés-Arnaut H, Casarrubias-Castillo K et al (2011) Transcriptomic analysis of grain amaranth (*Amaranthus hypochondriacus*) using 454 pyrosequencing: comparison with *A. tuberculatus*, expression profiling in stems and in response to biotic and abiotic stress. BMC Genomics 12:363. https://doi.org/10.1186/1471-2164-12-363

Esatbeyoglu T, Wagner AE, Motafakkerazad R et al (2014) Free radical scavenging and antioxidant activity of betanin: electron spin resonance spectroscopy studies and studies in cultured cells. Food Chem Toxicol 73:119–126. https://doi.org/10.1016/j.fct.2014.08.007

Escribano J, Cabanes J, Jiménez-Atiénzar M, Ibañez-Tremolada M, Gómez-Pando LR, García-Carmona F, Gandía-Herrero F (2017) Characterization of betalains, saponins and antioxidant power in differently colored quinoa (*Chenopodium quinoa*) varieties. Food Chem 234:285–294. https://doi.org/10.1016/j.foodchem.2017.04.187

Gandía-Herrero F, García-Carmona F (2020) The dawn of betalains. New Phytol 227:664–666. https://doi.org/10.1111/nph.16295

Gandía-Herrero F, García-Carmona F (2013) Biosynthesis of betalains: yellow and violet plant pigments. Trends Plant Sci 18:334–343. https://doi.org/10.1016/j.tplants.2013.01.003

Gandía-Herrero F, Escribano J, García-Carmona F (2016) Biological activities of plant pigments betalains. Crit Rev Food Sci Nutr 56:937–945. https://doi.org/10.1080/10408398.2012.740103

Gengatharan A, Dykes GA, Choo WS (2015) Betalains: natural plant pigments with potential application in functional foods. LWT-Food Sci Technol 64:645–649. https://doi.org/10.1016/j.lwt.2015.06.052

Gill M (1996) Pigments of fungi (macromycetes). Nat Prod Rep 13:513–528. https://doi.org/10.1039/np9961300513

Gliszczyńska-Świgło A, Szymusiak H, Malinowska P (2006) Betanin, the main pigment of red beet: molecular origin of its exceptionally high free radical-scavenging activity. Food AdditContam 23:1079–1087. https://doi.org/10.1080/02652030600986032

Griffiths AJ, Miller JH, Suzuki DT et al (2000) Reverse genetics. an Introduction to genetic analysis, 7th edn. W. H. Freeman, New York

Grewal PS, Modavi C, Russ ZN et al (2018) Bioproduction of a betalain color palette in *Saccharomyces cerevisiae*. MetabEng 45:180–188

Gurumurthy CB, Grati M, Ohtsuka M et al (2016) CRISPR: aversatile tool for both forward and reverse genetics research. Hum Genet 135:971–976. https://doi.org/10.1007/s00439-016-1704-4

Harris NN, Javellana J, Davies KM et al (2012) Betalain production is possible in anthocyanin-producing plant species given the presence of DOPA-dioxygenase and L-DOPA. BMC Plant Biol 12:34. https://doi.org/10.1186/1471-2229-12-34

HatlestadGJ ANA, Sunnadeniya RM et al (2015) The beet Y locus encodes an anthocyanin MYB-like protein that activates the betalain red pigment pathway. Nat Genet 47:92–96. https://doi.org/10.1038/ng.3163

HatlestadGJ SRM, Akhavan NA et al (2012) The beet R locus encodes a new cytochrome P450 required for red betalain production. Nat Genet 44:816–820. https://doi.org/10.1038/ng.2297

Heuer S, Richter S, Metzger JW et al (1994) Betacyanins from bracts of *Bougainvillea glabra*. Phytochemistry 37:761–767. https://doi.org/10.1016/S0031-9422(00)90354-6

Hilou A, Nacoulma OG, Guiguemde TR (2006) *In vivo* antimalarial activities of extracts from *Amaranthus spinosus L.* and *Boerhaaviaerecta* L. in mice. JEthnopharmacol 103:236–240. https://doi.org/10.1016/j.jep.2005.08.006

Hughes NM, Smith WK (2007) Attenuation of incident light in *Galax urceolata*(Diapensiaceae): concerted influence of adaxial and abaxial anthocyanic layers on photoprotection. Am J Bot 94:784–790

Ibdah M, Krins A, Seidlitz H, Heller W, Strack D, Vogt T (2002) Spectral dependence of flavonol and betacyanin accumulation in *Mesembryanthemum crystallinum* under enhanced UV radiation. Plant Cell Environ 25:1145–1154

Imamura T, Takagi H, Miyazato A et al (2018) Isolation and characterization of the betalain biosynthesis gene involved in hypocotyl pigmentation of the allotetraploid *Chenopodium quinoa*. BiochemBiophys Res Commun 496:280–286. https://doi.org/10.1016/j.bbrc.2018.01.041

Isah T, Isah T (2019) Stress and defense responses in plant secondary metabolites production. Biol Res. https://doi.org/10.1186/s40659-019-0246-3

Jain G, Gould KS (2015) Functional significance of betalain biosynthesis in leaves of *Disphymaaustrale* under salinity stress. Environ Exp Bot 109:131–140. https://doi.org/10.1016/j.envexpbot.2014.09.002

Jain G, Schwinn KE, Gould KS (2015) Betalain induction by l-DOPA application confers photoprotection to saline-exposed leaves of *Disphymaaustrale*. New Phytol 207:1075–1083. https://doi.org/10.1111/nph.13409

Joshi DC, Sood S, Hosahatti R et al (2018) From zero to hero: the past, present and future of grain amaranth breeding. Theor Appl Genet 131:1807–1823. https://doi.org/10.1007/s00122-018-3138-y

Kanner J, Harel S, Granit R (2001) Betalains—a new class of dietary cationized antioxidants. J Agric Food Chem 49:5178–5185. https://doi.org/10.1021/jf010456f

Khan MI, Giridhar P (2015) Plant betalains: chemistry and biochemistry. Phytochemistry 117:267–295. https://doi.org/10.1016/j.phytochem.2015.06.008

Khanam UKS, Oba S (2013) Bioactive substances in leaves of two amaranth species, *Amaranthus tricolor* and *A. hypochondriacus*. Can J Plant Sci 93:47–58. https://doi.org/10.4141/CJPS2012-117

Khandaker L, Ali MB, Oba S (2009) Influence of cultivar and growth stage on pigments and processing factors on betacyanins in Red Amaranth (*Amaranthus tricolor* L.). Food Sci Technol Int 15:259–265. https://doi.org/10.1177/1082013209341119

Koche D (2014) Role of secondary metabolites in plants' defense mechanism. In: Pawar SE (ed) MakdeKH. A glimpse of current vistas plant science research, Hislop College Publication Cell, Nagpur, pp 1–16

Kugler F, Stintzing FC, Carle R (2007) Characterisation of betalain patterns of differently coloured inflorescences from *Gomphrena globosa* L. and *Bougainvillea* sp. by HPLC-DAD-ESI-MSn. Anal Bioanal Chem 387:637–648. https://doi.org/10.1007/s00216-006-0897-0

Li G, Meng X, Zhu M, Li Z (2019) Research progress of betalain in response to adverse stresses and evolutionary relationship compared with anthocyanin. Molecules 24:3078. https://doi.org/10.3390/molecules24173078

Liu S, Zheng X, Pan J et al (2019) RNA-sequencing analysis reveals betalains metabolism in the leaf of *Amaranthus tricolor* L. PLoS ONE. https://doi.org/10.1371/journal.pone.0216001

Lloyd A, Brockman A, Aguirre L et al (2017) Advances in the MYB-bHLH-WD repeat (MBW) pigment regulatory model: addition of a WRKY factor and co-option of an anthocyanin MYB for betalain regulation. Plant Cell Physiol 58:1431–1441. https://doi.org/10.1093/pcp/pcx075

Miguel MG (2018) Betalains in some species of the amaranthaceae family: a review. Antioxidants 7:53. https://doi.org/10.3390/antiox7040053

Nakashima T, Araki T, Ueno O (2011) Photoprotective function of betacyanin in leaves of *Amaranthus cruentus* L. under water stress. Photosynthetica 49:497–506. https://doi.org/10.1007/s11099-011-0062-7

Nakatsuka T, Yamada E, Takahashi H et al (2013) Genetic engineering of yellow betalain pigments beyond the species barrier. Sci Rep 3:1970. https://doi.org/10.1038/srep01970

Niveyro SL, Mortensen AG, Fomsgaard IS, Salvo A (2013) Differences among five amaranth varieties (*Amaranthus* spp.) regarding secondary metabolites and foliar herbivory by chewing insects in the field. Arth-Plant Int 7:235–245. https://doi.org/10.1007/s11829-012-9219-y

Piattelli M, Minale L, Prota G (1964) Isolation and structure of amaranthine and isoamaranthine. Ann Chim 54:963–968

Polturak G, Aharoni A (2018) "La Vie en Rose": biosynthesis, sources, and applications of betalain pigments. Mol Plant 11:7–22. https://doi.org/10.1016/j.molp.2017.10.008

Polturak G, Aharoni A (2019) Advances and future directions in betalain metabolic engineering. New Phytol 224:1472–1478. https://doi.org/10.1111/nph.15973

Polturak G, Breitel D, Grossman N et al (2016) Elucidation of the first committed step in betalain biosynthesis enables the heterologous engineering of betalain pigments in plants. New Phytol 210:269–283. https://doi.org/10.1111/nph.13796

Polturak G, Grossman N, Vela-Corcia D et al (2017) Engineered gray mold resistance, antioxidant capacity, and pigmentation in betalain-producing crops and ornamentals. Proc Natl Acad Sci USA 114:9062–9067. https://doi.org/10.1073/pnas.1707176114

Sarker U, Oba S (2018a) Augmentation of leaf color parameters, pigments, vitamins, phenolic acids, flavonoids and antioxidant activity in selected *Amaranthus tricolor* under salinity stress. Sci Rep 8:12349. https://doi.org/10.1038/s41598-018-30897-6

Sarker U, Oba S (2019a) Nutraceuticals, antioxidant pigments, and phytochemicals in the leaves of *Amaranthus spinosus* and *Amaranthus viridis* weedy species. Sci Rep 9:1–10. https://doi.org/10.1038/s41598-019-50977-5

Sarker U, Oba S (2019b) Salinity stress enhances color parameters, bioactive leaf pigments, vitamins, polyphenols, flavonoids and antioxidant activity in selected amaranthus leafy vegetables. J Sci Food Agric 99:2275–2284. https://doi.org/10.1002/jsfa.9423

Schliemann W, Cai Y, Degenkolb T, Schmidt J, Corke H (2001) Betalains of *Celosia argentea*. Phytochemistry 58:159–165

Schwinn KE (2016) The dope on l-DOPA formation for betalain pigments. New Phytol 210:6–9. https://doi.org/10.1111/nph.13901

Sheehan H, Feng T, Walker-Hale N et al (2020) Evolution of l-DOPA 4,5-dioxygenase activity allows for recurrent specialisation to betalain pigmentation in Caryophyllales. New Phytol 227:914–929. https://doi.org/10.1111/nph.16089

SlimenIB NT, Abderrabba M (2017) Chemical and antioxidant properties of betalains. J Agric Food Chem 65:675–689. https://doi.org/10.1021/acs.jafc.6b04208

Spórna-Kucab A, Wróbel N, Kumorkiewicz-Jamro A, Wybraniec S (2020) Separation of betacyanins from *Iresineherbstii* Hook. ex Lindl. leaves by high-speed countercurrent chromatography in a polar solvent system. J Chromatogr A 1626:461370. https://doi.org/10.1016/j.chroma.2020.461370

Sreekanth D, Arunasree MK, Roy KR et al (2007) Betanin a betacyanin pigment purified from fruits of *Opuntia ficus-indica* induces apoptosis in human chronic myeloid leukemia Cell line-K562. Phytomedicine 14:739–746. https://doi.org/10.1016/j.phymed.2007.03.017

Stintzing FC, Carle R (2004) Functional properties of anthocyanins and betalains in plants, food, and in human nutrition. Trends Food Sci Technol 15:19–38. https://doi.org/10.1016/j.tifs.2003.07.004

Strack D, Vogt T, Schliemann W (2003) Recent advances in betalain research. Phytochemistry 62:247–269. https://doi.org/10.1016/S0031-9422(02)00564-2

Sunil M, Hariharan AK, Nayak S et al (2014) The draft genome and transcriptome of *Amaranthus hypochondriacus*: a C4 dicot producing high-lysine edible pseudo-cereal. DNA Res 21:585–602. https://doi.org/10.1093/dnares/dsu021

Sunnadeniya R, Bean A, Brown M et al (2016) Tyrosine hydroxylation in betalain pigment biosynthesis is performed by cytochrome p450 enzymes in beets (*Beta vulgaris*). PLoS ONE. https://doi.org/10.1371/journal.pone.0149417

Tang Y, Li X, Zhang B et al (2015) Characterisation of phenolics, betanins and antioxidant activities in seeds of three *Chenopodium quinoa*Willd, genotypes. Food Chem 166:380–388. https://doi.org/10.1016/j.foodchem.2014.06.018

The UniProt Consortium (2018) UniProt: a worldwide hub of protein knowledge. Nucleic Acids Res 47: D506–D515. https://doi.org/10.1093/nar/gky1049

Thulin M, Moore AJ, El-Seedi H et al (2016) Phylogeny and generic delimitation in Molluginaceae, new pigment data in Caryophyllales, and the new family Corbichoniaceae. Taxon 65:775–793. https://doi.org/10.12705/654.6

Timoneda A, Feng T, Sheehan H et al (2019) The evolution of betalain biosynthesis in Caryophyllales. New Phytol 224:71–85. https://doi.org/10.1111/nph.15980

Veličanski AS, Cvetković DD, Markov SL et al (2011) Antibacterial activity of *Beta vulgaris* L. pomace extract. Acta Period Technol 42:263–269. https://doi.org/10.2298/APT1142263V

Wang CQ, Xu C, Wei JG et al (2008) Enhanced tonoplast H+-ATPase activity and superoxide dismutase activity in the halophyte *Suaeda salsa* containing high level of betacyanin. J Plant Growth Regul 27:58–67. https://doi.org/10.1007/s00344-007-9031-6

Wink M (1997) Compartmentation of secondary metabolites and xenobiotics in plant vacuoles. Adv Bot Res 25:141–169. https://doi.org/10.1016/S0065-2296(08)60151-2

Zheng X, Liu S, Cheng C et al (2016) Cloning and expression analysis of betalain biosynthesis genes in *Amaranthus tricolor*. Biotechnol Lett 38:723–729. https://doi.org/10.1007/s10529-015-2021-z

Molecular Characterisation of Seed Storage Proteins (SSPs) in Grain Amaranth

5

Esaú Bojórquez-Velázquez,
Fátima Alejandra Talamantes-Herrera,
Selma Valles, Ivan Takeshi Cerritos-Castro,
and Ana Paulina Barba de la Rosa

Abstract

Amaranth is well recognized as a crop with the highest agri-food potential because its leaves and seeds are highly nutritious and, both can be used as food and feed resources. Amaranth seeds are the main vegetable source of squalene, an unsaturated hydrocarbon, which have been attributed to diverse beneficial health effects. It is also a source of compounds such as tocotrienols, phytosterols, and sterols that help lower blood cholesterol. Likewise, amaranth has been outlined as a promising system for food production since it provides seeds of excellent nutritional quality, mainly due to their proteins that contain an adequate balance of essential amino acids, and the negligible prolamin content. Besides, the identification and characterisation of bioactive peptides encrypted in amaranth seed storage proteins, such as peptides that function as inhibitors of dipeptidyl peptidase IV and angiotensin-converting enzyme, has increased the interest of these seeds as a nutraceutical food system with important biological functions to improve the consumer´s health. However, little is known about amaranth seed storage proteins genetics and the mechanisms that govern their accumulation in seeds. Future work is needed in this direction in order to be able to improve the functional properties of amaranth proteins that will permit their increased use in the food industry. This chapter summarizes the state of knowledge regarding the biochemical and molecular characteristics of amaranth proteins and offers a broad picture of their possible applications in food science and technology.

5.1 Amaranth Species Used for Grain Production

Three amaranth species are used for seed production: *Amaranthus hypochondriacus* L. and *A. cruentus* L., which are originated in Mexico and Central America, respectively, and *A. caudatus* L. with a centre of origin in Peru and Ecuador. Although these cultivated amaranth species have exceptional nutritional and nutraceutical characteristics, further studies are needed to characterize wild amaranth species, which are a potential source of genes with agrobiotechnological applicability for crop improvement. In Mexico, several wild species can be found, amongst them *A. hybridus* and *A. powellii* are considered to be ancestors of the cultivated species *A. hypochondriacus* and *A. cruentus* (Kietlinski et al. 2014; Sauer 1967).

E. Bojórquez-Velázquez · F. A. Talamantes-Herrera · S. Valles · I. T. Cerritos-Castro · A. P. B. de la Rosa (✉)
Instituto Potosino de Investigación Científica y Tecnológica A.C, San Luis Potosi, Mexico
e-mail: apbarba@ipicyt.edu.mx

D. Adhikary et al. (eds.), *The Amaranth Genome*, Compendium of Plant Genomes,
https://doi.org/10.1007/978-3-030-72365-1_5

5.2 Seeds Morphological Characteristics

Amaranth seeds are smooth, shiny, and slightly flattened with sizes between 0.9 and 1.7 mm in length and 1.0 and 1.3 mm in width, depending on the species. Seeds are the smallest in *A. powelli*, while *A. cruentus* produces the largest ones (Fig. 5.1). Seed colour is characteristic for each amaranth species; wild species are black while cultivated species were selected because of their cream light colour. *Amaranthus hypochondriacus* presents two phenotypes: opaque and translucent. The vitreous or translucent characteristic has been related to the type, degree of cross-linking, and molecular distribution of proteins and starch in seeds (Bojórquez-Velázquez et al. 2018; Konishi et al. 1985a; Okuno and Sakaguchi 1982; Park and Nishikawa 2012).

A cross-section of *A. cruentus* cv. Amaranteca seeds stained with fast green FCF (recognising mainly proteins) and Safranin O (recognizing mainly lignified, suberized or cutinized cell walls) are shown in Fig. 5.2. The largest part of the seed is the starchy perisperm located at the centre of the seed. Starchy perisperm is surrounded by the embryo, which is wound in a circle and consists of the radicle, hypocotyl, and cotyledons (the hypocotyl is not visible in Fig. 5.2 due to the cutting plane). Embryo and starchy perisperm are covered by the seed coat, which consists of a lignified/suberized monolayer with an average thickness of 7 μm with no detectable protein content.

Fig. 5.1 Size and colour of amaranth seeds from different species and cultivars

Starchy perisperm contains mainly starch inside of large polyhedral-like cells with sizes up to 80 μm. These cells have only cellulosic primary cell walls (green stained), and low protein content in their protoplasm. However, some cells near the perisperm edges showed higher protein content that those near the centre. The embryo is the more complex structure of the seed; it is composed of a radicle, hypocotyl, and cotyledons. Each organ is further composed of three kinds of tissues: (i) protoderm, which will become the epidermis, (ii) ground meristems, which will become ground tissues, and (iii) procambium, from which the vascular bundles will emerge (Evert and Eichhorn 2013).

In an amaranth seed, the embryo accumulates the highest protein content, as revealed by the intense green staining (Fig. 5.2), appearing as turquoise dots inside cells (i.e., protein bodies). On average, the ground meristems of the radicle possess between 10 and 25 spherical shaped protein bodies, with a diameter between 0.7 and 2.0 μm, depending on cell size. The protoderm possesses 3–5 "large" protein bodies with a diameter between 0.7 and 2.5 μm, and several smaller ones. Ground meristems of cotyledons have between 4 and 5 pleomorphic protein bodies, with a size between 1.1 up and 3.0 μm. Ground meristems and protoderm have a thin secondary lignified/suberized cell wall, the endosperm has a thicker one. Procambium cells seem to have one cylindrical protein body of about 5.5 μm length per cell and with no apparent secondary cell wall. On the other hand, the remaining endosperm also has protein bodies, with similar shapes and sizes than the ground meristem, but only around 40 protein bodies in large cells.

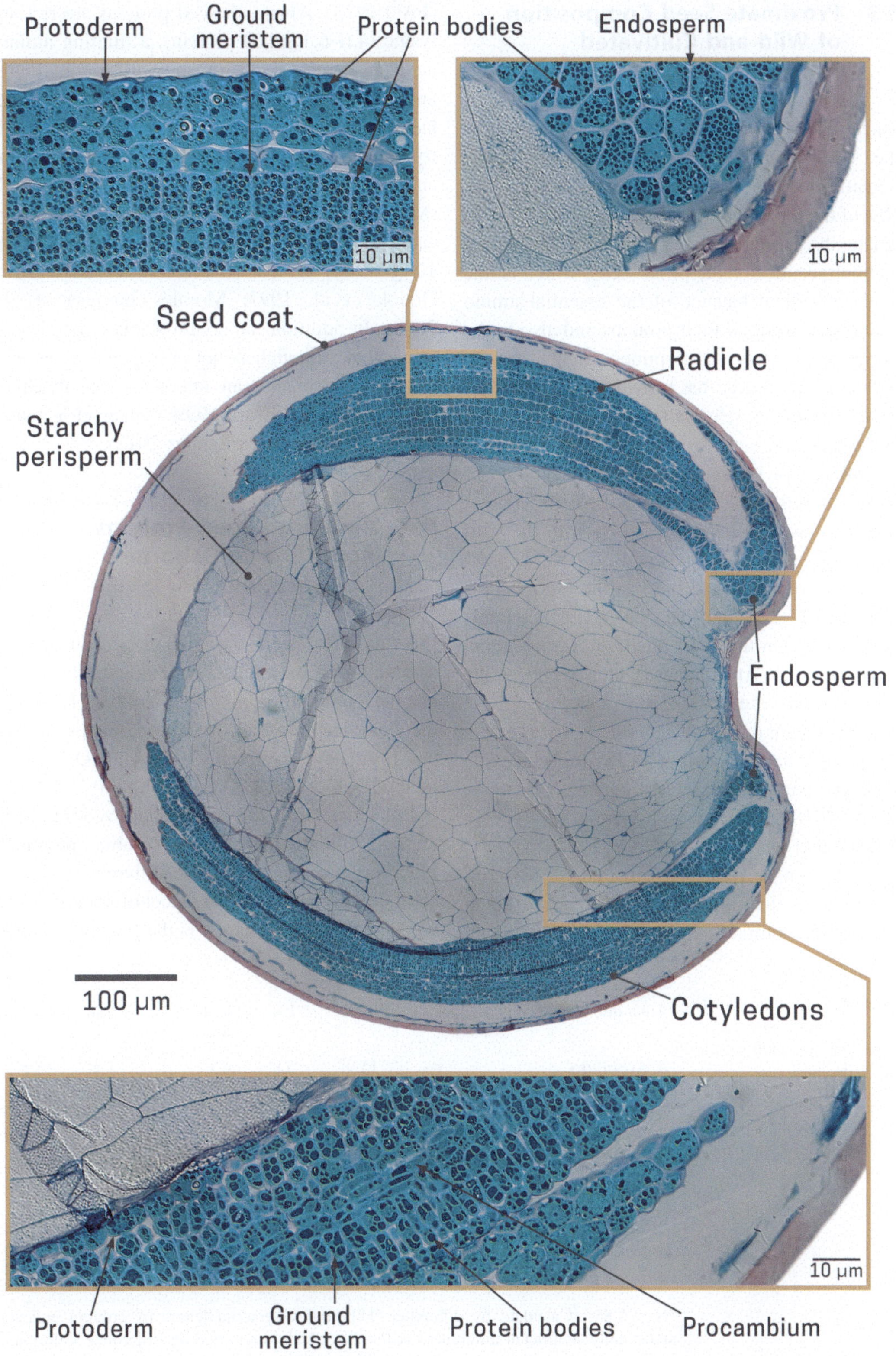

Fig. 5.2 Cross-section of an *Amaranthus cruentus* cv. Amaranteca seed. The upper panel shows an amplified image of the radicle (left) and the endosperm (right). The lower panel shows an amplified image of cotyledons. Protein bodies are observed as turquoise spots

5.3 Proximate Seed Composition of Wild and Cultivated Amaranth Species

Since the 70s, amaranth has been considered as the crop with the highest biotechnological potential for use as a source of functional food. One of the most important characteristics of grain amaranths is undoubtedly the high nutritional and nutraceutical value of its seeds, which is due to the excellent balance of the essential amino acid composition of their proteins and also to the presence of encrypted peptides having diverse biological functions that provide health benefits to the consumers (Huerta-Ocampo and Barba de la Rosa 2011).

Differences in the proximal composition are found amongst species. Total protein content seems to have an indirect relation with the seed size (Bojórquez-Velázquez et al. 2018). *A. powelli* is the smallest seed but has the highest protein (ca. 17.8%) and fat (ca. 8.2%) contents (Table 5.1). On the other hand, *A. caudatus* and *A. cruentus* cv. Amaranteca, which produces the largest seeds, are the species with the lowest protein content (ca. 14.6%), while *A. hybridus*, is the species having seeds with the highest crude fibre content (ca. 6.1%) (Bojórquez-Velázquez et al. 2018; Bressani and García-Vela 1990; Singhal and Kulkarni 1988).

As shown in Table 5.2, amaranth seeds contain proteins rich in essential amino acids, being close to the optimum required for the human diet (FAO 1973). Amaranth seed proteins are rich in lysine (4.6–6.1 g/100 protein), a limiting amino acid in cereals, and in the sulfur-containing amino acids, cysteine and methionine, which are lacking in legumes. Since the quantification of tryptophan (Trp) requires alkaline hydrolysis in anaerobic conditions, its content is not reported (Mota et al. 2016), but a few reports have indicated that Trp in amaranth ranges from 0.8 to 1.8 g/100 g protein (Betschart et al. 1981; Dodok et al. 1997; Morales-Guerrero et al. 2009). In addition, amaranth seeds contain histidine that, although is not considered an essential amino acid, has an important role in child development (Millward 1999; Valcárcel-Yamani and Caetano Da Silva Lannes 2012).

5.4 Amaranth Seed Proteins According to Osborne's Classification

Proteins have been classified in various ways, depending on their location, biological function, chemical composition, or solubility. However, the classification of seed storage proteins (SSPs) has been done essentially following Osborne´s method, which is based on a solubility criterion. In this sense, SSPs has been categorized into four classes: albumins or water-soluble proteins; globulins proteins soluble in different salt concentrations; prolamins or alcohol-soluble proteins, and glutelins, which is the fraction soluble

Table 5.1 Seed proximate composition of wild and cultivated amaranth species (in % dry basis)

Amaranth species	Protein[1]	Fat	Crude fibre	Ash	CH[2]
A. hybridus	15.9[b]	5.9[c]	6.1[a]	3.7[a]	68.5[d]
A. powellii	17.8[a]	8.2[a]	5.9[b]	3.6[a]	64.5[e]
A. cruentus cv. Amaranteca	14.8[c]	6.9[b]	2.5[e]	2.8[b]	73.0[a]
A. hypochondriacus cv. Opaca	16.7[b]	6.9[b]	3.5[d]	3.0[b]	69.9[c]
A. hypochondriacus cv. Cristalina	16.7[b]	5.7[c]	3.9[c]	2.9[b]	70.9[b]
A. hypochondriacus cv. Nutrisol	15.8[b]	6.9[b]	2.4[f]	3.6[a]	71.4[b]
A. caudatus[3]	14.6				

[1]N × 5.85; [2]Calculatd by difference. Different superscript letters by column indicate statistically significant differences at $P < 0.05$. Values reported by Bojórquez-Velázquez et al. (2018) and [3]Bressani and García-Vela (1990)

Table 5.2 Amino acid profile of seed proteins of edible grain *Amaranthus* species (grams/100 g protein)

Essential amino acid	Grain amaranth species			FAO[a,b]	
	A. caudatus[c]	*A. cruentus*	*A. hypochondriacus*	Adults	Children
Histidine	1.6	5.4	5.1	1.6	1.9
Cysteine	2.3	2.0–3.8	2.0–3.9		
Isoleucine	3.6–4.1	3.4–3.7	2.8–3.8	1.3	2.8
Leucine	5.9–6.3	4.8–5.9	5.0–5.8	1.9	4.4
Lysine	5.7–6.4	4.8–5.8	3.2–6.0	1.6	4.4
Threonine	3.8	3.2–4.2	2.6–4.3	0.9	2.8
Tryptophan[d,e]	1.1	0.81	1.1–4.0	0.5	0.9
Valine	4.1–4.7	3.9–4.3	3.2–4.2	1.3	2.5
Met-Cys[f]	4.7	4.6	4.7	1.7	2.2
Phe + Tyr[g]	6.2	7.1	6.2	1.9	2.2

[a]FAO (1973), [b]Millward (1999), [c]Valcárcel-Yamani and Caetano Da Silva Lannes (2012), [d]Dodok et al. (1997), [e]Morales-Guerrero et al. (2009), [f]Sulfur-containing amino acids. [g]Aromatic amino acids

in acidic or alkaline solutions or with solutions containing denaturant or chaotropic agents (Osborne 1924). Amaranth SSPs are characterized by a content of albumins (19–45%), closely followed by glutelins (22–41%) and globulins (16–35%), with little or negligible amounts of prolamins (0.7–1.3%).

However, the SSPs composition in amaranth seeds is variable, which could be related to the application of different extraction methods and/or to the species of seeds reported (Barba de la Rosa et al. 1992a, b; Bressani and García-Vela 1990; Duarte-Correa et al. 1986; Gorinstein et al. 1991; Konishi et al. 1985b; Paredes-López et al. 1988; Segura-Nieto et al. 1992). The electrophoretic profile of amaranth SSPs (Fig. 5.3) differs from the classical patterns of cereals and legumes, which are characterized by high amounts of prolamins and globulins, respectively.

5.4.1 Albumins

Albumins are the water-soluble or very low ionic strength protein fraction. In amaranth seeds, represent one of the major protein groups, ranging from 19 to 45% of the total seed protein content. Albumins have been classified into fractions according to their sedimentation patterns, namely 1.4S or 2.0S and the second 4.6S (Konishi et al. 1991). The electrophoretic pattern of amaranth seed albumins is represented by the main band, of around 35 kDa, and several bands ranging in size from 5 to 120 kDa (Fig. 5.4). Proteins rich in methionine and high levels of lysine, known as MRP-1 and MRP-2, respectively, were detected in the amaranth seed albumins fraction. Two antimicrobial peptides (i.e., Ac-AMP1 and Ac-AMP2), with unique properties as potent inhibitors of fungal growth, were isolated from this fraction (Segura-Nieto et al. 1994).

Reports regarding amaranth seed albumins have concentrated on deal mainly about the most abundant 34–35 kDa protein (Barba de la Rosa et al. 1992a, b, 2009), which was cloned and named AmA1 (Raina and Datta 1992). AmA1 is characterized by an appropriate balance and high levels of essential amino acids according to the FAO guidelines for human nutrition (FAO 1973). For this reason, the *AmA1* gene was inserted in potato and wheat crops leading to an increase in the total protein content of the tubers, as well as in the level of lysine and tyrosine essential amino acids (Chakraborty et al. 2010; Tamás et al. 2009). In addition, when mice were fed with this protein, did not stimulate their immune response, therefore discarding its potential as an allergen (Chakraborty et al. 2010). At the time of these studies, the role of AmA1 was unknown; however, through sequence

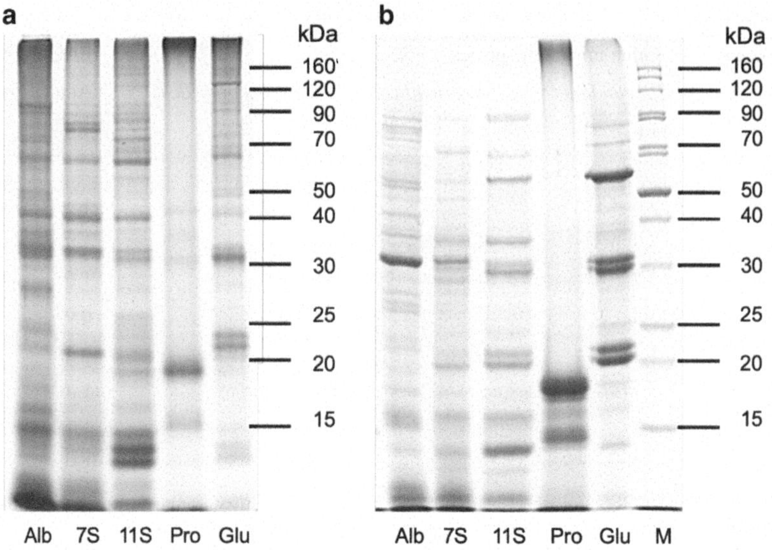

Fig. 5.3 Comparative electrophoretic profiles of Osborne protein fractions extracted from *A. hypochondriacus* cv. Nutrisol seeds. Alb = Albumins; 7S = 7S Globulins; 11S = 11S Globulins; Pro = Prolamins; Glu = Glutelins; M = Molecular weight marker (in kDa). An abrupt change is observed when contrasting separations performed under non-reducing (**a**) and reducing conditions (**b**). The high molecular weight aggregates disappear in the presence of a reducing agent, which indicates that a considerable number of disulfide bonds are involved in the stability of these proteins in their native state

analysis, it is now known that AmA1 has agglutinin domains and shares 97% identity with an *A. caudatus* orthologue (Fig. 5.5), which has been crystallized in the presence of the Gal-β1,3-GalNAc-α-O-benzyl ligand (Transue et al. 1997).

Some other important proteins isolated from the albumins fractions include α-amylase inhibitor and trypsin inhibitors, cystatin and a Cu/Zn superoxide dismutase (SOD). The latter was classified as an intrinsically disordered protein (IDP) having high thermal stability (Chagolla-López et al. 1994; Montero-Morán et al. 2015; Valdés-Rodríguez et al. 1993, 2007).

5.4.2 Globulins

In terms of functionality, globulins are an important protein fraction that is well represented in legumes. Soybean 11S globulins have a key role in the establishment of certain functional properties used in the preparation of diverse food products (Morr 1990; Swanson 1990). In amaranth seeds, as in legumes (e.g. soybean, peas, French beans) and some cereals (e.g. oats, rice), 11S globulins represent an important fraction comprising 13.7–18.1% of the total seed protein content (Barba de la Rosa et al. 1992b). On the other hand, amaranth 7S globulins are less abundant and, consequently, are less studied than the 11S globulins.

Native 7S globulins are typically large trimeric proteins of ~150–190 kDa, whose lack of cysteine residues impedes disulfide bond formation. Native 11S globulins consist of six monomers that interact in a non-covalent way, each monomer containing one acidic and one basic subunit linked by a disulfide bond, producing a hexameric protein of ~300–400 kDa. The electrophoretic pattern of amaranth seed 7S globulins, under reducing conditions, shows the main band of ~38 kDa (Fig. 5.6a), while the 11S globulins are characterized by the presence of the acidic (30–35 kDa) and basic (~20 kDa) subunits (Fig. 5.6b).

The 11S and 7S globulins have similar properties, including the ability to form both trimeric and hexameric structures and their similar tertiary

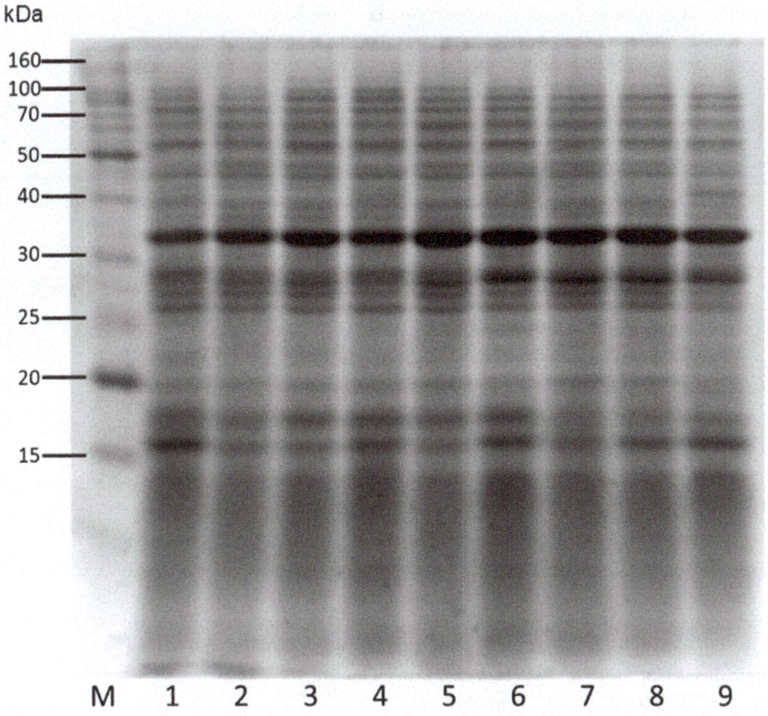

Fig. 5.4 Electrophoretic pattern of amaranth seed albumins extracted from different grain amaranth species, hybrids, and races. Albumins show a main a 35 kDa band. M = molecular weight marker (in kDa). Lane 1 = *A. cruentus* cv. Africana Cristalina Col 805-624; 2 = *A. hypochondriacus* cv. Mercado PSM_0232GA; 3 = *A. hypochondriacus* cv. Mercado PSM_0245GA; 4 = *A. hypochondriacus* cv. Nutrisol; 5 = *A. hypochondriacus* cv. Mercado Revancha; 6 = *A. hypochondriacus x hybridus* cv. Blanca Cruza 1; 7 = *A. hypochondriacus x hybridus* cv. Café Cruza 2; 8 = *A. hypochondriacus x hybridus* cv. Café Cruza 3; 9 = *A. hypochondriacus x hybridus* cv. Negra Cruza 4

```
1JLX_AMACA          -AGLPVIMCLKSNNHQKYLRYQSDNIQQYGLLQFSADKILDPLAQFEVEPSKTYDGLVHI  59
Q8S390_AMAHP        MAGLPVIMCLKSNNNQEYLRYQSDNIQQYGLLQFSADKILDPLAQFEVEPSKTYDGLVHI  60
                     *************:*:*******************************************

                                  ↓↓ ↓
1JLX_AMACA          KSRYTNKYLVRWSPNHYWITASANEPDENKSNWACTLFKPLYVEEGNMKKVRLLHVQLGH  119
Q8S390_AMAHP        KSRYTNKYLVRWSPNHYWITASANEPDENKSNWACTLFKPLYVEEGNMKKVRLLHVQLGH  120
                    ************************************************************

                        ↓          ↓
1JLX_AMACA          YTQNYTVGGSFVSYLFAESSQIDTGSKDVFHVIDWKSIFQFPKGYVTFKGNNGKYLGVIT  179
Q8S390_AMAHP        YTENYTVGGSFVSYLFAESSQIDTGSKDVFHVIDWKSIFQFPKTYVTFKGNNGKYLGVIT  180
                    **:**************************************** ****************

1JLX_AMACA          INQLPCLQFGYDNLNDPKVAHQMFVTSNGTICIKSNYMNKFWRLSTDDWILVDGNDPRET  239
Q8S390_AMAHP        INQLPCLQFGYDNLNDPKVAHQMFVTSNGTICIKSNYMNKFWRLSTDNWILVDGNDPRET  240
                    ***********************************************:************

1JLX_AMACA          NEAAALFRSDVHDFNVISLLNMQKTWFIKRFTSGKPGFINCMNAATQNVDETAILEIIEL  299
Q8S390_AMAHP        NEAAALFRSDVHDFNVISLLNMQKTWFIKRFTSGKPEFINCMNAATQIVDETAILEIIEL  300
                    ************************************ ********** ************

1JLX_AMACA          GQNN  303
Q8S390_AMAHP        GSNN  304
                    *.**
```

Fig. 5.5 Sequence alignment of AmA1 (Q8S390_AMAHP) and its orthologue present in *A. caudatus* (1JLX_AMACA). A high degree of identity between both proteins is evident. The amino acids conforming to the binding sites for the Gal-β1,3-GalNAc-α-O-benzyl substrate are indicated with an arrow (N74, H75, W77, Y124, and F135)

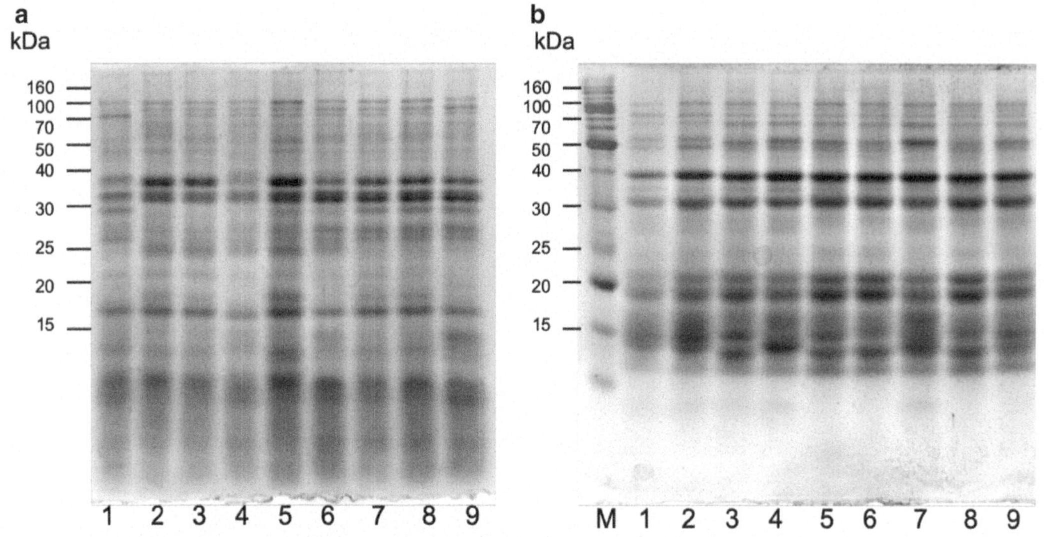

Fig. 5.6 Electrophoretic patterns, under reducing conditions, of **a** 7S and **b** 11S amaranth seed globulins extracted from different species, cultivars, and races. Lane 1 = *A. cruentus* cv. Guatemalteca Col 805-620; 2 = *A. cruentus* cv. Guatemalteca Col 805-545; 3 = *A. cruentus* cv. Guatemalteca Col 805-648; 4 = *A. cruentus* cv. guatemalteca Col SM 15-86; 5 = *A. cruentus* cv. Mexicana Col 161; 6 = *A. cruentus* cv. Mexicana Col 167; 7 = *A. cruentus* cv. Mexicana Café; 8 = *A. hypochondriacus* cv. Nepal Col 9; 9 = *A. hypochondriacus* cv. Nepal Col 160

structure. This led to the conclusion that they evolved from a common ancestral gene (Lawrence et al. 1994). Globulins are rich in essential amino acids, with high amounts of lysine (3.7–7.6%) and sulfur amino acids, Met + Cys (3.1–7.1%). They also have high levels of arginine (Arg), glutamine (Gln), glutamic acid (Glu), asparagine (Asn), and aspartic acid (Asp) (Barba de la Rosa et al. 1992b; Konishi et al. 1985b; Marcone and Yada 1991; Segura-Nieto et al. 1992).

In amaranth, as in other seeds, 7S globulins have been difficult to purify and characterize (Shewry et al. 1995). On the other hand, the amaranth 11S globulin (Ah11S) was the first amaranth cloned and characterized globulin (Barba de la Rosa et al. 1996). The crystallographic structure of Ah11S (Fig. 5.7) showed high similarity with 11S globulins of other plants, including soybean, pumpkin, pea, and rapeseed (Tandag-Silvas et al. 2012). High nutritional qualities have been attributed to this protein, considering that it contains higher levels of lysine and sulfur amino acids than other legumes and that, in general, it contains an excellent balance of essential amino acids, close to that required to meet human nutritional needs, as established by the FAO guidelines (FAO 1973). Because of its high nutritive value, the *Am11S* gene has been used for the genetic transformation of cultivated crops, such as corn, with the aim to increase its nutritional quality (Luna-Suárez et al. 2008; Rascón-Cruz et al. 2004; Valdez-Ortiz et al. 2005).

5.4.3 Prolamins

Prolamins are proteins associated with celiac toxicity (Ballabio et al. 2011; Valcárcel-Yamami and Caetano Da Silva Lannes 2012). In amaranth, this faction represents from 0.7 to 1.3% of the total protein seed content (Barba de la Rosa et al. 1992a). It is characterized by the presence of low molecular weight proteins <20 kDa and for the formation of very high molecular weight protein aggregates, which are not resolved by SDS-PAGE (Fig. 5.8). Immunological assays showed that amaranth prolamins do not have a cross-reaction with antibodies of celiac disease patients able to recognize wheat gliadin. Therefore, amaranth seed

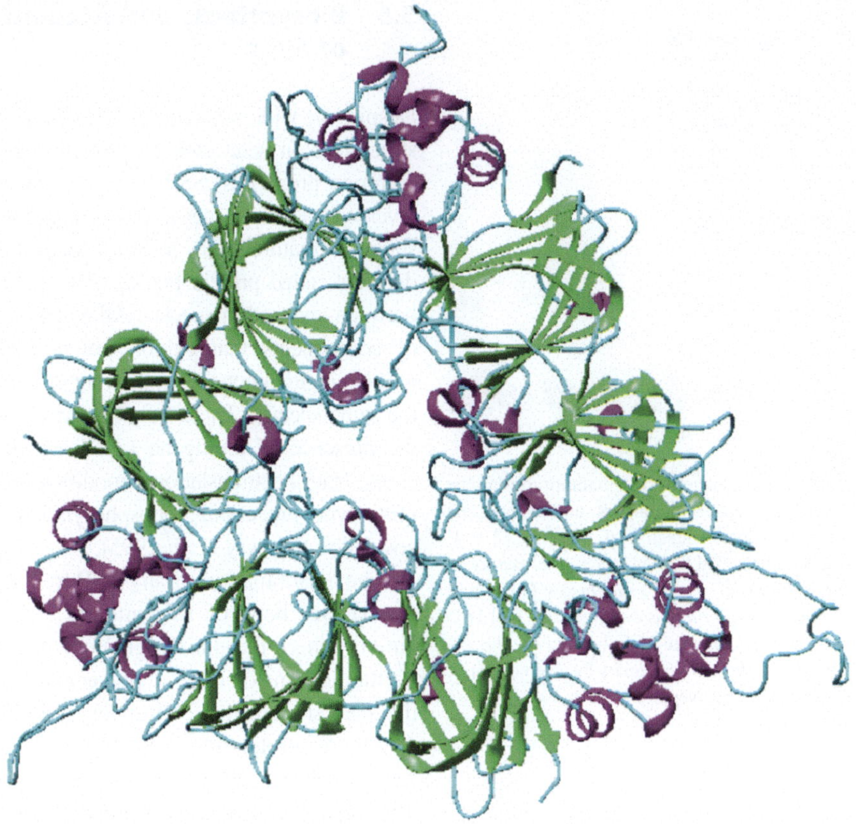

Fig. 5.7 3D model of the Ah11S protein structure. The *Ah11S* gene was cloned into *Escherichia coli* and the recombinant protein was subsequently purified and crystallized. Green and magenta segments represent the β-barrel α-helix domains, respectively (PDB accesion number 3QAC)

proteins have been considered as a safe and promising food due to its lack of gluten (Bergamo et al. 2011; Saturni et al. 2010).

5.4.4 Glutelins

Glutelins are the second most important fraction in amaranth SSPs. Their content in seeds is similar to globulins. Interestingly, glutelins and 11S globulins have a similar electrophoretic profile (Fig. 5.9). This suggests that amaranth seed glutelins could be 11S globulins present in a high aggregation state or with a higher proportion of hydrophobic amino acids, that is, genetic variants with diminished solubility in neutral polar conditions (Barba de la Rosa et al. 2009; Bojórquez-Velázquez et al. 2019). Another fraction, intermediate between globulins and glutelins, called globulin-P, has been reported (Abugoch et al. 2003). However, it appears likely that it was an experimental artefact produced by the protein extraction conditions.

Glutelins are rich in leucine (5.9–10.2%) as well as in aromatic amino acids, Phe + Tyr (6.7–10.3%). In this fraction, the lunasin-like peptide, having important cancer-preventive properties, was detected (Maldonado-Cervantes et al. 2010; Silva-Sánchez et al. 2008). Clear differences in glutelin fractions have been observed amongst *Amaranthus* species (Fig. 5.9), due to this fact, the utilisation of electrophoretic profiles of this particular protein fraction has been proposed for the identification of amaranth accessions (Dzunková et al. 2011).

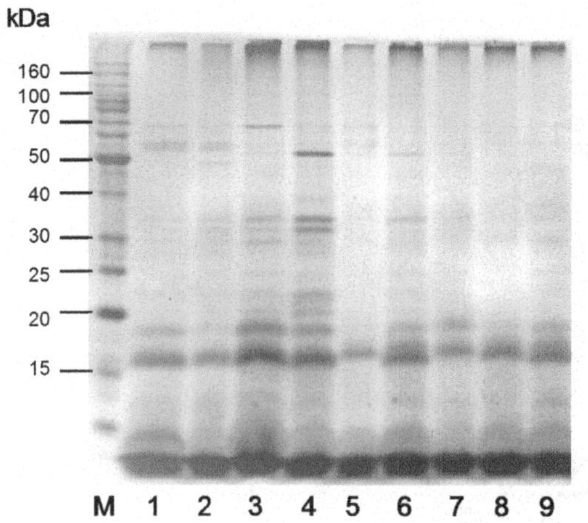

Fig. 5.8 Electrophoretic pattern of amaranth prolamins extracted from different species, cultivars, and races. Lane 1 = *A. cruentus* cv. Guatemalteca Col 805-620; 2 = *A. cruentus* cv Guatemalteca Col 805-545; 3 = *A. cruentus* cv Guatemalteca Col 805-648; 4 = *A. cruentus* cv Guatemalteca Col SM 15-86; 5 = *A. cruentus* cv. Mexicana Col 161; 6 = *A. cruentus* cv. Mexicana Col 167; 7 = *A. cruentus* cv. Mexicana Café; 8 = *A. hypochondriacus* cv. Nepal Col 9; 9 = *A. hypochondriacus* cv. Nepal Col 160

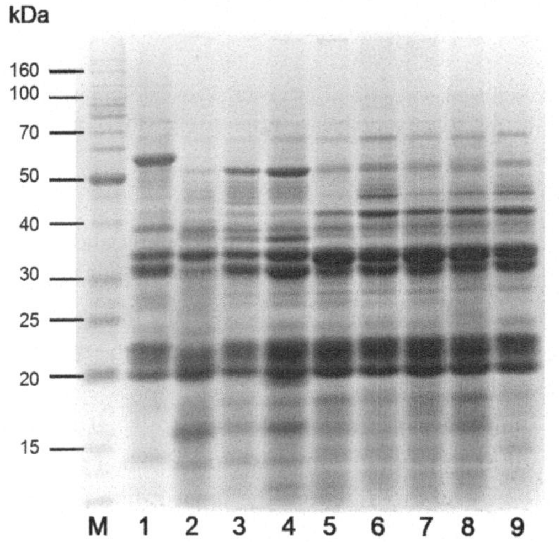

Fig. 5.9 Electrophoretic pattern of amaranth glutelins extracted from different species, cultivars, and races. M = molecular weight markers. Lane 1 = *A. cruentus* cv. Africana Cristalina Col 805-624; 2 = *A. hypochondriacus* cv. Mercado PSM_0232GA; 3 = *A. hypochondriacus* cv Mercado PSM_0245GA; 4 = *A. hypochondriacus* cv. Azteca Nutrisol; 5 = *A. hypochondriacus* cv. Mercado Revancha; 6 = *A. hypochondriacus x hybridus* cv. Blanca Cruza 1; 7 = *A. hypochondriacus x hybridus* cv. Café Cruza 2; 8 = *A. hypochondriacus x hybridus* cv. Café Cruza 3; 9 = *A. hypochondriacus x hybridus* cv. Negra Cruza 4

5.5 Biosynthesis and Accumulation of SSPs

Despite the wide variation of SSPs in their detailed structures, all seed storage proteins have several common properties. SSPs are synthesized at high levels in specific tissues and at certain development stages. In mature seeds, SSPs are located in discrete deposits called protein bodies (PB) that may vary between and within species. All SSPs are mixtures of components that exhibit polymorphism both within single genotypes and amongst genotypes of the same species. These polymorphisms arise from the presence of multigene families and, in some cases, from posttranslational modifications such as proteolytic processing, phosphorylation, and glycosylation (Shewry et al. 1995).

The accumulation of any SSP fraction reflects the balance between its synthesis and its degradation, where the main regulatory step seems to be the amount of mRNA present in the cells of the seed storage tissue (Bewley and Greenwood 1990). Accumulation of SSPs occurs during the last two-thirds of its development, beginning after cell division and completed embryo formation, and ending at the stage of seed maturation and drying. The time, speed, and extent of synthesis vary depending on each seed and in each type of SSPs. Although these variables are genetically defined, the environment also plays an important regulatory role (Higgins 1984).

SSPs are synthesized in the rough endoplasmic reticulum (RER) as large precursors called preproprotein and are transported to the RER lumen thanks to a signal peptide found at its N-terminal, which is proteolytically removed during this transition (Fig. 5.10) (Miernyk and Hajduch 2011). During SSPs permanence and transit in the RER, the pre-proteins are subjected to initial processing including the formation of disulfide bonds, incorporation of glycosylation and folding via chaperones, which prevent the incorrect formation of inter- and intramolecular interactions. The pro-proteins are transported through the secretory route to their corresponding storage or deposition organelles. According to the classification noted by Fukushima (1991), there are two

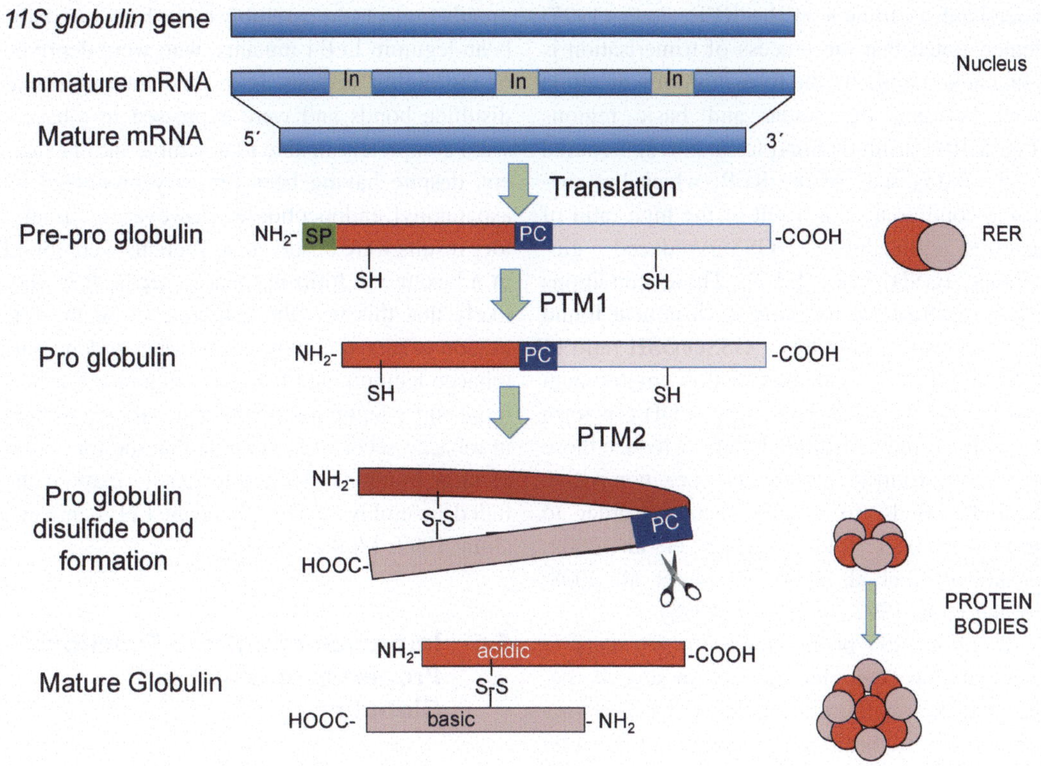

Fig. 5.10 Schematic representation of globulins biosynthesis and processing in legumes

main ways by which SSPs can be transported: the first is the prolamins route while the second corresponds to albumins and globulins. In the former, the organelles where prolamins are stored result from direct RER vesiculation and constitute the PBs, meaning that prolamins do not follow the classic secretion route. On the other hand, globulins are transported via the Golgi apparatus to Protein Storage Vacuoles (PSV), the site where pro-globulins are assembled and deposited as mature proteins (Fukushima 1991; Hara-Nishimura 1987; Mandal and Mandal 2000; Miernyk and Hajduch 2011; Shewry et al. 1995; Tandang-Silvas et al. 2011).

It is very common for pro-globulins to undergo glycosylation, regularly in the Asn-X-Ser/Thr sequence and to remain N-glycosylated during their passage through the RER. Occasionally, N-glycosylation is subsequently removed during proteolytic processing. However, these modifications may persist and be present in the fully processed and mature SSPs (Tandang-Silvas et al. 2011). In its mature form, 11S globulins are processed to form the acidic and the basic subunits covalently bound by a disulfide bond (Shewry et al. 1995), which regulates 11S globulins assembly (Fig. 5.10).

Once the disulfide bonds have been formed and the trimers assembled, 11S globulins are transported through the Golgi apparatus to the PSV, where each monomer is subjected to a proteolytic excision, which is a pre-prerequisite for the assembly of the protein in its mature hexameric form. It has been established that the Vacuolar Processing Enzyme (VPE) an asparaginyl endopeptidase, acts on a specific site of the pro-globulin, between highly conserved asparagine and glycine (Asn-Gly) residues. This contributes to the formation of the aforementioned acid and basic subunits (Chua et al. 2008; Jung et al. 1998; Müntz 1998; Shewry et al. 1995; Tandang-Silvas et al. 2010a).

After synthesis and removal of the signal peptide, the prolegumins monomers are assembled as trimers in the RER. Jung (1997) demonstrated that the process of trimerisation is stimulated mainly by the formation of a disulfide bond between the acidic and basic regions (Fig. 5.10). Disulfide bond formation is favoured by the redox state of the RER, which has oxidising conditions as a result of the high ratio of oxidized glutathione (GSSG)/reduced glutathione (GSH) (i.e., 1:4.7). These conditions highly contrast the reducing environment found in the cytosol, which has a GSSG/GSH ratio of 1:30 to 1:100. It is thanks to this environment that Protein Disulfide Isomerase (PDI) can form the native protein disulfide bonds or reduce those that have formed incorrectly, i.e. non-native disulfide bonds, to modify them in order to promote proper folding (Bass et al. 2004; Chakravarthi et al. 2006; Dixon et al. 2008; Hwang et al. 1992; Jung et al. 1998).

The transition phenomenon from trimers to hexamers has been characterized in several species and thoroughly elucidated by the 3D structure of the soybean pro-glycinin and glycinin (Adachi et al. 2001, 2003; Duranti et al. 2008). However, in vitro assays revealed that trimers that do not undergo a proteolytic rupture cannot be assembled as hexamers, unless they are treated with papain, a promiscuous protease able to cleave at several sites in the protein. Jung et al. (1998) exposed soybean pro-glycinin and faba bean pro-legumin, whose Asn-Gly site was changed of position or suppressed, to the activity of purified asparaginyl endopeptidase, and managed to reproduce only the specific cleavage known to occur in the in vivo process, thereby establishing that it is only when the trimers are treated in this way that they can be assembled as hexamers (Jung et al. 1998; Müntz 1998). This rupture may cause a conformal change that is favourable for the process, as reported by Adachi et al. (2003), from their analysis of the crystalline structure of the soybean glycinin A3B4 homohexamer. These authors concluded that exposure of the hydrophobic residues necessary for the face-to-face association of the two trimers, may take place only after the excision.

Once globulins are in the trimeric form, the subsequent correct formation of disulfide bonds greatly impacts the assembly of the hexamer. Faba bean legumin LeB4 mutants, who were deprived of the ability to form intra- and intermolecular disulfide bonds and were expressed in tobacco, were completely unable to assemble into hexamers, despite having been properly processed by asparaginyl endopeptidase. However, contradictory results were observed, as proteins were found in a hexameric form in tobacco seeds. It is very likely that this was the consequence of the formation of mixed complexes between endogenous tobacco legumin and faba bean mutants. Furthermore, since legumins are the main storage protein in tobacco seeds, it is possible that the formation of large amounts of these hexamers masked the failed assembly of less abundant LeB4 mutants (Jung 1997; Müntz 1998).

5.6 Improvement of the Functional Properties of Amaranth Globulins

Food proteins should have good nutritional qualities and adequate physicochemical functions to allow their proper incorporation into processed foods, in addition to health-promoting. The nutritional quality of proteins is determined by its amino acid content (Maruyama et al. 2011). In this sense, amaranth proteins are considered of high quality; however, their physicochemical properties, including gelation, emulsification, foaming, and water-absorption fare poorly when compared to soy proteins. Because functional properties depend on the protein structure, protein engineering should take into consideration that SSPs are in a folded conformation that facilitates packaging into the protein bodies, and that if severe changes in structure occur, their accumulation will not take place and will be consequently unavailable at the germination stage (Utsumi 1992).

Comparison of the amino acid sequences of several 11S globulins shows they have five variable (I–V) regions and four conserved domains (Fig. 5.11). The variable domains have a minor influence in the formation and maintenance of the globulin structure, a property that explains why

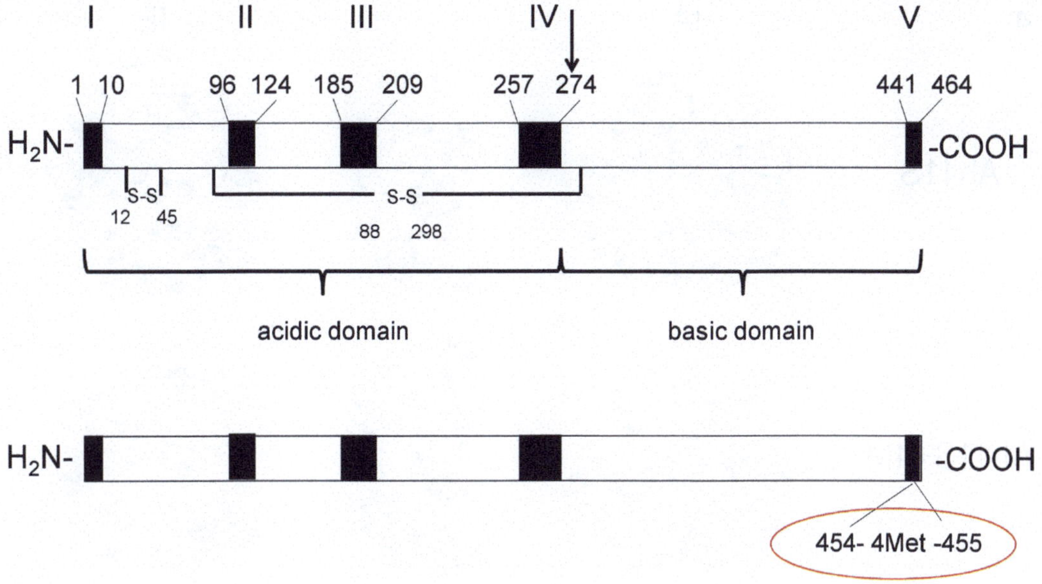

Fig. 5.11 Design of the grain amaranth globulins modifications that resulted in improved functional properties. The variable regions are named I to V, and the arrow indicates the proteolytic processing site that generates the acidic and basic subunits (Modified from Carrazco-Peña et al. 2013)

this region may tolerate modifications. For example, Carrazco-Peña et al. (2013) engineered a modification of the 11S amaranth gene that involved the inclusion of a cassette of our methionine residues designed to increase some functional properties. These results showed that this modification caused changes in the surface hydrophobicity and electrostatic potential (Fig. 5.12). This modification also increased the self-supporting gels formed, which had an intermediate hardness between the native and recombinant Ah11S. These functional properties are important in the food industry for the development of new products (Carrazco-Peña et al. 2013).

5.7 Phylogenetic Relationships of Amaranth Seed Storage Proteins

5.7.1 Albumins

The most representative SSPs of dicotyledonous plants are the 2S albumin and both the 7S and 11S globulins (Shewry 2015), and grain amaranth is no exception. As previously mentioned, AmA1 has high biochemical and nutritional relevance, due to its abundance in the seed and its applications for crop improvement; however, AmA1 does not belong to any of the storage protein categories, considering that the classification as albumin was based solely on its solubility in water. The canonical 2S albumins of amaranth have been scarcely studied since research on these proteins has not gone beyond the establishment of their electrophoretic profiles and the characterisation of their general basic biochemical properties such as assembly capacity and quaternary structure.

The low molecular weight monomeric units, of ~15 kDa, are constituted by two dissimilar subunits of 8–10 kDa and 4–5 kDa. These are originally synthesized as a single polypeptide chain that is subsequently proteolytically processed, but remain attached by two disulfide bonds (Barba de la Rosa et al. 2009; Janssen et al. 2017; Marcone et al. 1994; Shewry et al. 1995). With the publication of the amaranth genome and transcriptome (Clouse et al. 2016; Lightfoot et al. 2017), it is now possible to access the amino acid sequences of amaranth 2S albumins and analyse their relationships with

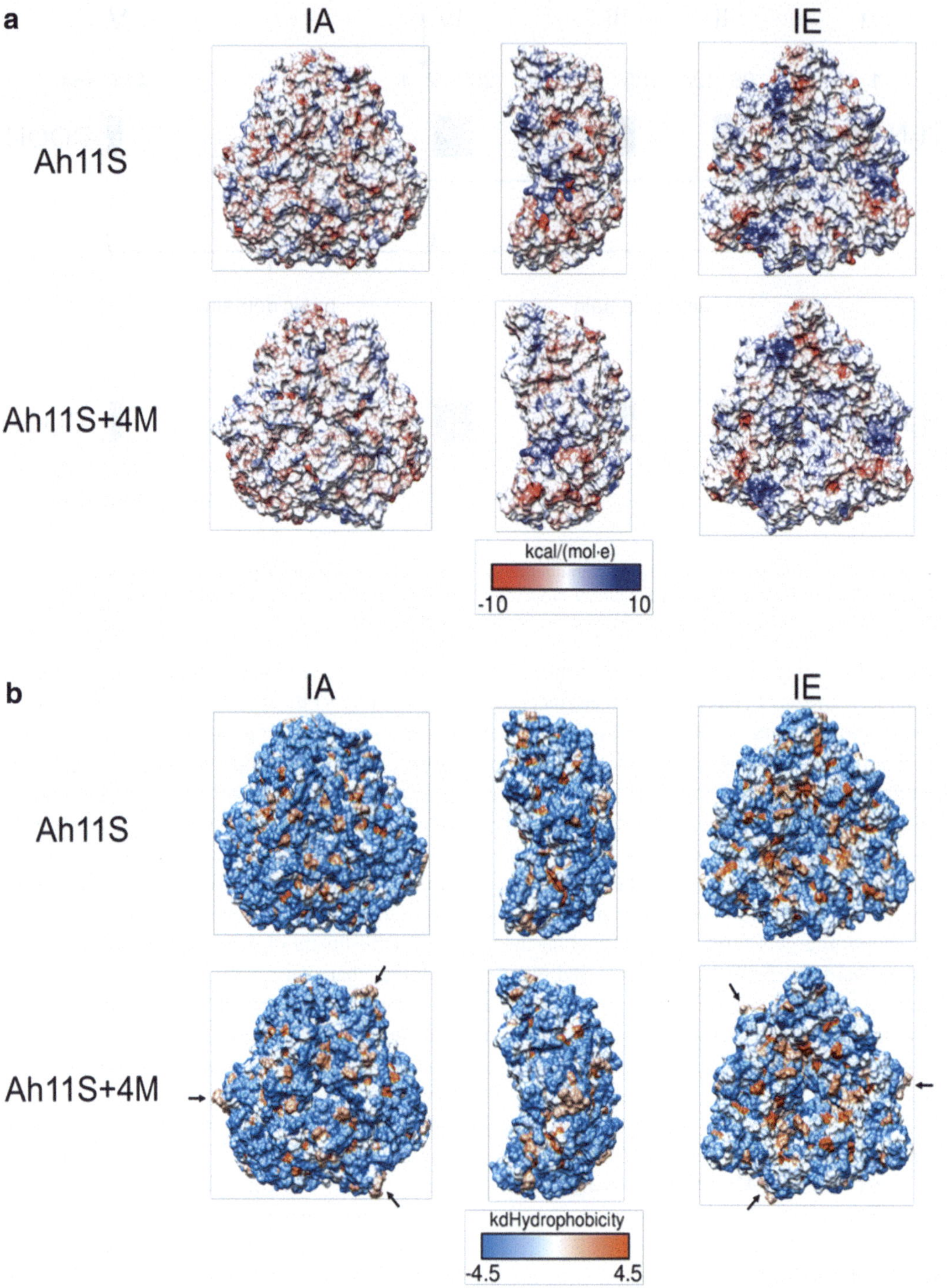

Fig. 5.12 Surface structural properties comparison of wild-type Ah11S and the Ah11S + 4Met modification. **a** Coulombic distribution and **b** topography of the hydrophobic surface. The incorporation of 4 methionine residues in the variable region V impacts not only the content of this amino acid but also the overall physicochemical properties of the protein. These changes result from the exposure of certain amino acids that can be observed in the bot, intra- and inter-chain disulphide bond containing faces (IA and IE), respectively. The location of the four additional methionine residues is indicated by arrows in the hydrophobic surface panel

orthologous proteins. In this sense, four 2S amaranth albumins sequences were detected by BLAST in the Phytozome databases (https://phytozome.jgi.doe.gov/pz/portal.html), with accession numbers AHYPO_006419-RA, AHYPO_006420-RA, AH009936-RA and AH009937-RA and lengths between 132 and 142 amino acids. The alignment shown (Fig. 5.13) shows that AHYPO_006419-RA and AH009936-RA have identical sequences, which is the same protein reported by different authors, while AHYPO_006420-RA and AH009937-RA differ only at the N-terminal end. All these proteins are rich in cysteine.

A phylogenetic tree of 2S albumins was constructed using MEGA-X (Kumar et al. 2018). The orthologous protein sequences of plant species belonging to four orders (Poales, Caryophyllales, Fabales, and Brassicales) were used. It was observed that amaranth 2S albumins are perfectly grouped within the Caryophyllales (Fig. 5.14). An interesting outcome was that the 2S albumins belonging to species of the Caryophyllales order grouped in a different clade from those in the Fabales (e.g., legumes such as soybean and peanuts) and Brassicales (e.g. Arabidopsis and canola) despite being dicotyledonous. The separation from albumins of the Poales order (e.g. cereals such as wheat and rice) was expected since these are evolutionarily more remote monocotyledonous plant species.

The 2S albumin from dicotyledons plants has been reported as highly allergenic proteins, including those from *Arachis hypogaea* (peanut), *Brassica napus* (canola), *Fagopyrum esculentum* (buckwheat), *Glycine max* (soybean) and *Sinapis alba* (mustard), among others (Moreno and Clemente 2008). Conversely, there are reports on allergenicity caused by similar proteins in species belonging to the Amaranthaceae or Chenopodiaceae families. As shown in Fig. 5.15, contrary to 2S albumins of allergy-causing species, amaranth albumin sequences were found to lack the allergy-causing IgE-binding.

5.7.2 Globulins

Both, 7S and 11S globulins, are SSPs belonging to the cupin superfamily since they have two barrel-shaped domains in their structures. In the case of 7S globulins, they may also contain vicilin-like domains, which give rise to peptides with antimicrobial activity once the protein is hydrolysed during germination. Probably due to its low abundance in seeds, there are few studies that have focused on amaranth 7S globulins, since it is difficult to obtain sufficient quantities of these proteins with an adequate degree of purity to implement extensive tests (Marcone 1999). However, the recent benefits of the omics sciences, which include sensitivity, permitted the

```
                                                                        ↓           ↓
AHYPO_006419-RA     MAGIKFMILAAAMAAMVVITHATIITTEVEIDDDSNPIFGRGKCERQLQSKWPNHCEEYM     60
AH009936-RA         MAGIKFMILAAAMAAMVVITHATIITTEVEIDDDSNPIFGRGKCERQLQSKWPNHCEEYM     60
AHYPO_006420-RA     MATKNVQLLLAAIATYLLI------STFKPTGATSNKGELSSQCERQLQWEKIDHCKNYL     54
AH009937-RA         MATKNVQLLLAAIATYLLI------STFKPTGATSNKGELSSQCERQLQWEKIDHCKNYL     54
                    **  :. :* **:*: ::*      :*    .  **     .:****** :  :**::*:

                                         ↓↓         ↓ ↓
AHYPO_006419-RA     MEEMR----PNQG--RGYFNQCCNELKMMDRECQCEAMKKMVQQMQ-----KSEREEMMD     109
AH009936-RA         MEEMR----PNQG--RGYFNQCCNELKMMDRECQCEAMKKMVQQMQ-----KSEREEMMD     109
AHYPO_006420-RA     RSEINMLMFPNSGNQACCLRECCEMLQQMDENCVCEVVESMAEEETEMMMKRGGEGRQMG     114
AH009937-RA         RSEINMLMFPNSGNQACCLRECCEMLQQMDENCVCEVVESMAEEETEMMMKRGGEGRQMG     114
                     .*:.    **.*     :.:**: *: **.:* **.::.*.::      :. . .  *.

                                 ↓
AHYPO_006419-RA     YEMMEQAMKLPMMCGTMQSMCPMRGYRRSPVV   141
AH009936-RA         YEMMEQAMKLPMMCGTMQSMCPMRGYRRSPVV   141
AHYPO_006420-RA     RVVLKKAMHLPQRCVVQRSNPRSKGKWK----   142
AH009937-RA         RVVLKKAMHLPQRCASIV--------------   132
                      ::::**:**  *
```

Fig. 5.13 BLAST alignment of grain amaranth 2S albumins. Arrows indicate conserved cysteines that are involved in the formation of disulfide bonds

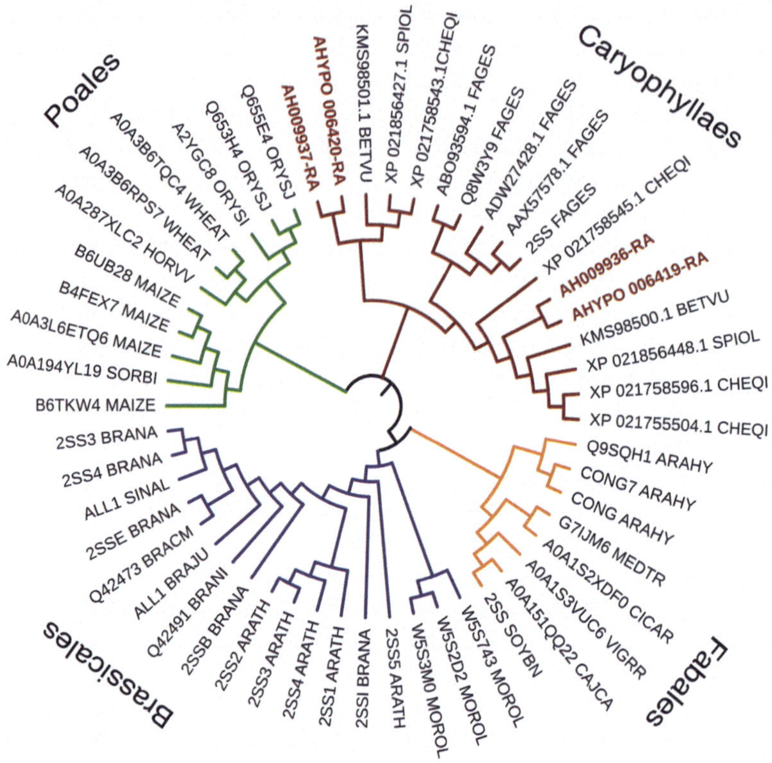

Fig. 5.14 Phylogenetic relationship of grain amaranth 2S albumins with those of other species. **Poales**; MAIZE = *Zea mays*; ORYS = *Oryza sativa*; SORBI = *Sorghum bicolor*; HORVV = *Hordeum vulgare*. **Caryophyllales**; AHYPO = *Amaranthus hypochondriacus*; BETVU = *Beta vuglaris*; CHEQI = *Chenopodium quinoa*; SPIOL = *Spinacia oleracea*; FAGES = *Fagopyrum esculentum*. **Fabales**: ARAHY = *Arachis hypogaea*; CICAR = *Cicer arietinum*; MEDTR = *Medicago truncatula*; VIGRR = *Vigna radiata*; SOYBN = *Glycine max*; CAJCA = *Cajanus cajan*. **Brassicales**: ARATH = *Arabidopsis thaliana*; BRANA = *Brassica rapa*; BRANI = *Brassica nigra*; SINAL = *Sinapis alba*; MOROL = *Moringa oleifera*. Amaranth sequences are highlighted in bold red. The sequences Red, Caryophyllales; Orange, Fabales; Purple, Brassicales; Green, Poales. The sequences were aligned with ClustalW and the phylogenetic tree was constructed using MEGA-X by the Neighbour-Joining method with 1000 Bootstrap replicates

identification of at least four canonical 7S globulins with cupin domains in gran amaranth seeds, two of which also contain vicilin-like domains (Bojórquez-Velázquez et al. 2019). Figure 5.16 represents the cladogram resulting from the phylogenetic analysis of these sequences, in which an interesting behaviour can be observed, since two of the sequences (AHYPO_007944-RA and AHYPO_018839-RA) cluster together with the 7S globulins of the Caryophyllales order, to which grain amaranth belongs, while the other two sequences (AHYPO_006304-RA and AHYPO_010140-RA) are grouped with members of the Brassicales order. Through a search in the Conserved Domains Database of the NCBI (Marchler-Bauer et al. 2005), it was observed that only the sequences of the species belonging to the order Caryophyllales contain a vicilin-like domain, which is present in AHYPO_007944-RA and AHYPO_018839-RA but are absent in AHYPO_006304-RA and AHYPO_010140-RA. This difference partially explains this behaviour; however, it is unknown why the latter two sequences are more closely related proteins belonging to the Brassicales order.

The 11S globulins are present through multigenic families containing three to five genes that codify for these proteins in a given species,

```
AHYPO_006419-RA    MAGIKFMILAAAMAAMVVITHATIITTEVEIDDDSNPIFGRGKCERQLQ-SKWPNHCEEY    59
AHYPO_006420-RA    MATKNVQLLLAAIATYLLI------STFKPTGATSNKGELSSQCERQLQ-WEKIDHCKNY    53
AH009937-RA        MATKNVQLLLAAIATYLLI------STFKPTGATSNKGELSSQCERQLQ-WEKIDHCKNY    53
CONG7_ARAHY        MAK---LTILVALALFLLAAHASARQQW--------ELQGDRRCQSQLE-RANLRPCEQH    48
ALL1_SINAL         -----------------------PAGPF--------R---IPKCRKEFQQAQHLRACQQW    26
ALL1_BRAJU         ------------------------AGPF--------R---FPRCRKEFQQAQHLRACQQW    25
                                                            :*. :::      *::

AHYPO_006419-RA    MMEEMR-------------------------------P----NQ-G--RGYFNQCCNELK    81
AHYPO_006420-RA    LRSEINM---------------------------LMFP----NS-GNQACCLRECCEMLQ    81
AH009937-RA        LRSEINM---------------------------LMFP----NS-GNQACCLRECCEMLQ    81
CONG7_ARAHY        LMQKIQRDEDSYGRDPYSPSQDPYSPSQDPDRRDPYSPSPYDRRGAGSSQHQERQCNELN    108
ALL1_SINAL         LHKQAMQ----SGSGPSWTLDDEFDFEDD--MENPQGP-------QQRPPLLQQCCNELH    73
ALL1_BRAJU         LHKQAMQ----SGSGP-----------------QPQGP-------QQRPPLLQQCCNELH    57
                   : .:                                *          ..**: *:

AHYPO_006419-RA    MMD--RECQCEAMKKMVQQMQ-----KSEREEMMDY-------EMMEQAMKLPMMCGTMQ    127
AHYPO_006420-RA    QMD--ENCVCEVVESMAEEETEMMMKRGGEGRQMGR-------VVLKKAMHLPQRCVVQR    132
AH009937-RA        QMD--ENCVCEVVESMAEEETEMMMKRGGEGRQMGR-------VVLKKAMHLPQRCASIV    132
CONG7_ARAHY        EFENNQRCMCEALQQIMENQSDRL--QGRQQ---EQ-------QFKRELRNLPQQCGLRA    156
ALL1_SINAL         QEE--PLCVCPTLKGASKAVKQQVRQQLGQQGQQGPHLQHVISRIYQTATHLPKVCNIRQ    131
ALL1_BRAJU         QEE--PLCVCPTLKGASKAVKQQIRQQGQQQGQQGQQLQHEISRIYQTATHLPRVCNIPR    115
                     :    * * .::   :       :  .        . .   :**  *
```

Fig. 5.15 Amino acid sequences alignment between grain amaranth 2S albumins and those known allergy-causing species. Black boxes indicate the IgE binding epitopes that are absent in amaranth albumins. AHYPO_006419-RA, AHYPO_006420-RA, and AH009937-RA are from *Amaranthus hypochondriacus*; CONG7_ARAHY, *Arachis hypogaea*; ALL1_SINAL, *Sinapis alba*; ALL1_BRAJU, *Brassica juncea*

which favours their presence as heterohexamers in their mature state. In soybeans, for example, five paralogues have been identified, which are classified into two groups based on their amino acid sequence. Group I comprise the monomers A1aB1b (53.6 kDa), A1bB2 (52.2 kDa) and A2B1a (52.4 kDa) monomers, while group II contains the A3B4 (55.4 kDa) monomers and A5A4B3 (61.2 kDa) monomers. The sequence similarity between monomers of the same group is 82–86%, and 42–45% between both groups (Chua et al. 2008; Mills et al. 2002; Nielsen et al. 1989; Tandang-Silvas et al. 2011). These are by far the most widely studied storage proteins in dicot plants, and amaranth is no exception. A large number of investigations have focused on the study of amaranth 11S globulin, including the isolation of the gene that codes for this protein, in addition to mutagenic and physicochemical analyses that have focused on its application in key aspects of food science and technology, and on the establishment of its three-dimensional structure through X-ray crystallography (Barba de la Rosa et al. 1996; Carrazco-Peña et al. 2013; Osuna-Castro et al. 2000; Tandang-Silvas et al. 2010b, 2012).

Only one 11S globulin sequence was known in amaranth (i.e., Accession number in UniProtKB Q38712_AMAHP), but recently other paralogues that differentially accumulated between wild and domesticated species have been identified through comparative proteomics (Bojórquez-Velázquez et al. 2019). There are five sequences for this type of proteins in the databases deposited in the Phytozome repository, whose accession numbers are AHYPO_006768-RA, AHYPO_001411-RA, AHYPO_021282-RA, AH002656-RA and AH017742-RA (Clouse et al. 2016; Lightfoot et al. 2017). These proteins, despite sharing specific characteristics common in 11S globulins, are divided into three groups when subjected to the phylogenetic analysis (Fig. 5.17). This cladogram shows three main branches, branch A, which is subdivided into two groups, representing the Poales order and the Fabales, Brassicales, and Caryophyllales orders, respectively.

Thereby, each clade in A represents mono- and dicotyledonous plants, whereas branches B and C contain only members of the order Caryophyllales. Amaranth sequences contained in branch A are those considered as canonical

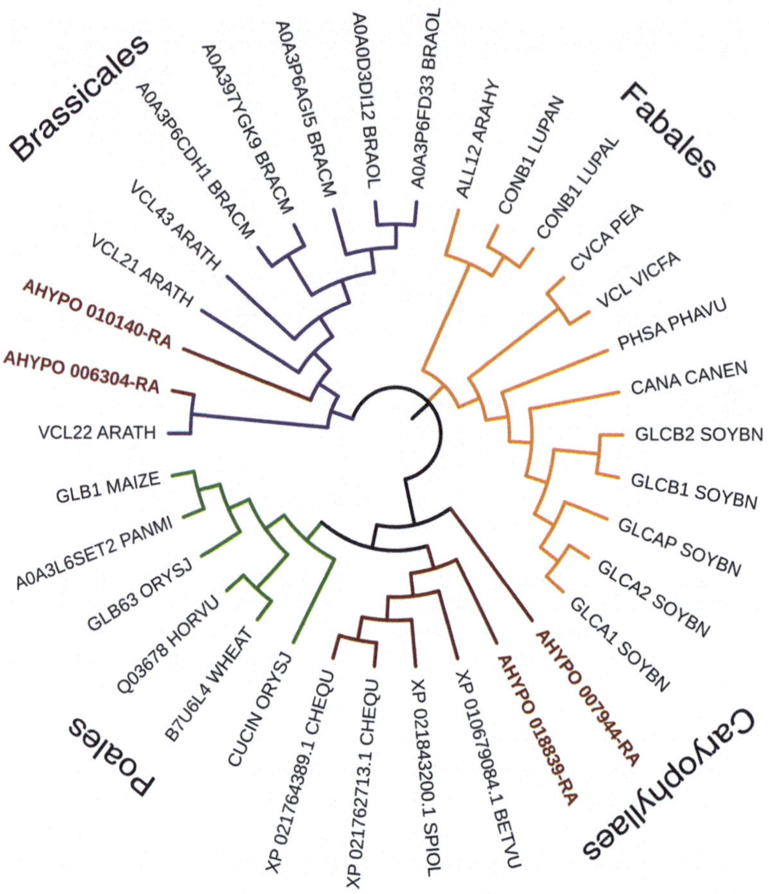

Fig. 5.16 Phylogenetic relationship of grain amaranth 7S globulins with their paralogues and orthologues. **Caryophyllales**: AHYPO = *Amaranthus hypochondriacus*; BETVU = *Beta vulgaris;* SPIOL = *Spinacia oleracea*; CHEQU = *Chenopodium quinoa*. **Poales**: ORYSJ = *Oryza sativa* japonica; HORVU = *Hordeum vulgare*; WHEAT = *Triticum aestivum*; MAIZE = *Zea mays*; PANMI = *Panicum miliaceum*. **Fabales**: ARAHY = *Arachis hypogaea*; LUPAN = *Lupinus angustifolius*; LUPAL = *Lupinus albus*; VICFA = *Vicia faba*; PEA = *Pisum sativum*; PHAVU = *Phaseolus vulgaris*; CANEN = *Canavalia ensiformis*; SOYBN = *Glycine max*. **Brassicales**: ARATH = *Arabidopsis thaliana*; BRACM = *Brassica campestris*; BRAOL = *Brassica oleracea*. Amaranth sequences are highlighted in bold red. As shown, the amaranth 7S globulins containing Vicilin-like domains are grouped in the Caryophyllales clade, while those in which these domains are absent are incorporated into the Brassicales order clade. The sequences were aligned with ClustalW and the phylogenetic tree was constructed using MEGA-X by the Neighbour-Joining method with 1000 Bootstrap replicas

11S globulins. Located in this branch is the widely studied Q38712_AMAHP sequence that has already been crystallized (PDB 3QAC), as well as and AHYPO_001411-RA and AH002656-RA, which were obtained from the genome of the *A. hypochondriacus* genome. The differences amongst them derive from the fact that they belong to different varieties or cultivars of the same species.

The separation of branches B and C suggests that there are other lineages for 11S globulins, all belonging to the Caryophyllales order. Branch B, groups representative of proteins of *A. hypochondriacus*, *B. vulgaris*, *C. quinoa.* and *S. oleracea*, as well as all reported the sequences of *F. esculentum* 11S globulins. The AHYPO_021282-RA amaranth sequence located in this branch is known as a high molecular mass

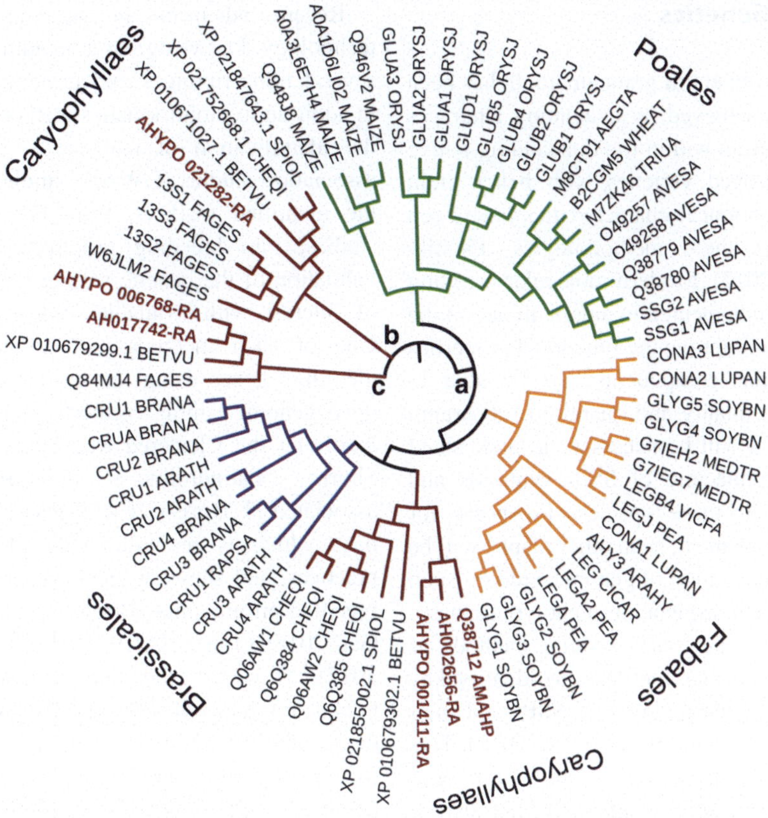

Fig. 5.17 Phylogenetic relationship of amaranth 11S globulins with their paralogues and orthologues. **a**, **b**, and **c**, denotes de different branches of segregation due to high variations in amino acid sequences, leading to different lineages of 11S globulins. **Poales**; MAIZE = *Zea mays*; ORYSJ = *Oryza sativa* japonica; AEGTA = *Aegilops squarrosa*; TRUA = *Triticum urartu*; WHEAT = *Triticum aestivum*; AVESA = *Avena sativa.* **Fabales**: LUPAN = *Lupinus angustifolius*; ARAHY = *Arachis hypogeae*; SOYBN = *Glycine max*; CICAR = *Cicer arietinum*; PEA = *Pisum sativum*; VICFA = *Vicia faba*; MEDTR = *Medicago truncatula*. **Caryophyllales**: AMAHP, AH, and AHYPO = *Amaranthus hypochondriacus*; BETVU = *Beta vulgaris;* SPIOL = *Spinacia oleracea*; CHEQI = *Chenopodium quinoa*; FAGES = *Fagopyrum esculentum*. **Brassicales**: ARATH = *Arabidopsis thaliana*; BRANA = Brasica napus; RAPSA = *Raphanus sativus*; BRACM = *Brassica campestris*; BRAOL = *Brassica oleracea*; Amaranth sequences are highlighted in bold red. The sequences were aligned with ClustalW and the phylogenetic tree was constructed using MEGA-X by the Neighbour-Joining method with 1000 Bootstrap replicates

11S globulin (Ah11SHMW) and contains 9 tandem repeats of G-S-E(Q)-W(R)-D(E)-P-R(S)-Y-P-G-H-G(E)-S-Q(E)-R-P-A(G/T)-H sequence not present in any other related protein. It has been reported that 11S globulins from *F. esculentum* also possess tandem repeats, although with a different sequence, which gives rise to variants of these proteins with different molecular masses (Bojórquez-Velázquez et al. 2019; Khan et al. 2012). The AHYPO_006768-RA and AH017742-RA 11S globulins that cluster in Clade C, are proteins with a length of 410 and 406 amino acids, respectively, and are approximately 100 residues smaller than the average size for proteins of this nature. AHYPO_006768-RA has also been reported to have high levels of phenylalanine, making it a highly hydrophobic protein (Bojórquez-Velázquez et al. 2019).

5.8 SSPs Genetics

Genetic improvement of grain amaranth has been systematically achieved by evaluating landraces and huge selection within the outstanding native varieties. Improved varieties with better grain yield and agronomical characteristics have been obtained using these methodologies (Espitia-Rangel et al. 2012). Modern molecular breeding tools will be implemented in amaranth due to the great interest to improve the functional properties of its molecular components. To be able to manipulate the quality and quantity of amaranth seed proteins, it will be necessary to understand the genetic architecture of SSPs synthesis and accumulation in these species. Prospects for amaranth improvement, as in other plants, will be based on the integration of genomic studies, such as whole-genome sequencing, genotyping, association mapping, and genome editing methods, as well as bioinformatics methods, to identify key genes regulating the quantity of SSPs and their amino acid composition (Gacek et al. 2018). This knowledge will serve as a guide for crop improvement, not only of amaranth, but also of legume and cereal grain crops (Bolon et al. 2010).

It has well documented that seed protein and oil contents are inversely correlated in soybean (Brim and Burton 1979; Cober and Voldeng 2000), whereas this relation appears to be different in grain (Bojórquez-Velázquez et al. 2018). Both seed protein content and seed size are inherited in a quantitative manner and several quantitative trait loci (QTLs) associated with these properties have been identified in different species including wheat (Uauy et al. 2006), rice (Zhang et al. 2008), pea (Timmerman-Vaughan et al. 2005), and barley (Jukanti et al. 2007). Candidate genes responsible for observable changes in seed protein and oil contents are alleles of the LG1 QTL (Bolon et al. 2010). The QTL associated with seed protein content, that mapped to soybean linkage group I (LGI), is of particular interest due that is associated with high protein accumulation and lower oil content, respectively (Brummer et al. 1997; Chung et al. 2003).

Recent advances in genome sequencing technology, knowledge of amaranth genome and several transcriptomes and proteome databases, in addition to bioinformatics methods, will allow the identification of markers in genome-wide association studies (GWAS). Implementation of the identified markers, from GWAS or QTL studies, into breeding programs will include validation of the genetic variant linked to a trait of interest with available molecular methods. One of such methods includes targeting the identified genes using the most recently developed genome editing methods, such as the one involving the Clustered Regulatory Interspaced Short Palindromic (CRISPR)/CRISPR-associated 9 (Cas9) (CRISPR/CAS9) technology, which is revolutionising plant breeding practices, and showing great potential for future breeding programmes (Bortesi and Fischer 2015; Gao 2018; Li et al. 2019; Rodriguez-Leal et al. 2017). However, the implementation of this potent strategy, will first require the development of an efficient grain amaranth genetic transformation method, a subject that recently has been explored and improved (Castellanos-Arévalo et al. 2020; Chap. 20 in this book).

5.9 Conclusions

Despite the molecular characterisation and nutritional quality studies of amaranth SSPs, genomic studies are needed to identify key genes regulating SSPs contents as well as the genetic and molecular pathways involved in SSPs deposition during seed formation. Recently developed technologies will shorten the length of the breeding process needed to obtain new varieties. For this reason, integration of modern bioinformatics techniques, sequencing technology methods and genetic mapping studies should be applied to grain amaranth breeding programs in order to further increase seed protein quantity and to improve their functionality without affecting the excellent amino acid balance of grain amaranth seed proteins.

Acknowledgements Authors thank Dr. E. Espitia-Rangel (Instituto Nacional de Investigaciones Forestales Agricolas y Pecuarias, Mexico) for his invaluable help in all our projects. Thanks to A. López-Covarrubias for her help in protein extraction and electrophoretic analyses and to M. L. Llamas-García for her support for the in silico analysis and elaboration of structural models.

References

Abugoch LE, Martínez EN, Añón MC (2003) Influence of the extracting solvent upon the structural properties of amaranth (*Amaranthus hypochondriacus*) glutelin. J Agric Food Chem 51:4060–4065. https://doi.org/10.1021/jf025949e

Adachi M, Kanamori J, Masuda T, Yagasaki K, Kitamura K, Mikami B, Utsumi S (2003) Crystal structure of soybean 11S globulin: Glycinin A3B4 homohexamer. Proc Natl Acad Sci U S A 100:7395–7400. https://doi.org/10.1073/pnas.0832158100

Adachi M, Takenaka Y, Gidamis AB, Mikami B, Utsumi S (2001) Crystal structure of soybean proglycinin A1aB1b homotrimer. J Mol Biol 305:291–305. https://doi.org/10.1006/jmbi.2000.4310

Ballabio C, Uberti F, Di Lorenzo C, Brandolini A, Penas E, Restani P (2011) Biochemical and immunochemical characterization of different varieties of amaranth (*Amaranthus* L. ssp.) as a safe ingredient for gluten-free products. J Agric Food Chem 59:12969–12974. https://doi.org/10.1021/jf2041824

Barba de la Rosa AP, Fomsgaard IS, Laursen B, Mortensen AG, Olvera-Martínez L, Silva-Sánchez C, Mendoza-Herrera A, González-Castañeda J, De León-Rodríguez A (2009) Amaranth (*Amaranthus hypochondriacus*) as an alternative crop for sustainable food production: phenolic acids and flavonoids with potential impact on its nutraceutical quality. J Cereal Sci 49:117–121. https://doi.org/10.1016/j.jcs.2008.07.012

Barba de la Rosa AP, Gueguen J, Paredes-LÓpez O, Viroben G (1992a) Fractionation procedures, electrophoretic characterization, and amino acid composition of amaranth seed proteins. J Agric Food Chem 40:931–936. https://doi.org/10.1021/jf00018a002

Barba de la Rosa AP, Paredes-López O, Gueguen J (1992b) Characterization of amaranth globulins by ultracentrifugation and chromatographic techniques. J Agric Food Chem 40:937–940. https://doi.org/10.1021/jf00018a003

Barba de la Rosa AP, Herrera-Estrella A, Utsumi S, Paredes-López O (1996) Molecular characterization, cloning and structural analysis of a cDNA encoding an amaranth globulin. J Plant Physiol 149:527–532. https://doi.org/10.1016/S0176-1617(96)80329-4

Bass R, Ruddock LW, Klappa P, Freedman RB (2004) A major fraction of endoplasmic reticulum-located glutathione is present as mixed disulfides with protein. J Biol Chem 279:5257–5262. https://doi.org/10.1074/jbc.M304951200

Bergamo P, Maurano F, Mazzarella G, Iaquinto G, Vocca I, Rivelli AR, De Falco E, Gianfrani C, Rossi M (2011) Immunological evaluation of the alcohol-soluble protein fraction from gluten-free grains in relation to celiac disease. Mol Nutr Food Res 55:1266–1270. https://doi.org/10.1002/mnfr.201100132

Betschart AA, Irving DW, Shepherd AD, Saunders RM (1981) *Amaranthus cruentus*: milling characteristics, distribution of nutrients within seed components, and the effects of temperature on nutritional quality. J Food Sci 46:1181–1187. https://doi.org/10.1111/j.1365-2621.1981.tb03019.x

Bewley JD, Greenwood JS (1990) Protein storage and utilization in seeds. In: Dennis, DT Turpin, DH (eds) Advanced plant biochemistry and molecular biology, 1st edn. Longman Harlow, UK, pp 456–469

Bojórquez-Velázquez E, Barrera-Pacheco A, Espitia-Rangel E, Herrera-Estrella A, Barba de la Rosa AP (2019) Protein analysis reveals differential accumulation of late embryogenesis abundant and storage proteins in seeds of wild and cultivated amaranth species. BMC Plant Biol 19:1–17. https://doi.org/10.1186/s12870-019-1656-7

Bojórquez-Velázquez E, Velarde-Salcedo AJ, De León-Rodríguez A, Jimenez-Islas H, Pérez-Torres JL, Herrera-Estrella A, Espitia-Rangel E, Barba de la Rosa AP (2018) Morphological, proximal composition, and bioactive compounds characterization of wild and cultivated amaranth (*Amaranthus spp.*) species. J Cereal Sci 83:222–228. https://doi.org/10.1016/j.jcs.2018.09.004

Bolon YT, Joseph B, Cannon SB, Graham MA, Diers BW, Farmer AD, May GD, Muehlbauer GJ, Specht JE, Tu ZJ, Weeks N, Xu WW, Shoemaker RC, Vance CP (2010) Complementary genetic and genomic approaches help characterize the linkage group I seed protein QTL in soybean. BMC Plant Biol 10:41. https://doi.org/10.1186/1471-2229-10-41

Bortesi L, Fischer R (2015) The CRISPR/Cas9 system for plant genome editing and beyond. Biotechnol Adv 33:41–52. https://doi.org/10.3389/fpls.2018.00890

Bressani R, García-Vela LA (1990) Protein fractions in amaranth grain and their chemical characterization. J Agric Food Chem 38:1205–1209. https://doi.org/10.1021/jf00095a010

Brim CA, Burton JW (1979) Recurrent selection in soybeans. II. Selection for increased percent protein in seeds. Crop Sci 19:494–498. https://doi.org/10.2135/cropsci1979.0011183x001900040016x

Brummer EC, Graef GL, Orf J, Wilcox JR, Shoemaker RC (1997) Mapping QTL for seed protein and oil content in eight soybean populations. Crop Sci 37:370–378. https://doi.org/10.2135/cropsci1997.0011183X003700020011x

Carrazco-Peña L, Osuna-Castro JA, De León-Rodríguez A, Maruyama N, Toro-Vazquez JF, Morales-Rueda JA, Barba de la Rosa AP (2013) Modification of solubility and heat-induced gelation of amaranth 11S globulin by protein engineering. J Agric Food Chem 61:3509–3516. https://doi.org/10.1021/jf3050999

Castellanos-Arévalo AP, Estrada-Luna AA, Cabrera-Ponce JL, Valencia-Lozano E, Herrera-Ubaldo H, de Folter S, Blanco-Labra A, Délano-Frier JP (2020) *Agrobacterium rhizogenes*-mediated transformation of grain (*Amaranthus hypochondriacus*) and leafy (*A. hybridus*) amaranths. Plant Cell Rep 39:1143–1160. https://doi.org/10.1007/s00299-020-02553-9

Chagolla-Lopez A, Blanco-Labra A, Patthy A, Sanchez R, Pongor SA (1994) Novel alpha amylase inhibitor from amaranth (*Amaranthus hypochondriacus*) seeds. J Biol Chem 269:23675–23680. https://www.jbc.org/content/269/38/23675.short

Chakraborty S, Chakraborty N, Agrawal L, Ghosh S, Narula K, Shekhar S, Naik PS, Pande PC, Chakrborti SK, Datta A (2010) Next-generation protein-rich potato expressing the seed protein gene Am A1 is a result of proteome rebalancing in transgenic tuber. Proc Natl Acad Sci USA 107:17533–17538. https://doi.org/10.1073/pnas.1006265107

Chakravarthi S, Jessop CE, Bulleid NJ (2006) The role of glutathione in disulphide bond formation and endoplasmic-reticulum-generated oxidative stress. EMBO Rep 7:271–275. https://doi.org/10.1038/sj.embor.7400645

Chua ACN, Hsiao ESL, Yang YC, Lin LJ, Chou WM, Tzen JTC (2008) Gene families encoding 11S globulin and 2S albumin isoforms of jelly fig (*Ficus awkeotsang*) achenes. Biosci Biotechnol Biochem 72:506–513. https://doi.org/10.1271/bbb.70588

Chung J, Babka HL, Graef GL, Staswick PE, Lee DJ, Cregan PB, Shoemaker RC, Specht JE (2003) The seed protein, oil, and yield QTL on soybean linkage Group I. Crop Sci 43:1053–1067. https://doi.org/10.2135/cropsci2003.1053

Clouse JW, Adhikary D, Page JT, Ramaraj T, Deyholos MK, Udall JA, Fairbanks DJ, Jellen EN, Maughan PJ (2016) The amaranth genome: genome, transcriptome, and physical map assembly. Plant Genome 9:1–14. https://doi.org/10.3835/plantgenome2015.07.0062

Cober ER, Voldeng H (2000) Developing high-protein, high-yield soybean populations and lines. Crop Sci 40:39–42. https://doi.org/10.2135/cropsci2000.40139x

Dixon BM, Heath SHD, Kim R, Suh JH, Hagen TM (2008) Assessment of endoplasmic reticulum glutathione redox status is confounded by extensive ex vivo oxidation. Antioxid Redox Sign 10:963–972. https://doi.org/10.1089/ars.2007.1869

Dodok L, Modhir AA, Buchtová V, Halásová G, Poláček I (1997) Importance and utilization of amaranth in food industry. Part 2. Composition of amino acids and fatty acids. Food/Nahrung 41:108–110. https://doi.org/10.1002/food.19970410211

Duarte-Correa A, Jokl L, Carlsson R (1986) Amino acid composition of some *Amaranthus sp.* grain proteins and of its fractions. Arch Latinoam Nutr 36:466–476

Duranti M, Consonni A, Magni C, Sessa F, Scarafoni A (2008) The major proteins of lupin seed: characterisation and molecular properties for use as functional and nutraceutical ingredients. Trends Food Sci Technol 19:624–633. https://doi.org/10.1016/j.tifs.2008.07.002

Džunková M, Janovská D, Čepková PH, Prohasková A, Kolář M (2011) Glutelin protein fraction as a tool for clear identification of amaranth accessions. J Cereal Sci 53:198–205. https://doi.org/10.1016/j.jcs.2010.12.003

Evert RF, Eichhorn S (2013) Raven biology of plants. W. H. Freeman and Company Publishers, USA, pp 526–533. ISBN-13:978-1464117800

Espitia-Rangel E, Escobedo Lopez D, Aguilar Delgado M (2012) Estrategy and methodology for genetic improvement of grain amaranth. Chapter XVI. In: Espitia-Rangel E (ed). Amaranto: Ciencia y Tecnología, pp 227–247. ISBN 978-607-425-874-4

FAO (1973) Energy and protein requirements. Nutrition meeting report series 52. Food and Agriculture Organization, Rome. https://apps.who.int/iris/handle/10665/41042

Fukushima D (1991) Structures of plant storage proteins and their functions. Food Rev Int 7:353–381. https://doi.org/10.1080/87559129109540916

Gao C (2018) The future of CRISPR technologies in agriculture. Nat Rev Mol Cell Biol 19:275–276. https://doi.org/10.1038/nrm.2018.2

Gacek K, Bartkowiak-Broda I, Batley J (2018) Genetic and molecular regulation of seed storage proteins (SSPs) to improve protein nutritional value of oilseed rape (*Brassica napus* L.) seeds. Front Plant Sci 9:890. https://doi.org/10.3389/fpls.2018.00890

Gorinstein S, Moshe R, Greene LJ, Arruda P (1991) Evaluation of four *Amaranthus* species through protein electrophoretic patterns and their amino acid composition. J Agric Food Chem 39:851–854. https://doi.org/10.1021/jf00005a006

Hara-Nishimura I (1987) Introduction of a disulfide bond in proglobulin molecules during the IIS globulin biosynthesis in endoplasmic reticulum of developing pumpkin cotyledons. Agric Biol Chem 51:2001–2008. https://doi.org/10.1080/00021369.1987.10868299

Higgins TJ V (1984) Synthesis and regulation of major proteins in seeds. Annu Rev Plant Physiol 35:191–221. https://doi.org/10.1146/annurev.pp.35.060184.001203

Huerta-Ocampo JA, Barba de la Rosa AP (2011) Amaranth: a pseudo-cereal with nutraceutical properties. Curr Nutr Food Sci 7:1–9. https://doi.org/10.2174/157340111794941076

Hwang C, Sinskey AJ, Lodish HF (1992) Oxidized redox state of glutathione in the endoplasmic reticulum. Science 257:1496–1502. https://doi.org/10.1126/science.1523409

Janssen F, Pauly A, Rombouts I, Jansens KJA, Deleu LJ, Delcour JA (2017) Proteins of amaranth (*Amaranthus spp.*), buckwheat (*Fagopyrum spp.*), and quinoa (*Chenopodium spp.*): a food science and technology perspective. Compr Rev Food Sci Food Saf 16:39–58. https://doi.org/10.1111/1541-4337.12240

Jukanti AK, Heidlebaugh NM, Parrott DL, Fischer IA, McInnerney K, Fischer AM (2007) Comparative transcriptome profiling of near-isogenic barley (*Hordeum vulgare*) lines differing in the allelic state of a major grain protein content locus identifies genes with possible roles in leaf senescence and nitrogen reallocation. New Phytol 177:333–349. https://doi.org/10.1111/j.1469-8137.2007.02270.x

Jung R (1997) Role of the sulfhydryl redox state and disulfide bonds in processing and assembly of 11S seed globulins. Plant Cell 9:2037–2050. https://doi.org/10.1105/tpc.9.11.2037

Jung R, Paul Scott M, Nam YW, Beaman TW, Bassüner R, Saalbach I, Müntz K, Nielsen NC (1998) The role of proteolysis in the processing and assembly of 11S seed globulins. Plant Cell 10:343–357. https://doi.org/10.1105/tpc.10.3.343

Khan N, Takahashi Y, Katsube-Tanaka T (2012) Tandem repeat inserts in 13S globulin subunits, the major allergenic storage protein of common buckwheat (*Fagopyrum esculentum* Moench) seeds. Food Chem 133:29–37. https://doi.org/10.1016/j.foodchem.2011.12.056

Kietlinski KD, Jimenez F, Jellen EN, Maughan PJ, Smith SM, Pratt DB (2014) Relationships between the weedy *Amaranthus hybridus* (Amaranthaceae) and the grain amaranths. Crop Sci 54:220–228. https://doi.org/10.2135/cropsci2013.03.0173

Konishi Y, Fumita Y, Ikeda K, Okuno K, Fuwa H (1985a) Isolation and characterization of globulin from seeds of *Amaranthus hypochondriacus* l. Agric Biol Chem 49:1453–1459. https://doi.org/10.1080/00021369.1985.10866912

Konishi Y, Horikawa K, Oku Y, Azumaya J, Nakatani N (1991) Extraction of two albumin fractions from amaranth grains: Comparison of some physicochemical properties and the putative localization in the grains. Agric Biol Chem 55:2745–2750. https://doi.org/10.1080/00021369.1991.10857150

Konishi Y, Nojima H, Okuno K, Asaoka M, Fuwa H (1985b) Characterization of starch granules from waxy, nonwaxy, and hybrid seeds of *Amaranthus hypochondriacus* L. Agric Biol Chem 49:1965–1971. https://doi.org/10.1080/00021369.1985.10867018

Kumar S, Stecher G, Li M, Knyaz C, Tamura K (2018) MEGA X: molecular evolutionary genetics analysis across computing platforms. Mol Biol Evol 35:1547–1549. https://doi.org/10.1093/molbev/msy096

Lawrence MC, Izard T, Beuchat M, Blagrove RJ, Colman PM (1994) Structure of phaseolin at 2.2 Å resolution: implications for a common vicilin/legumin structure and the genetic engineering of seed storage proteins. J Mol Biol 238:748–776. https://doi.org/10.1006/jmbi.1994.1333

Li C, Nguyen V, Liu J, Fu W, Chen C, Yu K, Cui Y (2019) Mutagenesis of seed storage protein genes in soybean using CRISPR/Cas9. BMC Res Notes 12:176. https://doi.org/10.1186/s13104-019-4207-2

Lightfoot DJ, Jarvis DE, Ramaraj T, Lee R, Jellen EN, Maughan PJ (2017) Single-molecule sequencing and Hi-C-based proximity-guided assembly of amaranth (*Amaranthus hypochondriacus*) chromosomes provide insights into genome evolution. BMC Biol 15:74. https://doi.org/10.1186/s12915-017-0412-4

Luna-Suárez S, Medina-Godoy S, Cruz-Hernández A, Paredes-López O (2008) Expression and characterization of the acidic subunit from 11S Amaranth seed protein. Biotechnol J 3:209–219. https://doi.org/10.1002/biot.200700146

Maldonado-Cervantes E, Jeong HJ, León-Galván F, Barrera-Pacheco A, De León-Rodríguez A, González de Mejia E, de Lumen BO, Barba de la Rosa AP (2010) Amaranth lunasin-like peptide internalizes into the cell nucleus and inhibits chemical carcinogen-induced transformation of NIH-3T3 cells. Peptides 31:1635–1642. https://doi.org/10.1016/j.peptides.2010.06.014

Mandal S, Mandal RK (2000) Seed storage proteins and approaches for improvement of their nutritional quality by genetic engineering. Curr Sci 79:576–589

Marchler-Bauer A, Anderson JB, Cherukuri PF, DeWeese-Scott C, Geer LY, Gwadz M, He S, Hurwitz DI, Jackson JD, Ke Z, Lanczycki CJ, Liebert CA, Liu C, Lu F, Marchler GH, Mullokandov M, Shoemaker BA, Simonyan V, Song JS, Thiessen PA, Yamashita RA, Yin JJ, Zhang D, Bryant SH (2005) CDD: a conserved domain database for protein classification. Nucl Acids Res 33:192–196. https://doi.org/10.1093/nar/gki069

Marcone MF, Yada RY (1991) Isolation, purification, and characterization of the oligomeric seed globulin from *Amaranthus hypochondriacus*. Agric Biol Chem 55:2281–2289. https://doi.org/10.1080/00021369.1991.10870949

Marcone MF (1999) Evidence confirming the existence of a 7S globulin-like storage protein in *Amaranthus hypochondriacus* seed. Food Chem 65:533–542. https://doi.org/10.1016/S0308-8146(98)00221-0

Marcone MF, Niekamp FK, Le Maguer M, Yada RY (1994) Purification and characterization of the physicochemical properties of the albumin fraction from the seeds of *Amaranthus hypochondriacus*. Food Chem 51:287–294. https://doi.org/10.1016/0308-8146(94)90029-9

Maruyama N, Mikami B, Utsumi S (2011) The development of transgenic crops to improve human health by advanced utilization of seed storage proteins. Biosci Biotechnol Biochem 75:823–828. https://doi.org/10.1271/bbb.100924

Miernyk JA, Hajduch M (2011) Seed proteomics. J. Proteomics 74:389–400. https://doi.org/10.1016/j.jprot.2010.12.004

Mills ENC, Jenkins J, Marigheto N, Belton PS, Gunning AP, Morris VJ (2002) Allergens of the cupin superfamily. In: Biochemical society transactions. Portland Press, pp 925–929. https://doi.org/10.1042/bst0300925

Millward DJ (1999) Meat or wheat for the next millennium? Plenary lecture. The nutritional value of plant-based diets in relation to human amino acid and protein requirements. Proc Nutr Soc 58:249–260. https://doi.org/10.1017/s0029665199000348

Montero-Morán GM, Sampedro JG, Saab-Rincón G, Cervantes-González MA, Huerta-Ocampo J, De León-Rodríguez A, Barba de la Rosa AP (2015) Biochemical and molecular characterization of a novel Cu/Zn superoxide dismutase from *Amaranthus hypochondriacus* L.: an intrinsically disordered protein. Appl Biochem Biotechnol 176:2328–2345. https://doi.org/10.1007/s12010-015-1721-0

Morales-Guerrero JC, Vázquez-Mata N, Bressani R (2009) El amaranto: características físicas, químicas, toxicológicas y funcionales y aporte nutricional. Instituto Nacional de Ciencias Médicas y Nutrición Salvador Zubirán, D.F., Mexico. ISBN 9786077797005

Moreno FJ, Clemente A (2008) 2S albumin storage proteins: what makes them food allergens? Open Biochem J 2:16–28. https://doi.org/10.2174/1874091x00802010016

Morr CV (1990) Current status of soy protein functionality in food systems. J Am Oil Chem Soc 67:265–271. https://doi.org/10.1007/BF02539674

Mota C, Santos M, Mauro R, Samman N, Matos AS, Torres D, Castanheira I (2016) Protein content and amino acids profile of pseudocereals. Food Chem 193:55–61. https://doi.org/10.1016/j.foodchem.2014.11.043

Müntz K (1998) Deposition of storage proteins. Plant Mol Biol 38:77–99. https://doi.org/10.1007/978-94-011-5298-3_4

Nielsen NC, Dickinson CD, Cho T-J, Thanh VH, Scallon BJ, Fischer RL, Sims TL, Drews GN, Goldberg RB (1989) Characterization of the glycinin gene family in soybean. Plant Cell 1:313–328. https://doi.org/10.2307/3869011

Okuno K, Sakaguchi S (1982) Inheritance of starch characteristics in perisperm of *Amaranthus hypochondriacus*. J Hered 73:467–467. https://doi.org/10.1093/oxfordjournals.jhered.a109700

Osborne T (1924) The vegetable proteins, 2nd edn. Longmans Green and Co., London

Osuna-Castro JA, Rascón-Cruz Q, Napier J, Fido RJ, Shewry PR, Paredes-López O (2000) Overexpression, purification, and in vitro refolding of the 11S globulin from amaranth seed in *Escherichia coli*. J Agric Food Chem 48:5249–5255. https://doi.org/10.1021/jf000795t

Paredes-López O, Mora-Escobedo R, Ordorica-Falomir C (1988) Isolation of amaranth proteins. LWT Food Sci Technol 27:59–61

Park YJ, Nishikawa T (2012) Characterization and expression analysis of the starch synthase gene family in grain amaranth (*Amaranthus cruentus* L.). Genes Genet Syst 87:281–289. https://doi.org/10.1266/ggs.87.281

Raina A, Datta A (1992) Molecular cloning of a gene encoding a seed-specific protein with nutritionally balanced amino acid composition from *Amaranthus*. Proc Natl Acad Sci USA 89:11774–11778. https://doi.org/10.1073/pnas.89.24.11774

Rascón-Cruz Q, Sinagawa-García S, Osuna-Castro JA, Bohorova N, Paredes-López O (2004) Accumulation, assembly, and digestibility of amarantin expressed in transgenic tropical maize. Theor Appl Genet 108:335–342. https://doi.org/10.1007/s00122-003-1430-x

Rodríguez-Leal D, Lemmon ZH, Man J, Bartlett ME, Lippman ZB (2017) Engineering quantitative trait variation for crop improvement by genome editing. Cell 171:470–480. https://doi.org/10.1016/j.cell.2017.08.030

Saturni L, Ferretti G, Bacchetti T (2010) The gluten-free diet: safety and nutritional quality. Nutrients 2:16–34. https://doi.org/10.3390/NU20100016

Sauer JD (1967) The grain amaranths and their relatives: A revised taxonomic and geographic survey. Ann Missouri Bot Gard 54:103–137. https://doi.org/10.2307/2394998

Segura-Nieto M, Barba de la Rosa AP, Paredes-Lopez O (1994) Biochemistry of Amaranth proteins. Amaranth biology, chemistry, and technology. CRC Press, pp 75–106. https://doi.org/10.1201/9781351069601-5

Segura-Nieto M, Vázquez-Sánchez N, Rubio-Velázquez H, Olguin-Martínez LE, Rodríguez-Nester CE, Herrera-Estrella L (1992) Caracterizarían of amaranth (*Amaranthus hypochondriacus* L.) seed proteins. J Agric Food Chem 40:1553–1558. https://doi.org/10.1021/jf00021a016

Shewry PR (2015) The protein chemistry of dicotyledonous grains. Encyclopedia of food grains, 2nd edn. Elsevier Inc., pp 109–114. https://doi.org/10.1016/B978-0-12-394437-5.00105-4

Shewry PR, Napier JA, Tatham AS (1995) Seed storage proteins: structures and biosynthesis. Plant Cell 7:945–956. https://doi.org/10.1105/tpc.7.7.945

Silva-Sánchez C, Barba de la Rosa AP, León-Galván MF, De Lumen BO, De León-Rodríguez A, González De Mejía E (2008) Bioactive peptides in amaranth (*Amaranthus hypochondriacus*) seed. J Agric Food Chem 56:1233–1240. https://doi.org/10.1021/jf072911z

Singhal RS, Kulkarni PR (1988) Composition of the seeds of some *Amaranthus* species. J Sci Food Agric 42:325–331. https://doi.org/10.1002/jsfa.2740420406

Swanson BG (1990) Pea and lentil protein extraction and functionality. J Am Oil Chem Soc 67:276–280. https://doi.org/10.1007/BF02539676

Tamás C, Kisgyörgy BN, Rakszegi M, Wilkinson MD, Yang MS, Láng L, Tamás L, Bedo Z (2009) Transgenic approach to improve wheat (*Triticum aestivum* L.) nutritional quality. Plant Cell Rep 28:1085–1094. https://doi.org/10.1007/s00299-009-0716-0

Tandang-Silvas MRG, Fukuda T, Fukuda C, Prak K, Cabanos C, Kimura A, Itoh T, Mikami B, Utsumi S, Maruyama N (2010a) Conservation and divergence on plant seed 11S globulins based on crystal structures. Biochim Biophys Acta Proteins Proteomics 1804:1432–1442. https://doi.org/10.1016/j.bbapap.2010.02.016

Tandang-Silvas MRG, Tecson-Mendoza EM, Mikami B, Utsumi S, Maruyama N (2011) Molecular design of seed storage proteins for enhanced food physicochemical properties. Annu Rev Food Sci Technol 2:59–73. https://doi.org/10.1146/annurev-food-022510-133718

Tandang-Silvas MR, Cabanos CS, Carrazco Peña LD, De La Rosa APB, Osuna-Castro JA, Utsumi S, Mikami B, Maruyama N (2012) Crystal structure of a major seed storage protein, 11S proglobulin, from *Amaranthus hypochondriacus*: Insight into its physico-chemical properties. Food Chem 135:819–826. https://doi.org/10.1016/j.foodchem.2012.04.135

Tandang-Silvas MR, Carrazco-Pẽa L, Barba de la Rosa AP, Osuna-Castro JA, Utsumi S, Mikami B, Maruyama N (2010b) Expression, purification and preliminary crystallization of amaranth 11S proglobulin seed storage protein from *Amaranthus hypochondriacus* L. Acta Crystallogr Sect F Struct Biol Cryst Commun 66:919–922. https://doi.org/10.1107/S1744309110021032

Timmerman-Vaughan GM, Mills A, Whitfield C, Frew T, Butler R, Murray S, Lakeman M, McCallum J, Russell A, Wilson D (2005) Linkage mapping of QTL for seed yield, yield components, and developmental traits in pea. Crop Sci 45:1336–1344. https://doi.org/10.2135/cropsci2004.0436

Transue TR, Smith AK; Mo H; I JG, Goldstein IJ, Saper MA (1997) Structure of benzyl T-antigen disaccharide bound to *Amaranthus caudatus* agglutinin. Nat Struct Biol 4:779–783. https://doi.org/10.1038/nsb1097-779

Uauy C, Distelfeld A, Fahima T, Blechl A, Dubcovsky J (2006) A NAC gene regulating senescence improves grain protein, zinc, and iron content in wheat. Science 314:1298–1301. https://doi.org/10.1126/science.1133649

Utsumi S (1992) Plant food protein engineering. Adv Food Nutr Res 36:89–208. https://doi.org/10.1016/S1043-4526(08)60105-9

Valcárcel-Yamani B, Lannes S (2012) Applications of quinoa (*Chenopodium quinoa* Willd.) and amaranth (*Amaranthus spp.*) and their influence in the nutritional value of cereal based foods. Food Publ Health 2:265–275. https://doi.org/10.5923/j.fph.20120206.12

Valdés-Rodríguez S, Guerrero-Rangel A, Melgoza-Villagómez C, Chagolla-López A, Delgado-Vargas F, Martínez-Gallardo N, Sánchez-Hernández C, Délano-Frier J (2007) Cloning of a cDNA encoding a cystatin from grain amaranth (*Amaranthus hypochondriacus*) showing a tissue-specific expression that is modified by germination and abiotic stress. Plant Physiol Biochem 45:790–798. https://doi.org/10.1016/j.plaphy.2007.07.007

Valdés-Rodriguez S, Segura-Nieto M, Chagolla-Lopez A, Verver Y Vargas-Cortina A, Martinez-Gallardo N, Blanco-Labra A (1993) Purification, characterization, and complete amino acid sequence of a trypsin inhibitor from amaranth (*Amaranthus hypochondriacus*) seeds. Plant Physiol 103:1407–1412. https://doi.org/10.1104/pp.103.4.1407

Valdez-Ortiz A, Rascón-Cruz Q, Medina-Godoy S, Sinagawa-García SR, Valverde-González ME, Paredes-López O (2005) One-step purification and structural characterization of a recombinant His-tag 11S globulin expressed in transgenic tobacco. J Biotechnol 115:413–423. https://doi.org/10.1016/j.jbiotec.2004.09.013

Zhang W, Bi J, Chen L, Zheng L, Ji S, Xia Y, Xie K, Zhao Z, Wang Y, Liu L, Jiang L, Wan J (2008) QTL mapping for crude protein and protein fraction contents in rice (*Oryza sativa* L.). J Cereal Sci 48:539–547. https://doi.org/10.1016/j.jcs.2007.11.010

Genetic Diversity in Amaranth and Its Close Relatives

6

Darshan T. Dharajiya, Anuj Kumar Singh, Kapil K. Tiwari, and N. N. Prajapati

Abstract

Amaranthus (Amaranthaceae) generally known as amaranth is an annual herbaceous genus, grown under diverse agro-climatic conditions. It has many species carrying nutraceutical properties with the potential to add values to the food, cosmetic, and pharmaceutical industries. According to their usage, amaranths are broadly classified into the grain, vegetable, weed, and wild type. This chapter helps in the understanding of intra- and inter-specific diversity present in amaranths, evaluated by morphological, biochemical, cytological, and molecular approaches. The understanding of genetic diversity in amaranth breeding programs support in identifying diverse parents to generate segregating population with genetic variability for introgression of desired agronomic traits and to develop high yielding elite varieties of amaranth.

D. T. Dharajiya (✉) · A. K. Singh · K. K. Tiwari
Bio Science Research Centre, Sardarkrushinagar Dantiwada Agricultural University (SDAU), Sardarkrushinagar 385506, Gujarat, India

N. N. Prajapati
Centre for Crop Improvement, SDAU, Sardarkrushinagar 385506, Gujarat, India

Abbreviations

AFLP	Amplified Fragment Length Polymorphism
APMC	Agricultural Produce Market Committee
ARS	Agricultural Research Service
AUJ	Agriculture University, Jodhpur
AVGRIS	AVRDC Vegetable Genetic Resources Information System
DBSKKV	Dr. Balasaheb Sawant Konkan Krishi Vidyapeeth
GBPUAT	Govind Ballabh Pant University of Agriculture and Technology
IARI	Indian Agricultural Research Institute
ICAR	Indian Council of Agricultural Research
IIHR	Indian Institute of Horticultural Research
INR	Indian Rupee
ISSR	Inter Simple Sequence Repeat
ITS	Internal Transcribed Spacer
KAU	Kerala Agricultural University
NBPGR	National Bureau of Plant Genetic Resources
OUAT	Odisha University of Agriculture and Technology
PCA	Principal Component Analysis
RAPD	Random Amplified Polymorphic DNA
RFLP	Restriction Fragment Length Polymorphism

D. Adhikary et al. (eds.), *The Amaranth Genome*, Compendium of Plant Genomes,
https://doi.org/10.1007/978-3-030-72365-1_6

SDAU	Sardarkrushinagar Dantiwada Agricultural University
SNP	Single Nucleotide Polymorphism
SSR	Simple Sequence Repeat
TNAU	Tamil Nadu Agricultural University
UAS	University of Agricultural Sciences
USDA	United States Department of Agriculture
VPKAS	Vivekananda Parvatiya Krishi Anusandhan Sansthan

6.1 Introduction

Genus *Amaranthus*, commonly known as pigweed or amaranth includes cosmopolitan and herbaceous plants of Amaranthaceae family. "Amaranth" word was originated from the Greek word "amarantos" meaning "unwithering" indicating its hearty characteristics symbolizing immortality (Martinez-Lopez et al. 2020). Amaranth carries a C4 photosynthesis pathway; and has a certain advantage over cereals. It has high water use efficiency, can adapt in tropical environment withstanding hot and dry conditions, short cultivation period, phenotypic plasticity, tolerant to low soil fertility, and has a wide range of sexuality from gynomonoecy, monoecy to dioecy (Das 2016). The cosmopolitan pseudocereal can be cultivated under various soil and agro-climatic conditions and has tolerance to abiotic stresses (heat, drought, and salinity) (Rastogi and Shukla 2013; Martinez-Lopez et al. 2020). It is a promising crop that has the potential to support food security for our future generations.

The taxonomy and species number of amaranths have no general agreement. Some natural hybrids or ecotypes of amaranths have been considered as separate new species due to their substantial morphological variation from the existing species (Das 2016). There is still confusion in the number of species in the *Amaranthus* genus, as approximately 50 species (Brenan and Townsend 1980), 60 species (Brenner et al. 2000), 70 species (Iamonico 2012), 75 species (Sauer 1993) and 77 species (USDA, ARS, NPGS 2021) have been mentioned previously in separate reports. As per another report, the genus *Amaranthus* includes 87 species among which 56 are found in America, 17 in Europe, and 14 in Australia (Jacobsen and Mujica 2003) and about 25 *Amaranthus* spp. are found in Asia (Das 2016).

Some species of grain amaranths are native to South and Central America while some are native to Asia, Europe, Africa, and Australia (Becker et al. 1981; Martinez-Lopez et al. 2020). Although the origin of grain species has been in Central and South America, India has a great distribution and diversity of amaranths (Rastogi and Shukla 2013; Das 2016). The grain amaranth is cultivated in some distinct areas of the world and recently, it has been mentioned that China is the largest producer of amaranth (grain and vegetable). Other major producers of amaranth are the USA, Canada, India, and Argentina (Coelho et al. 2018). India is one of the countries where the cultivation of amaranth was initiated since a very long time; however, no official data on the area and production of amaranths are available. We have collected the data on year-wise grain amaranth price and quantity income at Agricultural Produce Market Committee (APMC), Palanpur, Gujarat, India which are represented in Fig. 6.1. It shows that the price of amaranth grains has been increased drastically in the last three years and the yield has been reduced. During 2019–20, a total of 84,445 quintals of amaranth grains has been submitted to APMC, Palanpur, India (APMC-Palanpur 2020).

6.2 Classification of Amaranths on the Bases of Their Use

Most of the species of *Amaranthus* are annual weeds; few are used as pseudocereals, vegetables, and ornamentals. Hence, they have been classified according to their uses (Sauer 1967; Das 2016; Singh et al. 2019). Amaranths have been widely cultivated for vegetable and grain purpose. It has been reported that due to their high lysine and protein-rich grains, *A. hypochondriacus*, *A. cruentus*, and *A. caudatus* have been classified as grain amaranths (Das

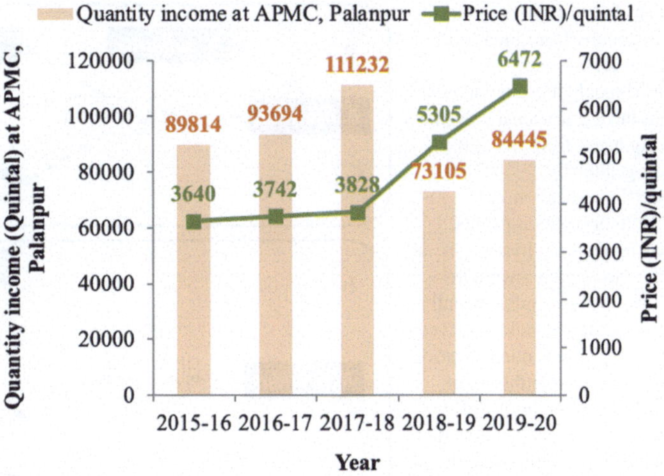

Fig. 6.1 Year-wise grain amaranth price (INR) and quantity income (quintal) in APMC, Palanpur (Banaskantha), India

2016; Pagi et al. 2017). Some species, namely *A. tricolor*, *A. blitum*, *A. dubius*, *A. viridis*, *A. graecizans*, *A. lividus*, and *A. spinosus* have been classified as vegetable amaranths. *A. tricolor* and *A. caudatus* were classified as an ornamental species. There are many species, namely *A. viridis, A. retroflexus*, *A. hybridus*, *A. gracilis*, *A. gangeticus*, *A. paniculatus*, and *A. graecizans* which are considered as wild relatives of some cultivated or domesticated *Amaranthus* spp. Many weedy species have been documented and some of them are monoecious (*A. retroflexus*, *A. hybridus*, *A. powellii*, *A. albus*, *A. blitoides*, and *A. spinosus*) and some are dioecious (*A. rudis*, *A. tuberculatus*, *A. palmeri*, and *A. arenicola*) (Sauer 1967; Das 2016; Singh et al. 2019).

6.3 Genetic Diversity in Amaranth and Their Wild Relatives

The *Amaranthus* genus has a wide range of morphological diversity. Genetic diversity is the variation of individual genotypes within and among species. It empowers a population in adapting to new conditions brought by environmental changes and it is important for the existence of species for a long time (Hamrick and Godt 1996). Genetic variability and phylogenetic relationships are assessed by data derived from experiments on morphological, biochemical, cytological, and molecular characters (Das 2016). The existence of several admixed amaranth genotypes indicated regular hybridization or introgression events (Das 2012a, b). Two different groups of amaranths, namely vegetable and grain amaranths have evolved from their particular weed ancestor through distinctive domestication events in diverse areas of the world (Das 2012a, b). The origin of grain amaranths from their wild ancestor has been explained by two hypotheses—monophyletic (all three grain amaranths have originated from a single progenitor (*A. hybridus*)) and polyphyletic (all the three grain amaranths have evolved independently from different wild relatives (*A. powellii, A. hybridus*, and *A. quitensis*)) (Sauer 1967, 1993; Das 2012a, 2016) (Fig. 6.2). The third hypothesis proposes that all the three grain amaranths have originated from genetically dissimilar populations of *A. hybridus* by independent domestication events (Das 2016) (Fig. 6.2). The confusion in evolutionary relations of grain amaranths and other amaranth species can be solved by evaluating genetic diversity analysis using molecular markers and next-generation sequencing (Khaing et al. 2013; Chaney et al. 2016; Wu and Blair 2017; Viljoen et al. 2018; Su-Young et al. 2019).

Evaluation of genetic diversity in crop breeding programs supports identifying diverse parents to generate segregating population having maximum genetic variability. It also helps in

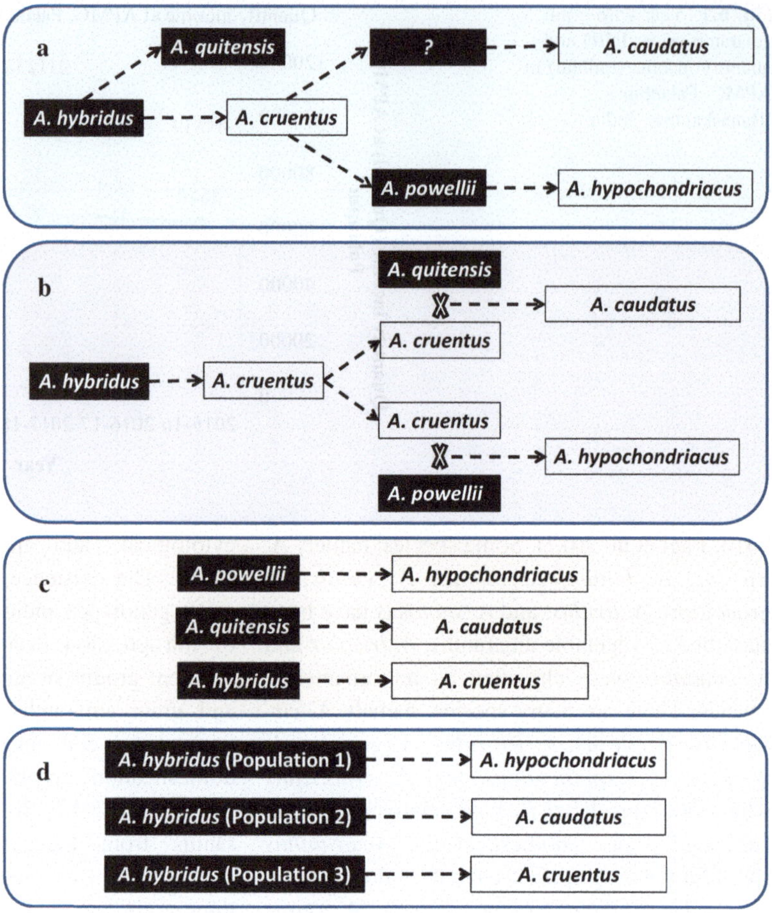

Fig. 6.2 Origin of grain amaranths from their wild ancestors. **a** and **b** Monophyletic hypothesis: all three grain amaranths have originated from a single progenitor by (**a**) domestication or (**b**) hybridization events, **c** Polyphylatic hypothesis: all the three grain amaranths have evolved independently from different wild relatives, **d** Independent domestication of grain amaranths from different populations of *A. hybridus* (Figure is drawn based on the information presented in Sauer 1967, 1993; Kietlinski et al. 2014; Das 2012a, Das 2016.)

the introgression of desired genes from diverse genotypes into the existing genetic base. Cluster analysis has been proven to be an effective way of grouping genotypes for effective management and use in crop-breeding programs (Shukla et al. 2010). The diverse genotypes from different clusters have been identified and utilized in different hybridization programs to develop high yielding and nutritionally rich varieties (Shukla et al. 2010).

6.3.1 Morphological Approach

Assessment of morphological diversity and characterization of plant genetic resources of crop species provides essential information for breeding programs of crops (Gerrano et al. 2017). Amaranths are rich in morphological diversity and through morphological characterization, taxonomic obscurities in amaranths have been resolved. Morphological characters of inflorescence, flowers, seed, stem, leaves, phyllotaxy, and pollen morphology have played an important part in distinguishing taxa in many circumstances (Das 2016).

On the bases of morphological characters of grain, vegetable, ornamental, and weed amaranths, intra-specific genetic diversity has been assessed in a few species, namely *A. tricolor* (Shukla et al. 2010; Ahammed et al. 2013), *A. hybridus* (Oboh 2007), and *A. hypochondriacus* (Venkatesh et al. 2014), whereas inter-specific diversity has been assessed among a set of species from *A. blitum, A.*

Table 6.1 Morphological and nutritional parameters considered for diversity analysis by Shankar et al. (2012)

Type of parameters	No. of parameters	Name of parameters
Morphological	18	Plant height, number of branches per plant, internode length, number of leaves per plant, node at which flower appeared, average leaf length, average leaf width, average petiole length, average plant weight, average leaf weight, average root weight, average petiole weight, average stem weight, average shoot weight, shoot to plant ratio, root to shoot ratio, leaf to shoot ratio, and yield per plant
Nutritional	4	Protein content, calcium content, Fe content, and vitamin-C content

dubius, A. tricolor, A. tricolor var. *tristis, A. caudatus, A. viridus, A. graecizans, A. gracilis, A. spinosus, A. hypochondriacus, A. hybridus*, and *A. cruentus* in separate studies (Anuja and Mohideen 2007; Erum et al. 2012; Shankar et al. 2012; Andini et al. 2013; Gerrano et al. 2015, 2017; Gueco et al. 2016).

Intra-specific diversity in vegetable amaranth (*A. tricolor*) has been assessed among thirty-nine genotypes by studying eight morphological and seven quality characters. The genotypes have been grouped into six clusters and a wide range of diversity has been reported for many characters (Shukla et al. 2010). Similarly, intra-specific genetic diversity among 22 genotypes of *A. tricolor* for 13 morphological characters was analyzed using Mahalanobis' D^2 statistics and principal component analysis (PCA) analysis (Ahammed et al. 2013). The genetic diversity among 16 genotypes of *A. hybridus* (Oboh 2007) and 100 genotypes of grain amaranth has been elucidated by morphological characters and cluster analysis (Venkatesh et al. 2014).

Inter-species genetic diversity for yield and yield components in hundred genotypes belonging to four different species, viz., *A. blitum, A. dubius, A. tricolor,* and *A. tricolor* var. *tristis* has been evaluated by cluster analysis. The clusters were heterogeneous with regard to the geographical origin and species of the genotypes (Anuja and Mohideen 2007). Thirty-two amaranth genotypes belonging to *A. caudatus, A. viridus, A. graecizans, A. tricolor, A. cruentus*, and unknown *Amaranthus* spp. have been used in field evaluation for 14 quantitative morphological traits to determine genetic diversity which showed the occurrence of an extensive genetic diversity (Gerrano et al. 2015, 2017).

Based on 18 morphological parameters and four nutritional parameters (Table 6.1), a set of 28 genotypes belonging *A. cruentus, A. dubius, A. hybridus*, and *A. tricolor* has been used for the assessment of genetic diversity by cluster analysis (Shankar et al. 2012). The first four principal components were for plant height, the number of branches per plant, internode length, and the number of leaves per plant which cumulatively contributed 68% of the variability (Shankar et al. 2012). The clusters of genotypes were formed on the basis of phenotypic characters and their geographic origin, but co-occurrence of genotypes belonging to different geographic regions was also reported (Shankar et al. 2012). Thirteen genotypes of *A. hypochondriacus* and *A. tricolor* were used for genetic diversity analysis on the bases of seven morphological characters and two distinct clusters according to their species have been observed except the China variety of *A. tricolor*, which has been grouped in the cluster of *A. hypochondriacus* (Erum et al. 2012). The floral variation and distinguishing morphological features of individual species of the *A. hybridus* species complex has been studied by Adhikary and Pratt (2015) which provided taxonomic information.

The variation in morphology and protein content of 53 Indonesian amaranth genotypes belonging to seven species, viz., *A. blitum, A. caudatus, A. dubius, A. hybridus, A. spinosus, A. tricolor*, and *A. viridis* of vegetable, ornamental, or weedy types has been assessed. Cluster and scatter plot analyses indicated that *A. viridis* and

A. dubius genotypes could be used as valuable parental lines in breeding programs for protein content and yield improvement, respectively (Andini et al. 2013). The clustering of the genotypes will be useful in the identification and selection of genetically diverse parents having the greatest inter-cluster distance in the amaranth breeding program as they may give high levels of heterosis for the desired traits (Anuja and Mohideen 2007).

6.3.2 Biochemical Approach

Biochemical evidences were used to confirm the elucidations based on morphological data and mostly those were concomitant with morphological data. Electrophoresis of seed proteins, isozyme polymorphism, leaf phenolic chromatogram, and other biochemical parameters have been applied to get knowledge about relationships among different species of amaranths (Das 2016). Genetic variability within individual *Amaranthus* species has been evaluated in *A. retroflexus* using isoenzyme analysis, whereas genetic variability among different *Amaranthus* species has been evaluated in a group of species from *A. acutilobus, A. albus, A. australis, A. caudatus, A. crassipes, A. cruentus, A. deflexus, A. dubius, A. fimbriatus, A. floridanus, A. hybridus, A. hypochondriacus, A. lividus, A. palmeri, A. powellii, A. quitensis, A. retroflexus, A. spinosus, A. standleyanus, A. standleyanus, A. tricolor*, and *A. viridis* in different studies (Chan and Sun 1997; Erum et al. 2012; Shankar et al. 2012; Shukla et al. 2018).

Genetic diversity and relationships of 60 genotypes belonging to 23 wild and cultivated grain species of *Amaranthus* have been evaluated using 30 polymorphic loci encoding 15 enzymes (Chan and Sun 1997). The evolutionary relationships between grain amaranths (*A. hypochondriacus, A. cruentus*, and *A. caudatus*) and their putative ancestors were studied, and the origin of grain amaranths was suggested to be a monophyletic with *A. hybridus* as the common ancestor (Chan and Sun 1997). The isozymes information generated truthful estimates of genetic diversity and relationships among cultivated species and their wild relatives (Chan and Sun 1997). Genetic diversity was evaluated in twenty populations of *A. retroflexus* collected from different habitats using isoenzyme analysis by Mandák et al. (2011). Twenty-eight genotypes belonging to *A. cruentus, A. dubius, A. hybridus*, and *A. tricolor* has been used for the assessment of genetic diversity by analyzing four nutritional parameters, Ca, Fe, protein, and vitamin-C content (Shankar et al. 2012). Likewise, the genetic diversity among thirteen genotypes belonging to *A. tricolor* and *A. hypochondriacus* has been analyzed on the bases of four nutritional parameters, total protein, total carbohydrate, total fats, and moisture content (Erum et al. 2012). Genetic diversity was studied in 218 genotypes of *A. hypochondriacus* and its seven wild relatives (*A. caudatus, A. cruentus, A. dubius, A. hybridus, A. spinosus, A. retroflexus*, and *A. tricolor*) for 24 nutritional parameters which include test weight, total protein content, total oil content, amino acid profile (Methionine, lysine, leucine, half cysteine, valine, isoleucine, tyrosine, arginine + threonine, histidine, phenylalanine, alanine, aspartatic acid, glutamine, proline, glycine, and serine), and fatty acid profile (Palmitic acid, stearic acid, oleic acid, linoleic acid, and lenolenic acid) (Shukla et al. 2018). The results showed the existence of significant diversity for nutritional parameters in amaranths (Shukla et al. 2018).

6.3.3 Cytological Approach

The karyological study is difficult in the *Amaranthus* genus due to the small size of the chromosomes. Three haploid numbers ($n = 14$, 16, and 17) have been reported in *Amaranthus* spp. (Das 2016). The reason behind the tribasic nature of the *Amaranthus* genus is probably resulted due to dysploidy or aneuploidy. Knowledge of cytological and karyotypic analysis is very important in the study of genetic diversity and phylogenetic relationship among the species. Alteration in chromosome structure is very crucial in the evolution of species as the wild relatives of amaranths have more symmetrical karyotype compared with the cultivated species. Chromosome number and ploidy level of various *Amaranthus* spp. on the basis of chromosome number are represented in Table 6.2 (Grant 1959a, b; Desai 1971; Behera and Patnaik 1974; Queirós 1989; Pal et al. 2000; Bao-Hua et al. 2002; Milan 2008; Reema and Roy 2014; Das 2016). Most of the species are diploid with chromosome numbers (2n) of 28, 32 or 34. There is still confusion in the chromosome number of some species which showed different numbers in different reports. Few species (*A. caturus, A. polygamous*, and *A. giganteus*) showed diploid and tetraploid sets with haploid number 16 or 17. The only tetraploid species is *A. dubius* with 2n = 64 (Grant 1959a).

Cytogenetic analysis of 12 populations from ten *Amaranthus* species, namely *A. blitum* subsp. *blitum, A. blitum* subsp. *emarginatus, A. albus, A. retroflexus, A. spinosus, A. cruentus, A. hybridus, A. powellii, A. viridis*, and *A. blitoides* along with two varieties, namely *A. graecizans* var. *graecizans* and *A. graecizans* var. *sylvestris* showed the presence of 32 or 34 chromosome number (2n) in them (Sheidai and Mohammadzadeh 2008). Cytogenetic characteristics were used to analyze genetic diversity and their evolutionary relationships. Significant variations in cytogenetic characteristics, like chiasma frequency, chiasma distribution, and chromosome pairing have been revealed among the species which indicated their genomic dissimilarities. The dendrogram grouped the species in four major clusters which almost agrees with the dendrogram based on morphological and protein data (Sheidai and Mohammadzadeh 2008).

The genetic diversity and evolutionary relationships between 24 genotypes from *A. hybridus, A. hypochondricus, A. palmeria, A. quitensis, A. retroflexus, A. spinosus, A. powellii*, and *A. caudatus* have been assessed based on the morphological features of the chromosomes and numerical characterization of the karyotypes by Radwan et al. (2014). The variation in the karyotype of the genotypes of the same species was reported. *A. powellii* and *A. hybridus* have been considered as a progenitor of *Amaranthus* species on the bases of cluster analysis. *A. powellii* has been considered as the most advanced species of *Amaranthus* due to its smallest chromosome length (Radwan et al. 2014).

Table 6.2 Classification of amaranth species on the basis of chromosome number

Ploidy level	No. of chromosomes	*Amaranthus* species
Diploid	28	*A. tenuifolius*
	32	*A. acutilobus; A. albus; A. angusifolius; A. arenicola; A. aureus; A. blitoides; A. caudatus; A. chlorostachys; A. edulis; A. flavus; A. graecizans* ssp. *sylvestris; A. hybridus; A. hypochondriacus; A. leucocarpus; A. mantegazzianus; A. quitensis; A. sylvestris; A. tuberculatus*
	34	*A. asplundii; A. deflaxus; A. gangeticus; A. graecizans* ssp. *graecizans; A. lividus; A. polygonoides; A. roxburghianus; A. salicifolius; A. spinosus; A. tricolor; A. viridis*
	28/34	*A. blitum*
	32/34	*A. cruentus; A. mangostanus; A. powellii; A. paniculatus; A. palmeri; A. retroflexus*
Diploid and Tetraploid	32 and 64	*A. caturtus*
	34 and 68	*A. polygamous*
	32/34 and 64	*A. giganteus*
Tetraploid	64	*A. dubius*

6.3.4 Molecular Approach

Knowledge of phylogenetic relations gained through molecular techniques would empower plant breeders to distinguish genotypes by assessing the amount of dissimilarity within and between species for suitable breeding methods (Gelotar et al. 2019). Molecular markers, random amplified polymorphic DNA (RAPD), restriction fragment length polymorphism (RFLP), amplified fragment length polymorphism (AFLP), simple sequence repeats (SSR), inter simple sequence repeats (ISSR), internal transcribed spacer (ITS), and single nucleotide polymorphisms (SNP) are highly valued in the field of phylogenetics and genetics due to their ability to distinguish between genotypes (Grover and Sharma 2016). Like biochemical markers, molecular markers suggested that all three grain amaranths have evolved from the weedy progenitor, *A. hybridus*. Among cultivated grain amaranth species, *A. hypochondriacus* and *A. caudatus* share more similarity with each other than to *A. cruentus* (Das 2016). Intra-specific diversity in *Amaranthus* has been evaluated in some species, viz., *A. hypochondriacus, A. palmeri,* and *A. caudatus* by ISSR, SNP, and AFLP, respectively (Chandi et al. 2013; Jimenez et al. 2013; Raut et al. 2014) (Table 6.3).

The use of RAPD markers in the evaluation of intra- and inter-specific genetic diversity and relationships among grain amaranth genotypes

Table 6.3 Genetic diversity among *Amaranthus* species using molecular markers

S.N.	Species	Type	No. of genotypes	Type of molecular marker	References
1	*A. cruentus, A. caudatus,* and *A. hypochondriacus*	Grain	33	RAPD	Transue et al. (1994)
2	*A. acutilobus, A. albus, A. australis, A. caudatus, A. crassipes, A. cruentus, A. deflexus, A. dubius, A. fimbriatus, A. floridanus, A. hybridus, A. lividus, A. hypochondriacus, A. palmeri, A. powellii, A. quitensis, A. retroflexus, A. spinosus, A. standleyanus, A. standleyanus, A. tricolor, A. viridis, Amaranthus hybrid* (unknown), and *Amaranthus* spp. (unknown)	Grain, wild	60	RAPD	Chan and Sun (1997)
3	*A. cruentus, A. caudatus, A. hypochondriacus, A. hybridus, A. quitensis, A. powellii,* and *A. tricolor*	Grain, wild	30	ITS, ISSR, AFLP	Xu and Sun (2001)
4	*A. palmeri, A. powellii, A. retroflexus, A. arenicola, A. hybridus, A. spinosus, A. albus,* and *A. tuberculatus*	Wild	98	AFLP	Wassom and Tranel (2005)
5	*A. gangeticus, A. paniculatus, A. viridis, A. hypochondriacus, A. caudatus,* and *A. cruentus*	Grain, wild	30	RAPD	Ray and Roy (2009)
6	*A. hypochondriacus, A. caudatus,* and *A. cruentus*	Grain, wild	103	RAPD	Legaria (2010)
7	*A. cruentus, A. hypocondriacus,* and *A. chlorostachys*	Grain	6	RAPD	Popa et al. (2010)
8	*A. hybridus, A. caudatus, A. cruentus, A. hypochondriacus, A. powellii, A. retroflexus,* and *A. tuberculatus*	Grain, wild	46	SNP	Maughan et al. (2011)
9	*A. palmeri*	Wild	159	AFLP	Chandi et al. (2013)
10	*A. caudatus*	Ornamental	48	SNP	Jimenez et al. (2013)

(continued)

Table 6.3 (continued)

S.N.	Species	Type	No. of genotypes	Type of molecular marker	References
11	*A. acanthochiton, A. acutilobus, A. albus, A. arenicola, A. australis, A. blitum, A. blitoides, A. blitum* var. *oleraceus, A. bouchonii, A. cannabinus, A. viridis, A. caudatus, A. crassipes, A. cruentus, A. dubius, A. fimbriatus, A. graecizans* ssp. *silvestris, A. graecizans* ssp. *thellungianus, A. greggii, A. hybridus, A. hypochondriacus, A. lividus, A. muricatus, A. palmeri, A. polygamous, A. powellii, A. powellii* ssp. *bouchonii, A. powellii* ssp. *powellii, A. quitensis, A. spinosus, A. tricolor, A. tuberculatus,* and *Amaranthus* spp. (unknown)	Grain, wild, vegetable, ornamental	63	SSR	Khaing et al. (2013)
12	*Amaranthus* spp. (unknown)	–	31	RAPD, ISSR, SSR	Singh et al. (2013)
13	*A. blitum, A. deflexus, A. graecizans* subsp. *sylvestris, A. mantegazzianus, A. standleyanus, A. viridis, A. quitensis, A. caudatus, A. hybridus, A. tricolor,* and *Amaranthus* spp. (unknown)	Wild, vegetable, ornamental	75	SSR	Oo and Park (2013)
14	*A. blitum, A. caudatus,* and *A. hypochondriacus,*	Grain	70	SSR	Wang and Park (2013)
15	*A. hybridus, A. hypochondriacus, A. quitensis, A. caudatus,* and *A. cruentus*	Grain, wild	110	SSR	Kietlinski et al. (2014)
16	*A. caudatus, A. cruentus, A. hypochondriacus, A. hybridus, A. quitensis, A. powellii,* and *A. tricolor*	Grain, wild	88	SSR	Park et al. (2014)
17	*A. hypochondriacus*	Grain	28	ISSR	Raut et al. (2014)
18	*A. acutilobus, A. albus, A. arenicola, A. australis, A. blitoides, A. blitum, A. blitum* var. *oleraceus, A. bouchonii, A. caudatus, A. caudatus* var. *albiflorus, A. crassipes, A. crispus, A. cruentus, A. deflexus, A. dubius, A. fimbriatus, A. floridanus, A. gangeticus* var. *melancholicus, A. hybridus, A. hypochondriacus, A. lividus, A. mangostanus, A. mantegazzianus, A. palmeri, A. powellii, A. powellii* subsp. *bouchonii, A. quitensis, A. retroflexus, A. spinosus, A. standleyanus, A. tricolor, A. tuberculatus, A. viridis,* and *Amaranthus* spp. (unknown)	Grain, wild, vegetable	348	SSR	Suresh et al. (2014)
19	*A. caudatus, A. cruentus,* and *A. hypochondriacus*	Grain	55	RAPD, ISSR	Štefúnová et al. (2015)
20	*A. edulis, A. caudatus, A. hypochondriacus,* and *A. cruentus*	Grain, wild	19	ISSR	Gelotar et al. (2019)
21	*A. tricolor* and *A. hypochondriacus*	Vegetable, grain	300	SSR	Nguyen et al. (2019)

ITS Internal transcribed spacer, *AFLP* Amplified fragment length polymorphism, *ISSR* Inter simple sequence repeat, *RAPD* Random amplified polymorphic DNA, *SSR* Simple sequence repeats, *SNP* Single nucleotide polymorphism for the first time has been described by Transue et al. (1994). The RAPD markers have also been utilized for the evaluation of genetic diversity and relationships among 60 genotypes belonging to 23 wild and cultivated grain amaranth species (Table 6.3) has been first time reported by Chan and Sun (1997) distinctive species groups were formed in the dendrogram. The *A. hybridus* species complex, comprising all three cultivated grain amaranth species, namely *A. hypochondriacus, A. caudatus*, and *A. cruentus* along with their putative wild progenitors, namely *A. quitensis, A. powellii*, and *A. hybridus* is the most economically important group of the genus amaranth. As these are very closely related taxa, taxonomic confusion occurs among them. Xu and Sun (2001) used ITS, ISSR, and AFLP markers to understand the taxonomic status and phylogenetic relationships among grain amaranths (*A. cruentus, A. caudatus,* and *A. hypochondriacus*) and their wild relatives (*A. hybridus, A. quitensis,* and *A. powellii*) along with *A. tricolor*. As a result, phylogenetic tree having species specific clades has been generated and a close similarity between *A. caudatus* and *A. quitensis* was found with *A. powellii* as the most divergent taxon (Xu and Sun 2001).

Intra- and inter-specific genetic diversity among core collection of 63 genotypes of 32 different amaranth species (Table 6.3) was elucidated using SSR markers. The results of structure and phylogenetic analysis showed that the core set comprising of 63 genotypes was grouped in three subpopulations and some admixtures (some genotypes of *A. caudatus* and *A. tricolor* scattered in diverse groups) (Khaing et al. 2013). Similarly, intra-specific and inter-specific genetic diversity and relationships have been studied among 348 genotypes belonging to 33 *Amaranthus* spp. using SSR markers by Suresh et al. (2014). The clustering of genotypes and structure analysis did not strictly follow their geographic affiliations and species type. The results showed that *A. blitum* var. *oleraceus* (vegetable type), *A. caudatus* (grain type), *A. cruentus* (grain type), *A. hypochondriacus* (grain type), and *A. mangostanus* (grain type) were very closely related to *A. dubius* (weedy type), *A. crassipes* (weedy type), *A. crispus* (weedy type), *A. spinosus* (weedy type), and *A. hybridus* (weedy type), respectively (Suresh et al. 2014).

Different accessions from *A. hybridus, A. hypochondriacus, A. quitensis, A. caudatus,* and *A. cruentus* belonging to diverse geographic locations were used to evaluate genetic diversity using SSR markers (Kietlinski et al. 2014). It was revealed that the grain amaranths appear to be cultivars of *A. hybridus* and developed by two separate domestication events. *A. caudatus* and *A. hypochondriacus* were supposed to be closely related based on a single domestication with translocation event or distinct domestications from a single *A. hybridus* ancestry. Presumably, *A. caudatus* was crossed with weedy *A. quitensis*, which was the contradictory result as compared to earlier studies (Chan and Sun 1997; Xu and Sun 2001). So, the large number of loci and samples from extended geographic regions can be useful in accurate investigation of the domestication events of grain amaranths (Kietlinski et al. 2014). With regard to that, a diversity analysis of the cultivated *Amaranthus* spp. and their putative ancestor species (*A. hybridus*) has been performed using highly polymorphic SNP loci (Maughan et al. 2011). The results suggested that *A. hybridus* is possibly the progenitor species of all three grain amaranth species which was in agreement with previous reports (Chan and Sun 1997; Kietlinski et al. 2014). It was also revealed that grain amaranths are monophyletic, whereas *A. hybridus* is polyphyletic as its genotypes were present in each of the three grain amaranth monophyletic clades (Maughan et al. 2011).

Genetic diversity among grain amaranth species using ISSR markers has been studied by Gelotar et al. (2019). Few genotype specific amplicons have been produced which can be helpful in the identification of specific genotypes (Gelotar et al. 2019). Similarly, intra- and inter-specific genetic diversity among 55 accessions of grain amaranths were assessed using ISSR and RAPD markers (Štefúnová et al. 2015). There are some reports on the utilization of SSRs (Wang and Park 2013; Park et al. 2014) and RAPDs (Ray and Roy 2009; Legaria 2010; Popa et al.

2010) in the study of genetic diversity among amaranth species.

Leafy amaranths (mainly *A. tricolor*) are used in traditional food in many Asian and African countries still *A. tricolor* has not been focused on its systematic improvement. Therefore, there is a need to improve leafy amaranth which can be done by an assessment of the genetic diversity for the betterment of parents' selection in breeding programs. A set of 300 genotypes of *A. tricolor* and *A. hypochondriacus* has been used for phylogenetic analysis using SSR markers (Nguyen et al. 2019). The information on the genetic variability and population structure of amaranth could be helpful in selecting typical genotypes and managing *Amaranthus* breeding programs (Suresh et al. 2014; Gelotar et al. 2019).

6.4 Amaranth Germplasm Collection

The maximum amaranth germplasms have been collected and conserved by the Indian Council of Agricultural Research-National Bureau of Plant Genetic Resources (ICAR-NBPGR) National Gene Bank, India with 5232 grain amaranth and 504 vegetable amaranth accessions, followed by North Central Regional Plant Introduction Station, Ames, Iowa with 3502 accessions. There are other institutes from Peru, Ecuador, Mexico, Taiwan, and Bolivia possesses an amaranth germplasm collection (Table 6.4) (Jacobsen and Mujica 2003; AVGRIS 2020; GENESYS 2020; ICAR-NBPGR 2020).

6.5 Amaranth Varieties of India

In India, various research organizations and universities are actively involved in the amaranth improvement and development of promising varieties. The list of varieties of vegetable and grain amaranth for different regions is given in Table 6.5 (Dua et al. 2009; TNAU 2017; KAU 2020).

6.6 Importance and Applications of Amaranth

The seeds of the grain amaranths have a tremendous nutritional value, and they are rich in protein, starch, oil, vitamins (A, C, and B), minerals (Ca, Fe, Mn, and Zn), and cholesterol-reducing soluble fiber (Peter and Gandhi 2017). Amaranth grains have more quantity and quality protein compared with other cereals (Coelho et al. 2018). Moreover, the proteins of amaranth have high digestibility ($\sim$90%) and they are rich in amino acid lysine (6.2–6.1 g Lys/100 g protein) (limiting amino acid in cereals) and sulfur-containing amino acids (limiting in the pulses) (Das 2016; Joshi et al. 2018). The seeds on average are composed of 15% protein, 7% fat, 67% carbohydrate, 6.7% dietary fiber, and 2.6% ash (Cole 1979; Das 2016; Jimoh et al. 2018). The amaranth grains are ground into gluten-free flour and to prepare bread by mixing proportion of wheat or rye flour. Amaranth grains are used in making pastes, custards, and salad dressings; grain flour in cookies, bread and bakery products; and starch in thickening of soups, sauces, and gravies (Jimoh et al. 2018). Grain amaranth is called the "third millennium grain" due to its high nutritional properties (Das 2016; Peter and Gandhi 2017).

Amaranth (especially *A. cruentus*) leaves are a good source of protein, dietary fiber, minerals (Mn, Fe, Cu, Ca, Mg, K, and P) and vitamins (pro-vitamin A carotenoid) (Peter and Gandhi 2017). The leaves, shoots, and tender stems of several cultivated vegetable amaranth and young leaves of grain amaranth species are used as food (a potherb in soups or sauces and as a leafy vegetable) and as an animal feed. Vegetable amaranth has many beneficial roles in human health (Das 2016).

Amaranthus species have been used in the treatment of numerous diseases since a very long time, but they have seen attention in recent decades. In vitro or in vivo studies on different pharmacological activities of various species of *Amaranthus* (mainly *A. spinosus*, *A. caudatus,* and *A. viridis*) have proved that these species

Table 6.4 Amaranth germplasm collections in the world

Institute	Country	No. of accessions	References
ICAR-NBPGR National GeneBank	India	6309	ICAR-NBPGR (2020)
North Central Regional Plant Introduction Station (NCRPIS), Ames, Iowa (USDA-ARS)	USA	3502	GENESYS (2020)
World Vegetable Centre	Taiwan	896	AVGRIS (2020)
Universidad Nacional San Antonio Abad de Cusco, Centro de	Peru	740	Jacobsen and Mujica (2003)
Universidad Nacional del Altiplano (UNAP) Escuela de	Peru	440	Jacobsen and Mujica (2003)
Instituto de Investigaciones Agropecuarias (INIAP) EE. Santa	Ecuador	420	Jacobsen and Mujica (2003)
Instituto Nacional de Investigaciones Forestales y Agropecuarias (INIFAP)	Mexico	500	Jacobsen and Mujica (2003)
Centro de Investigaciones Fitoecogenéticas de Pairumani (CIFP), Cochabamba	Bolivia	106	Jacobsen and Mujica (2003)

Table 6.5 Amaranth varieties released in India

Name of variety	Developed by
Vegetable type varieties	
CO-1, CO-2, CO-3, CO-4 (Vegetable cum grain type), CO-5, Sirukeerai, PLR 1	TNAU, Coimbatore
Pusa Chotti Chaulai, Pusa Badi Chaulai, Pusa Kirti, Pusa Kiran, Pusa Lal Chaulai	IARI, New Delhi
Arka Suguana, Arka Arunima, Arka Samraksha, Arka Varna	IIHR, Bangalore
Konkan Durangi	DBSKKV, Dapoli
Arun, Renusree, Krishnasree	KAU, Kerala
Grain type varieties for hills	
PRA-1, PRA-2, PRA-3, PRA-4	GBPUAT, Ranichauri
VL Chua-44	ICAR-VPKAS, Almora
Annapurna, Durga	NBPGR regional station, Shimla
Grain type varieties for plains	
Gujarat Amaranth-1 (GA-1), GA-2, GA-3	SDAU, S.K. Nagar
BGA-2 (Kapilasa), Suvadra (BGA 4–9)	OUAT, Bhubaneshwar
Suvarna	UAS, Bangalore
RMA-4, RMA-7	AUJ, Jodhpur

possess one or more pharmacological activities among analgesic, antigastric ulcer, anticancer, antidiabetic, antifertility, antihyperglycemic, antianxiety, anti-inflammatory, antimalarial, antimicrobial, antinociceptive, antioxidant, cardiovascular protective, cytotoxic, gastroprotective, immunomodulatory, neuroprotective, and wound healing activities (Peter and Gandhi 2017;

Martinez-Lopez et al. 2020). Most of the species have strong antioxidant activity due to the presence of flavonoids and some phenolic acids. Few of the results offer details of various biological activities of major bioactive components present in crude extracts of particular plant parts. Still, there are no specific reports on the bioavailability and principal mechanism of action of bioactive constituents isolated from *Amaranthus* spp. Thus, there is a necessity for dedicated research in this area (Peter and Gandhi 2017).

Some *Amaranthus* species possess the potential phytoremediation ability to the uptake of heavy metals in soil (Ziarati and Alaedini 2014). *Amaranthus* species are used to extract natural nutritive pigments (red-violet betacyanins) for the food industry (Cai et al. 1998). Amaranth is an underexploited plant source of squalene (from seed oil), a highly important compound in the food, cosmetic, and pharmaceutical industries (Krulj et al. 2016). The diversity analysis can be useful for the improvement of amaranth for a particular character to enhance its applicability in various industrial applications. Promising genotype for the specific character can be utilized in the development of new varieties with desired characters.

6.7 Conclusion

Amaranthus genus has many species with great importance in the food, cosmetic, and pharmaceutical industries. Amaranth is mainly cultivated for grain and vegetable purpose, whereas many species are economically important weeds. The taxonomy and species number of amaranth are not in agreement. Despite its tremendous nutraceutical value, research activities in *Amaranthus* spp. are limited. The knowledge of taxonomy and intra-specific as well as inter-specific genetic diversity among various amaranth species can support in a breeding program identifying diverse parents for introgression of desired genes to develop high yielding and nutritionally rich varieties of amaranth.

References

Adhikary D, Pratt DB (2015) Morphologic and taxonomic analysis of the weedy and cultivated *Amaranthus hybridus* species complex. Syst Bot 40(2):604–610. https://doi.org/10.1600/036364415X688376

Ahammed AU, Rahman MM, Mian MAK (2013) Mutivariate analysis in stem amaranth (*Amaranthus tricolor*). Bangladesh J Plant Breed Genet 26(1):11–17. https://doi.org/10.3329/bjpbg.v26i1.19979

Andini R, Yoshida S, Yoshida Y, Ohsawa R (2013) *Amaranthus* genetic resources in Indonesia: morphological and protein content assessment in comparison with worldwide amaranths. Genet Resour Crop Evol 60(7):2115–2128. https://doi.org/10.1007/s10722-013-9979-y

Anuja S, Mohideen MK (2007) Genetic diversity for green yield characteristics in vegetable amaranthus (*Amaranthus* spp.). Asian J Hort 2(1):158–160

APMC-Palanpur (2020) The agricultural produce market committee: Amaranth income-price, 2020

AVGRIS (2020) The AVRDC vegetable genetic resources information system (AVGRIS). http://seed.worldveg.org/search/characterization/amaranthus. Accessed 15 Aug 2020

Bao-Hua S, Xue-Jie Z, Fa-Zeng L, Peng W (2002) Chromosome numbers of 14 species in *Amaranthus* from China. J Syst Evol 40(5):428–432

Becker R, Wheeler EL, Lorenz K, Stafford AE, Grosjean OK, Betschart AA, Saunders RM (1981) A compositional study of amaranth grain. J Food Sci 46 (4):1175–1180

Behera B, Patnaik SN (1974) Cytotaxonomic studies in the family Amaranthaceae. Cytologia 39(1):121–131

Brenan JPM, Townsend CC (1980) Proposal to reject *Amaranthus blitum* L. (Amaranthaceae) under Art. 69 in favour of *A. lividus* L. Taxon 29:695–696

Brenner DM, Baltensperger DD, Kulakow PA, Lehmann JW, Myers RL, Slabbert MM, Sleugh BB (2000) Genetic resources and breeding of *Amaranthus*. Plant Breed Rev 19:227–285

Cai Y, Sun M, Wu H, Huang R, Corke H (1998) Characterization and quantification of betacyanin pigments from diverse *Amaranthus* species. J Agric Food Chem 46(6):2063–2070

Chan KF, Sun M (1997) Genetic diversity and relationships detected by isozyme and RAPD analysis of crop and wild species of *Amaranthus*. Theor Appl Genet 95 (5–6):865–873

Chandi A, Milla-Lewis SR, Jordan DL, York AC, Burton JD, Zuleta MC, Whitaker JR, Culpepper AS (2013) Use of AFLP markers to assess genetic diversity in Palmer amaranth (*Amaranthus palmeri*) populations from North Carolina and Georgia. Weed Sci 61(1):136–145. https://doi.org/10.1614/WS-D-12-00053.1

Chaney L, Mangelson R, Ramaraj T, Jellen EN, Maughan PJ (2016) The complete chloroplast genome sequences for four *Amaranthus* species (Amaranthaceae). Appl Plant Sci 4(9):1600063. https://doi.org/10.3732/apps.1600063

Coelho LM, Silva PM, Martins JT, Pinheiro AC, Vicente AA (2018) Emerging opportunities in exploring the nutritional/functional value of amaranth. Food Funct 9(11):5499–5512. https://doi.org/10.1039/C8FO01422A

Cole JN (1979) Amaranth from the past for the future. Rodale Press

Das S (2012a) Domestication, phylogeny and taxonomic delimitation in underutilized grain *Amaranthus* (Amaranthaceae)—a status review. Feddes Repert 123 (4):273–282. https://doi.org/10.1002/fedr.201200017

Das S (2012b) Systematics and taxonomic delimitation of vegetable, grain and weed amaranths: a morphological and biochemical approach. Genet Resour Crop Evol 59(2):289–303. https://doi.org/10.1007/s10722-011-9684-7

Das S (2016) *Amaranthus:* a promising crop of future. Springer, Singapore

Desai SR (1971) Morphological and cytological studies on *Amaranthaceae*. Cytologia 36(2):349–353

Dua RP, Raiger HL, Phogat BS, Sharma SK (2009) Underutilized crops: Improved varieties and cultivation practices. New Delhi, NBPGR, p 66

Erum S, Naeemullah M, Masood S, Qayyum A, Rabbani MA (2012) Genetic divergence in *Amaranthus* collected from Pakistan. J Anim Plant Sci 22(3):653–658

Gelotar MJ, Dharajiya DT, Solanki SD, Prajapati NN, Tiwari KK (2019) Genetic diversity analysis and molecular characterization of grain amaranth genotypes using inter simple sequence repeat (ISSR) markers. Bull Natl Res Cent 43(1):103. https://doi.org/10.1186/s42269-019-0146-2

GENESYS (2020) GENESYS: global portal on plant genetic resources. https://www.genesys-pgr.org/wiews/USA020. Accessed 15 Aug 2020

Gerrano AS, Van Rensburg WSJ, Adebola PO (2015) Genetic diversity of *Amaranthus* species in South Africa. S Afr J Plant Soil 32(1):39–46. https://doi.org/10.1080/02571862.2014.973069

Gerrano AS, Van Rensburg WJ, Mavengahama S, Bairu M, Venter S, Adebola PO (2017) Qualitative morphological diversity of *Amaranthus* species. J Trop Agric 55(1):12–20

Grant WF (1959a) Cytogenetic studies in *Amaranthus*: II. Natural interspecific hybridization between *Amaranthus dubius* and *A. spinosus*. Can J Bot 37(5):1063–1070

Grant WF (1959b) Cytogenetic studies in *Amaranthus*: III. Chromosome numbers and phylogenetic aspects. Can J Genet Cytol 1(4):313–328

Grover A, Sharma PC (2016) Development and use of molecular markers: past and present. Critic Rev Biotechnol 36(2):290–302. https://doi.org/10.3109/07388551.2014.959891

Gueco LS, Borromeo T, De Guzman C (2016) Diversity in the morphology of Amaranth (*Amaranthus* sp.) germplasm collection in the Philippines. Asian J Agric Food Sci 4(2):73–79

Hamrick JL, Godt MJW (1996) Conservation genetics of endemic plant species. In: Avise JC, Hamrick JL (eds) Conservation genetics: case histories from nature. Chapman and Hall, New York, pp 281–304

Iamonico D (2012) *Amaranthus powellii* subsp. *cacciatoi* comb. et stat. nov. (Amaranthaceae). Nord J Bot 30 (1):12–16

ICAR-NBPGR (2020) ICAR-NBPGR: Status of base collections in National GeneBank. http://www.nbpgr.ernet.in/Research_Projects/Base_Collection_in_NGB.aspx. Accessed 15 Aug 2020

Jacobsen SE, Mujica A (2003) The genetic resources of Andean grain amaranths (*Amaranthus caudatus* L., *A. cruentus* L. and *A. hipochondriacus* L.) in America. Plant Genet Resour Newsl 133:41–44

Jimenez FR, Maughan PJ, Alvarez A, Kietlinski KD, Smith SM, Pratt DB, Elzinga DB, Jellen EN (2013) Assessment of genetic diversity in Peruvian amaranth (*Amaranthus caudatus* and *A. hybridus*) germplasm using single nucleotide polymorphism markers. Crop Sci 53(2):532–541. https://doi.org/10.2135/cropsci2012.07.0413

Jimoh MO, Afolayan AJ, Lewu FB (2018) Suitability of *Amaranthus* species for alleviating human dietary deficiencies. S Afr J Bot 115:65–73. https://doi.org/10.1016/j.sajb.2018.01.004

Joshi DC, Sood S, Hosahatti R, Kant L, Pattanayak A, Kumar A, Yadav D, Stetter MG (2018) From zero to hero: the past, present and future of grain amaranth breeding. Theor Appl Genet 131(9):1807–1823. https://doi.org/10.1007/s00122-018-3138-y

KAU (2020) Kerala Agricultural University: varieties released. http://www.kau.in/basic-page/varieties-released. Accessed 15 Aug 2020

Khaing AA, Moe KT, Chung JW, Baek HJ, Park YJ (2013) Genetic diversity and population structure of the selected core set in *Amaranthus* using SSR markers. Plant Breed 132(2):165–173. https://doi.org/10.1111/pbr.12027

Kietlinski KD, Jimenez F, Jellen EN, Maughan PJ, Smith SM, Pratt DB (2014) Relationships between the weedy *Amaranthus hybridus* (Amaranthaceae) and the grain amaranths. Crop Sci 54(1):220–228. https://doi.org/10.2135/cropsci2013.03.0173

Krulj J, Brlek T, Pezo L, Brkljača J, Popović S, Zeković Z, Bodroža Solarov M (2016) Extraction methods of *Amaranthus* sp. grain oil isolation. J Sci Food Agric 96(10):3552–3558. https://doi.org/10.1002/jsfa.7540

Legaria S (2010) Genetic diversity in some species of amaranth (*Amaranthus* spp.). Rev Fitotec Mex 33 (2):89–95

Mandák B, Zákravský P, Dostál P, Plačková I (2011) Population genetic structure of the noxious weed *Amaranthus retroflexus* in Central Europe. Flora 206 (8):697–703. https://doi.org/10.1016/j.flora.2011.01.010

Martinez-Lopez A, Millan-Linares MC, Rodriguez-Martin NM, Millan F, Montserrat-de la Paz S (2020) Nutraceutical value of kiwicha (*Amaranthus caudatus* L.). J Funct Foods 65:103735. https://doi.org/10.1016/j.jff.2019.103735

Maughan PJ, Smith SM, Fairbanks DJ, Jellen EN (2011) Development, characterization, and linkage mapping of single nucleotide polymorphisms in the grain amaranths (*Amaranthus* sp.). The Plant Genome 4 (1):92–101. https://doi.org/10.3835/plantgenome2010.12.0027

Milan PR (2008) Chromosome behaviour and fertility in induced polyploids of grain amaranths. Caryologia 61 (3):199–205. https://doi.org/10.1080/00087114.2008.10589630

Nguyen DC, Tran DS, Tran TTH, Ohsawa R, Yoshioka Y (2019) Genetic diversity of leafy amaranth (*Amaranthus tricolor* L.) resources in Vietnam. Breed Sci 19050. https://doi.org/10.1270/jsbbs.19050

Oboh BO (2007) Multivariate analysis of the diversity among some Nigerian accessions of *Amaranthus hybridus*. Int J Plant Breed Genet 1(2):89–94

Oo WH, Park YJ (2013) Analysis of the genetic diversity and population structure of amaranth accessions from South America using 14 SSR markers. Korean J Crop Sci 58(4):336–346

Pagi N, Prajapati N, Pachchigar K, Dharajiya D, Solanki SD, Soni N, Patel P (2017) GGE biplot analysis for yield performance of grain amaranth genotypes across different environments in western India. J Exp Biol Agric Sci 5(3):368–376. https://doi.org/10.18006/2017.5(3).368.376

Pal M, Ohri D, Subramaniyam GV (2000) A new basic chromosome number for *Amaranthus* (Amaranthaceae). Cytologia 65(1):13–16. https://doi.org/10.1508/cytologia.65.13

Park YJ, Nishikawa T, Matsushima K, Minami M, Tomooka N, Nemoto K (2014) Molecular characterization and genetic diversity of the starch branching enzyme (SBE) gene from *Amaranthus*: the evolutionary origin of grain amaranths. Mol Breed 34(4):1975–1985. https://doi.org/10.1007/s11032-014-0156-6

Peter K, Gandhi P (2017) Rediscovering the therapeutic potential of *Amaranthus* species: a review. Egypt J Basic Appl Sci 4(3):196–205. https://doi.org/10.1016/j.ejbas.2017.05.001

Popa G, Cornea CP, Ciuca M, Babeanu N, Popa O, Marin D (2010) Studies on genetic diversity in *Amaranthus* species using the RAPD markers. Analele Univ din Oradea Fasc Biol 17(2):280–285

Queirós M (1989) Estudos citotaxonómicos em *Amaranthus* de Portugal. Lazaroa 11:9–17

Radwan SAA, Badr S, Mira MM, Helmy R (2014) Genetic variability and phylogenetic relationships in *Amaranthus* using karyotype. J Biotechnol Res 14:53–70

Rastogi A, Shukla S (2013) Amaranth: a new millennium crop of nutraceutical values. Crit Rev Food Sci Nutr 53(2):109–125. https://doi.org/10.1080/10408398.2010.517876

Raut VR, Dodake SS, Chimote VP (2014) Evaluation of genetic diversity in grain amaranth (*Amaranthus hypochondriacus*) at molecular level using ISSR markers. Ind J Agric Biochem 27(1):60–65

Ray T, Roy SC (2009) Genetic diversity of *Amaranthus* species from the Indo-Gangetic plains revealed by RAPD analysis leading to the development of ecotype-specific SCAR marker. J Hered 100(3):338–347. https://doi.org/10.1093/jhered/esn102

Reema S, Roy BK (2014) A new chromosome number for *Amaranthus blitum*. J N Biol Rep 3(2):111–114

Sauer JD (1967) The grain amaranths and their relatives: a revised taxonomic and geographic survey. Ann Missouri Bot Gard 54(2):103–137

Sauer JD (1993) Historical geography of crop plants: a select roster. CRC Press

Shankar R, Lal A, da Silva JAT, More TA (2012) Diversity analysis of fleshy leaf type *Amaranthus* for semi-arid ecosystems. Int J Plant Breed 6:27–33

Sheidai M, Mohammadzadeh Z (2008) Cytogenetic study of *Amaranthus* L. species in Iran. Cytologia 73(1):1–7. https://doi.org/10.1508/cytologia.73.1

Shukla S, Bhargava A, Chatterjee A, Pandey AC, Mishra BK (2010) Diversity in phenotypic and nutritional traits in vegetable amaranth (*Amaranthus tricolor*), a nutritionally underutilised crop. J Sci Food Agric 90(1):139–144. https://doi.org/10.1002/jsfa.3797

Shukla A, Srivastava N, Suneja P, Yadav SK, Hussain Z, Rana JC, Yadav S (2018) Untapped amaranth (*Amaranthus* spp.) genetic diversity with potential for nutritional enhancement. Genet Resour Crop Evol 65(1):243–253. https://doi.org/10.1007/s10722-017-0526-0

Singh B, Pandey S, Kumar J (2013) A comparative study of inter simple sequence repeat (ISSR), random amplified polymorphic DNA (RAPD) and simple sequence repeat (SSR) loci in assessing genetic diversity in *Amaranthus*. Indian J Genet Plant Breed 73:411–418. https://doi.org/10.5958/j.0975-6906.73.4.062

Singh N, Singh P, Shevkani K, Virdi AS (2019) Amaranth: potential source for flour enrichment. In: Preedy V, Watson R (eds) Flour and breads and their fortification in health and disease prevention. Academic Press, pp 123–135. https://doi.org/10.1016/b978-0-12-814639-2.00010-1

Štefúnová V, Bežo M, Žiarovská J, Ražná K (2015) Detection of the genetic variability of *Amaranthus* by RAPD and ISSR markers. Pak J Bot 47(4):1293–1301

Suresh S, Chung JW, Cho GT, Sung JS, Park JH, Gwag JG, Baek HJ (2014) Analysis of molecular genetic diversity and population structure in *Amaranthus* germplasm using SSR markers. Pl Biosyst 148 (4):635–644. https://doi.org/10.1080/11263504.2013.788095

Su-Young H, Cheon KS, Yoo KO, Hyun-Oh L, Mekapogu M, Cho KS (2019) Comparative analysis of the complete chloroplast genome sequences of three

Amaranthus species. Plant Genet Resour 17(3):245–254. https://doi.org/10.1017/S1479262118000485

TNAU (2017) Lec 31: Origin, area, production, varieties, package of practices for amaranthus, palak and gogu. In: HORT 281—production technology of vegetables and flowers. Development of e-Course for B.Sc. (Agriculture). TNAU, Tamil Nadu, India, pp 283–293

Transue DK, Fairbanks DJ, Robison LR, Andersen WR (1994) Species identification by RAPD analysis of grain amaranth genetic resources. Crop Sci 34(5): 1385–1389

USDA, ARS, NPGS (2021) USDA, Agricultural Research Service, National Plant Germplasm System (NPGS), Germplasm Resources Information Network (GRIN-Taxonomy). National Germplasm Resources Laboratory, Beltsville, Maryland. https://npgsweb.ars-grin.gov/gringlobal/taxon/taxonomysearch. Accessed 19 Jan 2021

Venkatesh L, Murthy N, Nehru SD (2014) Analysis of genetic diversity in grain amaranth (*Amaranthus* spp.). Indian J Genet 74(4):522–525

Viljoen E, Odeny DA, Coetzee MP, Berger DK, Rees DJ (2018) Application of chloroplast phylogenomics to resolve species relationships within the plant genus *Amaranthus*. J Mol Evol 86(3–4):216–239. https://doi.org/10.1007/s00239-018-9837-9

Wang XQ, Park YJ (2013) Comparison of genetic diversity among amaranth accessions from south and Southeast Asia using SSR markers. J Korean Med Sci 21(3):220–228. https://doi.org/10.7783/KJMCS.2013.21.3.220

Wassom JJ, Tranel PJ (2005) Amplified fragment length polymorphism-based genetic relationships among weedy *Amaranthus* species. J Hered 96(4):410–416. https://doi.org/10.1093/jhered/esi065

Wu X, Blair MW (2017) Diversity in grain amaranths and relatives distinguished by genotyping by sequencing (GBS). Front Plant Sci 8:1960. https://doi.org/10.3389/fpls.2017.01960

Xu F, Sun M (2001) Comparative analysis of phylogenetic relationships of grain amaranths and their wild relatives (*Amaranthus*; Amaranthaceae) using internal transcribed spacer, amplified fragment length polymorphism, and double-primer fluorescent inter simple sequence repeat markers. Mol Phylogenet Evol 21(3):372–387. https://doi.org/10.1006/mpev.2001.1016

Ziarati P, Alaedini S (2014) The phytoremediation technique for cleaning up contaminated soil by *Amaranthus* sp. J Environ Anal Toxicol 4(208):2161-0525. https://doi.org/10.4172/2161-0525.1000208

7 Searching for an Identity: Functional Characterization of Taxonomically Restricted Genes in Grain Amaranth

Gabriela Cabrales-Orona and John P. Délano-Frier

Abstract

Taxonomically restricted genes, or TRGs, are specific to a particular taxon that can be found only in the genomes of single species or are represented as orthologs in closely related genera. Despite being regarded with a mixture of skepticism and awe by the scientific community, progress has been gradually attained in the understanding of their presumed origin and function in most, if not all, forms of life. Grain amaranth is not an exception, as shown by the numerous unknown function TRGs that were unveiled by a recent transcriptomic analysis undergone under different (a)biotic stress conditions. True to their nature, amaranth TRGs appear to be mostly stress-related genes that may offer a clue to better understand the ability of these remarkable plants to thrive under unfavorable ambient conditions. This chapter will concentrate on the description of what has gradually emerged from the incipient study of TRGs in grain amaranth and will place this knowledge in the context of what is known about these enigmatic genes in other organisms.

G. Cabrales-Orona · J. P. Délano-Frier (✉)
Departamento de Biotecnología y Bioquímica, Centro de Investigación y de Estudios Avanzados del I.P.N. (Cinvestav), Unidad Irapuato. Km 9.6 del Libramiento Norte Carretera Irapuato-León. Apartado Postal 629, C.P. 36824 Irapuato, Gto., México
e-mail: john.delano@cinvestav.mx

7.1 Introduction

Once, in what could be considered gentler, less frenzied times, Fred Sanger′s intellectually innovative and gargantuan effort to decipher the sequence of insulin was rewarded with the Nobel Prize in Chemistry, in 1953. Now, entire genomes are sequenced in a matter of days, with a tendency towards even faster procedures as the technologies become more efficient and less costly. This continuously improved capacity has created a deluge of genomic information that has gradually become a daunting challenge to those seeking to understand it. One of the emerging enigmas resulting from this technological revolution has been the discovery of a massive number of taxonomically unique, or restricted, gene sequences whose numbers keep increasing every time a newly sequenced species genome is uncovered. The percentage of genes that are unique in various species ranges widely, with 5–15% being typical (Wilson et al. 2005; Arendsee et al. 2014).

These genes are also known as taxonomically restricted genes, orphan genes, or "ORFans" (Dujon 1996; Fischer and Eisenberg 1999; Wilson et al. 2005, 2007; Khalturin et al. 2009; Nelson and Buggs 2016). The term "orphan"

D. Adhikary et al. (eds.), *The Amaranth Genome*, Compendium of Plant Genomes,
https://doi.org/10.1007/978-3-030-72365-1_7

(Dujon 1996), was first employed with reference to the yeast genome. At present, this term includes a subset of the more precise "taxonomically restricted genes" definition (or TRGs, which is synonymous to "lineage-specific genes") that are specific to a particular taxon. These sequences can be found only in the genomes of single species (i.e., autapomorphic genes), or are represented as orthologs in closely related genera (i.e., synapomorphic genes) (Narra et al. 2008; Arendsee et al. 2014; Nelson and Buggs 2016). Other definitions for these genes, such as "singleton ORFans", "orthologous ORFans" and "paralogous ORFans" (Siew and Fischer 2003), has contributed to the present lack of a unifying terminology.

It is generally considered that the use of the TRGs term represents a more orderly definition of "orphan" or "novel" genes, since it not only considers their lack of sequence similarity to genes or proteins in other organisms, but also their narrow phylogenetic distribution (Wilson et al. 2005). TRGs have at times been suggested to be non-functional sequences, or annotation artifacts (Skovgaard et al. 2001; Clamp et al. 2007). However, support for their functionality and biological significance is supported by rapidly accumulating experimental evidence (see below). For the sake of clarity, TRGs will be employed as a single inclusive term throughout this chapter.

The evolutionary origin of TRGs is poorly understood. Many models have been proposed to date to explain the origin and evolution of some TRGs. They include scenarios where the relaxed selection of one gene copy following duplication gradually leads to novel function(s) (a.k.a. duplication and divergence model), the formation of hybrid genes either from segments or complete sequences of separate genes (a.k.a. gene fusion model), the separation of a single gene to form two distinct genes via differential degeneration of the two copies (a.k.a. gene fission model), the relocation of genes from other species undergo divergence and neo-functionalization (a.k.a. horizontal transfer model) and integration of reverse-transcribed genes into a different site in the genome followed by divergence (a.k.a. retroposition model) (van Oss and Carvunis 2019). A perceived weakness of these models is that their progression necessarily leaves behind traceable putative-progenitor sequences that can be detectable by similarity searches (Zhou et al. 2008; Donoghue et al. 2011). This implies that if TRGs arose by any one of these mechanisms, they must have rapidly diverged from their progenitor sequences, beyond the threshold of detection of these scrutinies (Tautz and Domazet-Lošo, 2011; Zhou et al. 2008), thereby contradicting the gradual mutation/selection mechanism of evolution (Fisher 1930; Wright 1931). An alternative to these scenarios is the so-called de novo gene evolution model (Carvunis et al. 2012; Ding et al. 2012; Neme and Tautz 2013), in which TRGs are proposed to have evolved from non-coding sequences. The de novo gene evolution concept has been supported by several studies reporting the finding of sequences orthologous to TRGs in the non-coding DNA of other species (Levine et al. 2006; Zhou et al. 2008; Knowles and McLysaght 2009; Li et al. 2010a, b; Wu et al. 2011; Yang and Huang 2011; Zhao et al. 2015). One particularly informative report showed that *Poldi*, a mouse gene expressed in the testes, has orthologous regions in human and rat that are not expressed (Heinen et al. 2009). Critics of the de novo gene evolution theory are many. Perhaps the most sobering observation concerning de novo gene origination is that it represents a highly unlikely evolutionary process because of the almost negligible probability that a functional protein sequence may emerge from a random sequence after acquiring essential characteristics. (Jacob 1977; Siepel 2009). An alternative, although still contentious model, proposes that that de novo genes may emerge from short "proto-genes", or non-genic ORFs. These can result from mutations in non-genic sequences in the genome that subsequently undergo a continual transformation from short, limitedly expressed, unannotated ORFs with restricted taxonomic distribution to longer, frequently expressed, well-annotated ORFs with broad taxonomic distribution (Carvunis et al. 2012). This model coincides with the fact that short length is often a defining feature of TRGs (see below; Lipman et al. 2002; Daubin and Ochman 2004; Wolf et al. 2009).

7.2 Identification of TRGs

Two major approaches are used for the systematic identification of novel genes: genomic phylostratigraphy and synteny-based methods. The former is contingent on the mapping of the gene distribution of fully-sequenced genomes onto taxonomic (or phylogenetic) categories, such that genes are arranged in hierarchic relation to each other (Domazet-Lošo and Tautz 2007, 2008, 2010). It permits to infer the presence or absence of ancestral homologs of a given TRG and to identify the so-called "founder genes", defined as the phylogenetically oldest genes present at the base of a new gene lineage, protein domain or gene family (Tautz and Domazet-Lošo 2011). One noted disadvantage of this method is that its reliance on sequence similarity is often an impediment to determine whether a novel gene, particularly one evolving more rapidly, emerged de novo or diverged from an ancestral gene beyond recognition (Elhaik et al. 2006).

The synteny-based approaches are based on the analysis of syntenic sequences, that is, genomic blocks or sections in which the order and relative positioning of particular features has been maintained (Tautz and Domazet-Lošo 2011; Ghiurcuta and Moret 2014; McLysaght and Hurst 2016; Saripella et al. 2016). By such means, a de novo origin can be justified with increasing certainty when a high-quality syntenic alignment demonstrates that specific mutation(s) in a given syntenic region created coding potential that is lacking in closely related out-group species (McLysaght and Hurst 2016). Acquired coding potential, ideally absent in the out-group species, can be demonstrated using diverse molecular techniques that are usually complemented with genetic approaches potentially able to infer a possible biological function (Kellis et al. 2014).

7.3 Features of TRGs

TRGs or de novo genes have several distinctive characteristics that distinguish them from established genes: they usually have shorter lengths, evolve at a more accelerated pace, undergo a more relaxed purifying selection and tend to have lower levels of expression, which is usually highly developmental-, tissue- or condition-specific-dependent. Their specialization into enzyme catalytic activity is very rare (Arendsee et al. 2014; van Oss and Carvunis 2019). Also, various predicted TRG structural features, including intrinsic structural disorder (Wilson et al. 2017), and transmembrane domains, show a strong GC dependency. Chromatin configurations and DNA methylation levels have been linked with the formation of novel TRGs and/ or with their regulated expression in *Arabidopsis thaliana*, *Magnaporthe oryzae* and the nematode *Pristionchus pacificus* (Jeon et al. 2015; Li et al. 2016; Vakirlis et al. 2018; Werner et al. 2018; van Oss and Carvunis 2019).

As mentioned already, every sequenced genome has revealed a sizeable number of TRGs. Many have demonstrated to have widespread biological functions, particularly designed to adapt to different and/ or adverse environmental conditions (Khalturin et al. 2009; Arendsee et al. 2014; Nelson and Buggs 2016; Johnson 2018). Curiously, the highest TRG frequency is found in viral genomes. Apart from their ubiquity, it appears that many of these sequences are functionally significant or even essential, and may be factors defining growth on certain bacterial hosts (Ang and Georgopoulos 2012). In bacteria, the analysis of complete genomes of *Streptococcus agalactiae* or Group B *Streptococcus and Escherichia coli* revealed that each newly sequenced strain contained genes not previously seen in any other strain (Tettelin et al. 2005; Rasko et al. 2008; Touchon et al. 2009). This suggested that no single strain could be regarded as highly representative of the species, while the ever-increasing profusion of the strain-specific sequences was interpreted as evidence of the presence of the so-called "accessory" or "dispensable" bacterial genomes that were needed for survival in specific environments (Mira et al. 2010). Several other representative examples in diverse taxonomic groups further support the role that TRGs play in biotic and abiotic interactions such as defense, stress response, predation, signaling and diverse interactions with other organisms (Arendsee et al. 2014).

For instance, a fraction of the thousands of novel secreted proteins found to be codified in the gall-producing insects genome was proposed to be responsible for inducing the formation of the plant structures where these insects find shelter and sustenance (Zhao et al. 2015). Another study reported that multiple aphid-specific salivary proteins functioned as virulence effectors able to silence plant defenses and facilitate infestation (Elzinga and Jander 2013), whereas a TRG, identified as *Meloidogyne, a virulence protein*-1 (or *MAP-1*), encoded a secreted protein used as a virulence factor by *Meloidogyne incognita*, an endo-parasitic plant nematode (Tomalova et al. 2012). Polyp *Hydra* embryos utilize short proteins possessing high bactericidal activity to selectively incorporate a protective bacterial microbiota that confers protection from pathogens (Fraune et al. 2010). Similarly, TRGs encoding small secreted proteins were proposed to promote the retention of beneficial endosymbiotic nitrogen-fixing bacteria in unique differentiated cells, or bacteriocytes, in aphids (Shigenobu and Stern 2013).

TRGs are also involved in developmental processes. Included are their participation in nacre formation during shell construction in the bivalve *Pinctada maxima* and the gastropod *Haliotis asinine* (Jackson et al. 2010), limb regeneration following injury in salamanders, (da Silva et al. 2002; Fröbisch and Shubin 2011; Brockes and Gates 2014), developmental phenotypic plasticity observed in several nematode species subjected to stressful growing conditions (Sommer and Ogawa 2011; Mayer et al. 2015), the definition of octopus-specific morphological structures and nervous system (Albertin et al. 2015; Foote et al. 2015) and the establishment of developmental differences in tentacle formation in *Hydra vulgaris* and *H. oligactis* (Khalturin et al. 2008). TRGs have also been considered to have played an important role in the evolution of social insects from their solitary ancestors (Johnson and Tsutsui 2011; Jasper et al. 2015).

In plants, the importance of TRGs was demonstrated in a study showing that more than 80% of the knockout mutants of genes having unknown function in *A. thaliana*, many of which were TRGs, generated an altered phenotype when subjected to stress conditions (Luhua et al. 2013). This finding was preceded by the characterization of the *QUA-QUINE STARCH* or *QQS* gene, an Arabidopsis TRG encoding a small protein of only 59 amino acids (aa) with no known functional/structural motifs and no homology with proteins/genes of very closely species, such *A. lyrata* and *A. halerieii* (Li et al. 2009; Li and Wurtele 2015). QQS is a regulator of carbon and nitrogen partition into starch, lipids, and proteins in leaves and seeds. It is also responsive to an ample array of environmental, developmental, genetic and epigenetic conditions (Seo et al. 2011a, b, 2013). Up-regulation of this gene was also promoted in susceptible to disease *pen3* mutant plants affected in an ABC transporter associated with non-host resistance against fungal and oomycete pathogens (Stein 2006), a response later associated with syntaxin proteins involved in the fusion of transport vesicles to target membranes (Zhang et al. 2008; Maekawa et al. 2014). QQS has also been shown to interact with the Nuclear Factor Y subunit C4 protein (NFYC4) (Li et al. 2015), a subunit of the heterotrimeric NF-YA/B/C transcription factor, which is conserved across eukaryotes (Laloum et al. 2013; Nardini et al. 2013; Li et al. 2015; Li and Wurtele 2015). In this respect, the overexpression of QQS and NF-YC4 in Arabidopsis and soybean plants reduced their susceptibility to viruses, bacteria, fungi, aphids and soybean cyst nematodes (Qi et al. 2019). *QQS* down-regulation is suggested to occur concomitantly with an increased expression of *EXORDIUM LIKE 1*, known to be important for plant adaptation to depleted carbon, energy and/ or oxygen conditions (Schroder et al. 2011). In this respect, the *Senescence-Associated and QQS-Related* gene was proposed to participate in the QQS network, specifically in the senescence process (Jones et al. 2016).

Some particular TRGs are characterized by not being of recent de novo origin. They conserve a highly stable secondary structure and general function that is usually possible by the concerted binding of conserved cysteines that establish a strongly inter-connected backbone of the hydrophobic core. Outside this compact and impervious structure, duplications, rearrangements, and mutations tend to occur quite frequently. They are known as small, cysteine-rich, secreted proteins (or CRISPs) (Marshall et al. 2011). CRISPs are found in almost all pathogenic fungi (Stergiopoulos and de Wit 2009) and also in plants. For example, an antifungal CRISP possessing a unique fold was recently discovered in the garden pea, *Pisum sativum* (Mandal et al. 2013). Other plant TRGs having a role in defense have been described. These include *OsDR10*, which codes for a suppressor of pathogen-induced defense responses in rice (Xiao et al. 2009) and the 49 aa ENHANCER OF VASCULAR WILT RESISTANCE 1, a Brassicaceae-specific antifungal peptide shown to confer fungal resistance when transferred to *Nicotiana benthamiana* (Yadeta et al. 2014). Similarly, a wheat-specific TRG denominated *Triticum aestivum* Fusarium Resistance Orphan Gene (or TaFROG) increased the resistance to the Fusarium head blight disease caused by *Fusarium graminearum* (Perochon et al. 2015). TaFROG was found to be an intrinsically disordered protein and was, therefore, suggested to participate in cellular signaling or function as a hub in signaling networks through protein–protein interactions (Tompa et al. 2015; Wright and Dyson 2015). This hypothesis was supported by data showing that TaFROG formed distinct protein complexes with the TaNACL-D1 transcription factor and with TaSnRK1a, a master regulator of metabolism, in planta (Perochon et al. 2015, 2019).

In other stress-related aspects, several unannotated genes were thought to contribute to desiccation tolerance in the resurrection plant *Craterostigma plantagineum*. This led to the characterization of two TRGs encoding for a cysteine-rich rehydration-responsive protein 1 (CpCRP1) and an early dehydration-responsive protein 1 (CpEDR1), respectively. Expression profiles suggested that CpCRP1 is active in normal conditions and during rehydration, whereas CpEDR1 is needed to achieve desiccation tolerance partially through the protection of photosynthetic structures during both dehydration and rehydration (Giarola et al. 2015). Finally, other family-specific proline-rich tandem repeat cell wall-related secreted proteins may be required to generate specialized structures needed for bacterial symbiosis in legumes and seed storage vacuoles in grasses (Newman and Cooper 2011), while a regulatory role in development was ascribed to the *Poaceae specific and BR responsive gene 1* coding for a brassinosteroid-regulated mitochondrial protein in bamboo, which inhibited plant growth when overexpressed in *Arabidopsis* (Guo et al. 2020).

7.4 TRGs in Grain Amaranth

The genus Amaranthus consists of 60–70 species. Some, (i.e., *Amaranthus hypochondriacus*, *A. cruentus*, and *A. caudatus*) are semi-domesticated cultivated crops that yield highly nutritional grains and can readily adapt to drought and poor soils (Caselato-Sousa and Amaya-Farfán 2012). Numerous mechanisms have been identified to be related to (a)biotic stress tolerance in grain amaranths. These include C4 photosynthesis, high water use efficiency, an indeterminate flowering habit, the ability to develop long tap-roots together with an extensive web of lateral roots (Miller et al. 1984; Johnson and Henderson 2002; Omamt et al. 2006). Other determinants of stress resistance have also been identified, including osmotic adjustment linked to higher photosynthetic capacity under duress, the expression of stress-related genes and transcription factors, ABA-dependent signaling responses and the implementation of a sugar starvation response (Lal and Edwards 1996; Aguilar-Hernández et al. 2011; Huerta-Ocampo et al. 2009, 2011, 2014;

González-Rodríguez et al. 2019). Grain amaranths are also tolerant to severe defoliation (Castrillón Arbeláez et al. 2012; Vargas-Ortiz et al. 2013), a trait that was associated with efficient utilization of abundant carbon reserves in stem and roots. They are fairly resistant to insect pests and pathogens (Brenner et al. 2000), although increased insect protection against certain insect herbivores or pathogenic bacteria can be induced by chemical elicitors of defense responses, such as jasmonic acid (JA), and benzothiadiazole (BTH), a synthetic salicylic acid analog used to induce plant defense responses against biotrophic pathogens (Délano-Frier et al. 2004; Sánchez-Hernández et al. 2004; Casarrubias-Castillo et al. 2014). A subsequent transcriptomic analysis of *A. hypochondriacus* plants subjected to diverse biotic stress conditions supported this information and further revealed that a substantial number of the genes whose expression was modified by these conditions were unknown (Délano-Frier et al. 2011; Fig. 7.1). Several of these genes were expressed in multiple stress conditions. The authors hypothesized that the study of many of these genes could unravel potentially novel mechanisms utilized by *Amaranthus* plant species to thrive in unfavorable environmental conditions.

The first of these stress-responsive genes of unknown function to be characterized in more detail was the so-called *Ah24* gene (Massange-Sánchez 2015; Massange-Sánchez et al. 2015). It was shown to be TRG, since Ah24 homologs were present almost exclusively in plants belonging to the Caryophyllales. Initially, the *Ah24* gene was reported to encode for a predicted cytosolic protein of 164 aa residues with a predicted molecular mass of 17.6 kDa and pI of 5.71. However, the release of the *A. hypochondriacus* genome (Clouse et al. 2016), revealed that it is actually 100 aa longer, with a predicted molecular mass of 29.8 kDa and pI of 5.34 (Fig. 7.2c). A fluorescence microscopy analysis of cells near the root tip of transgenic *A. thaliana* plants revealed that the localization of the Ah24-GFP fusion protein tended to concentrate in the periphery of the nucleus and, apparently, in the nucleolus as well. This suggests that the biological activity of Ah24 could be associated, at least partially, with this particular localization in the cell. This proposal was in agreement with several studies reporting the significance that the nuclear surface/ envelope has on cell division and cell elongation through its participation as a microtubule-organizing center from which a plasma membrane-cell wall continuum is generated. This network has been proposed to play a role in plant growth and in the regulation of plant responses to environmental stress conditions, including immune system activation and/ or sensing of biotroph pathogens, cell wall mechanical tension or osmotic stress (Stoppin et al. 1994; Zhou et al. 2007; Ambrose and Wasteneys 2014; Liu et al. 2015). Intriguingly, some of the enriched gene onthology (GO) categories detected in the microarray data obtained from *Ah24* overexpressing (OE) *A. thaliana* transgenic plants included biological process and molecular function and cellular component categories associated with this particular intracellular communication, such as microtubule polymerization or depolymerization, nucleologenesis, regulation of cell communication, actomyosin structure organization in organelle envelope, and constituents of membranes, microtubules and/or cell walls (Massange-Sánchez 2015; Massange-Sánchez et al. 2015).

Bio-informatic analyses (https://zhanglab.ccmb.med.umich.edu/I-TASSER/output/S553256/) revealed that the secondary structure of the Ah24 protein includes 9 α helices and 10 β-sheets. Also predicted were one possible N glycosylation site ($N_{41}L_{42}$) and four putative protein kinase C tyrosine phosphorylation sites, at positions T_{43}, T_{76}, T_{98}, and T_{162} (Figs. 7.2a–c).

As mentioned above, an in silico comparison with predicted protein sequences available in the 1000 Plants Initiative public database, or 1KP (www.onekp.com/), which includes medicinal plants and extremophytes, showed that the highest homology of the Ah24 protein was shared with other predicted Ah24 proteins in *Amaranthus* and *Alternanthera* and less so with predicted plant proteins in species belonging almost exclusively to the Caryophyllales order, characterized by an abundance of plants able to thrive in extreme

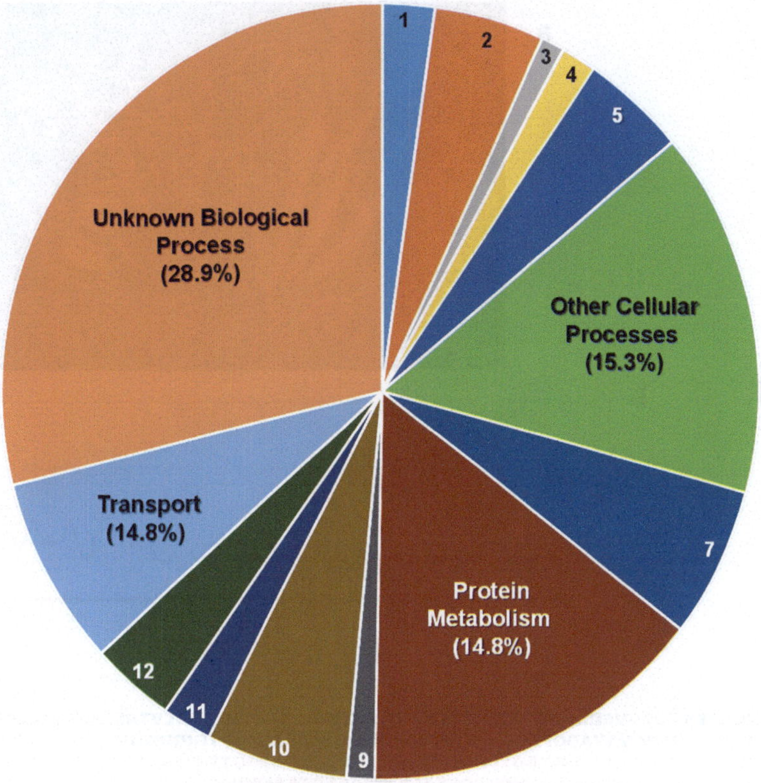

	Biological Process	%
1	Cell organization and biogenesis	2.2
2	Development process	4.7
3	DNA or RNA metabolism	1.0
4	Electron transport and energy pathway	1.5
5	Other biological process	4.5
6	Other cellular process	15.3
7	Other metabolic process	6.3
8	Protein metabolism	14.8
9	Response to (a)biotic stimulus	1.2
10	Response to stress	6.0
11	Signal transduction	2.1
12	Transcription	3.6
13	Transport	7.9
14	Unknown biological process	28.9

Fig. 7.1 Functional Gene Ontology (GO) assignment for Biological Process. The figure indicates that 28.9% of the contigs/isotigs annotated within this category in the *Amaranthus hypochondriacus* transcriptome had no known function (original image, adapted from Délano-Frier et al. 2011)

habitats (Flowers and Colmer 2008; Oh et al. 2012). Thus, Ah24 homologs are also present in *Salicornia europaea*, an halophyte widely distributed in coastal and inland salt marshes worldwide (Fan et al. 2013; Ma et al. 2013) and in other exceptionally stress-tolerant plants such as *Kochia scoparia* (Kalinina et al. 2012), *Spergularia media* (syn. *S. maritime*; Flowers and Colmer 2008), *Dianthus* sp. and various *Atriplex* species. Many of the latter are renowned for their capacity to remediate saline or heavy-metal contaminated soils (Kachout et al. 2010).

In agreement with the role of TRGs in development and/ or stress responses, *Ah24* was expressed in young or developing tissues of *A. hypochondriacus* plants. Moreover, its

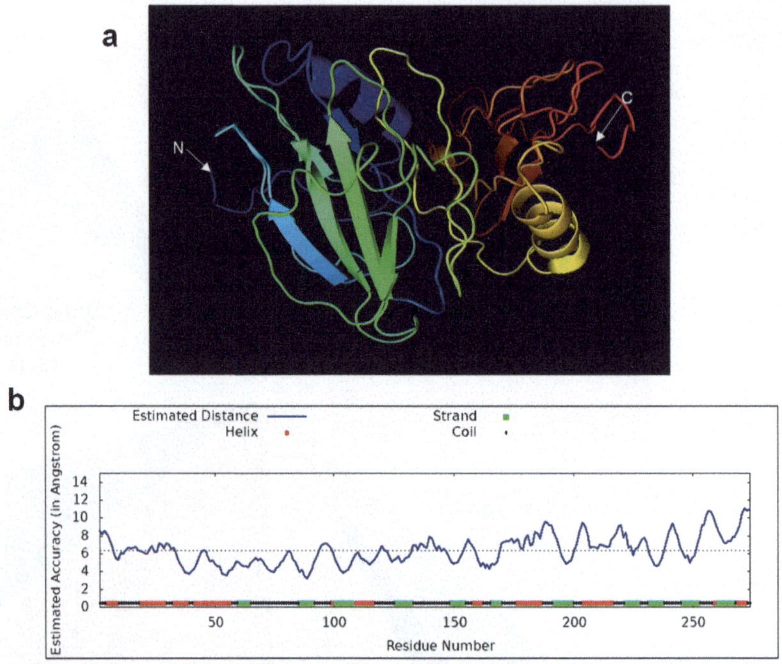

c

**MAEIEAQVFGVEDKNADHDKIKSDLKAKASNTVDLLTSCTNLTTKLVTRLEREHFWAGRAVTVLPAIAPWTYGVPTQRG
ELPAGVKWGVAYADGPDSTARKWVVAFDTAAQKAYAEAGPTGPIDWNVVEVKLGLSSSKSEVEDPILGGKCTAEIDGL
NAWATFPQVYKVEDKNADYDKIMSDLKTQASNGNTVDMQTTFTNLTPKALWRLDREHFWAGQGVNFLFGFASGTYGV
VTQRGELPAGVKWGVAYVDGPDSTARKWVVAFDTAAQKVN**

Fig. 7.2 **a** Predicted structural conformation of the Ah24 protein. **b** The blue line represents the B-factor, which indicates the extent of the inherent thermal mobility of residues/atoms in the Ah24 protein. Segments of red or green cylinders represent the distribution of α-helixes and β-strands, respectively, while the black lines represent random coils. **c** Up-dated amino acid sequence of the Ah24 protein (original images modified from Massange-Sánchez et al. 2015; generated using https://zhanglab.ccmb.med.umich.edu/I-TASSER)

expression was strongly induced in meristems and newly emerging leaves of severely defoliated plants. Its identification in a proteomic analysis performed in leaves of grain amaranth plants subjected to chewing insect herbivory (Navarro-Meléndez 2009), was later corroborated by data (Massange-Sánchez 2015; Massange-Sánchez et al. 2015) revealing that leaves exposed to mechanical damage, insect herbivory or treated with exogenous methyl jasmonate (MeJA) underwent a fast and time-dependent accumulation of *Ah24* transcripts (Fig. 7.3). This was corroborated by a study in which the regulation of *Ah24* gene expression was examined by fusing a 1500 bp promoter region upstream of the start codon to a GUS reporter gene, and using this recombinant transgene to generate *A. thaliana* transgenic lines (Hernández-Mateo 2019). Histochemical analysis of these plants revealed that the *Ah24* promoter was inactive in untreated control plants and plants transformed with an empty vector (Fig. 7.4). However, in concordance with the gene expression data, a strong response to MeJA and mechanical wounding was detected in leaves of the transgenic *A. thaliana* lines. MeJA activated the *Ah24* promoter predominantly in the vascular tissue, whereas mechanical wounding stress produced a strong reaction only around the damaged area (Fig. 7.4). Further, a possible role in heavy metal tolerance was suggested by a recent study showing that *Ah24* was strongly induced in the roots of the "Plainsman" cultivar (*A. hypochondriacus* × *A. hybridus*; Baltensperger et al. 1992), previously classified as a cadmium hyper-tolerant grain amaranth (Lancíková et al. 2020). Accordingly,

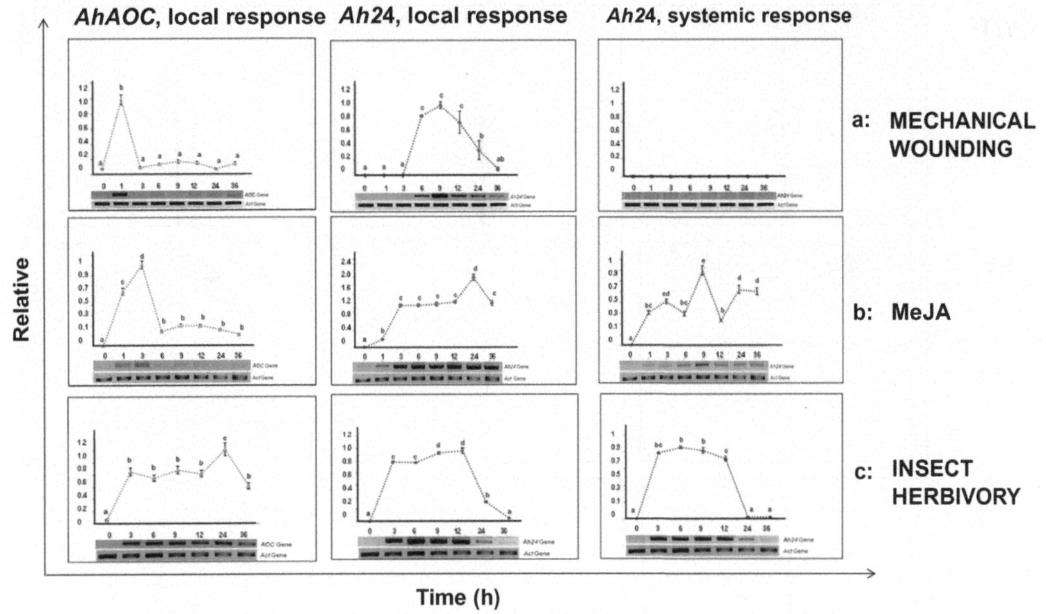

Fig. 7.3 *Ah24* transcript accumulation in leaves of *A. hypochondriacus* plants subjected to **a** mechanical wounding, **b** foliar methyl jasmonate (MeJA) application and **c** chewing insect herbivory. The response in both damaged/ treated leaves (local response; middle row) and undamaged/ untreated leaves (systemic response; right end row) was determined periodically during a 36 h period. The results were compared to the local response of the jasmonic acid biosynthetic gene, *AhAOC* (left end row), which was used as a positive control for JA-regulated wound responses (original images, adapted from Massange-Sánchez 2015 and Massange-Sánchez et al. 2015)

an in silico analysis of a 1500 bp segment of its promoter region unveiled the prominence of boxes involved in development and (a)biotic stress responses (Fig. 7.5; Table 7.1).

Further characterization of this gene *in planta* was examined in transgenic *A. thaliana* and *Nicotiana tabacum Ah24* OE plants. Compared to controls, transformed *A. thaliana* plants were larger and produced more leaves without suffering a fitness penalty, measured as seed yield (Fig. 7.6a). They also showed reduced stomatal opening (Fig. 7.6b), a characteristic that probably contributed to their increased water-stress tolerance (Fig. 7.6c).

Additional microarray analysis of *Ah24* OE transgenic plants kept under normal growing conditions yielded a gene expression pattern that was in accordance with the phenotypes observed. Highly representative changes were the modified expression of various *MIR*-encoding genes whose targets affect plant hormone signaling (Schommer et al. 2008; Siré et al. 2009; Zhu et al. 2011; Barah et al. 2013; Turner et al. 2013; Curaba et al. 2014), *CYP714A2* and *AGF1*, involved in the maintenance of gibberellin homeostasis (Zhu et al. 2006; Matsushita et al. 2007; Zhang et al. 2011), *SLEEPY1*, coding for an F-box family protein gene that positively regulates gibberellin signaling and *CRF2* and *SEUSS-like*, coding for an ethylene-responsive transcription factor that controls cytokinin signal transduction, and for a transcriptional adaptor of floral and embryonic development and of auxin-related responses, respectively (Bao et al. 2010; Ariizumi et al. 2011; Cutcliffe et al. 2011). The expression of the *RHE1a* gene, coding for a RING finger E3 ligase, which is considered to be part of the overall signal involved in plant development and defense (Schwechheimer et al. 2009), was also affected in *Ah24* OE *A. thaliana*

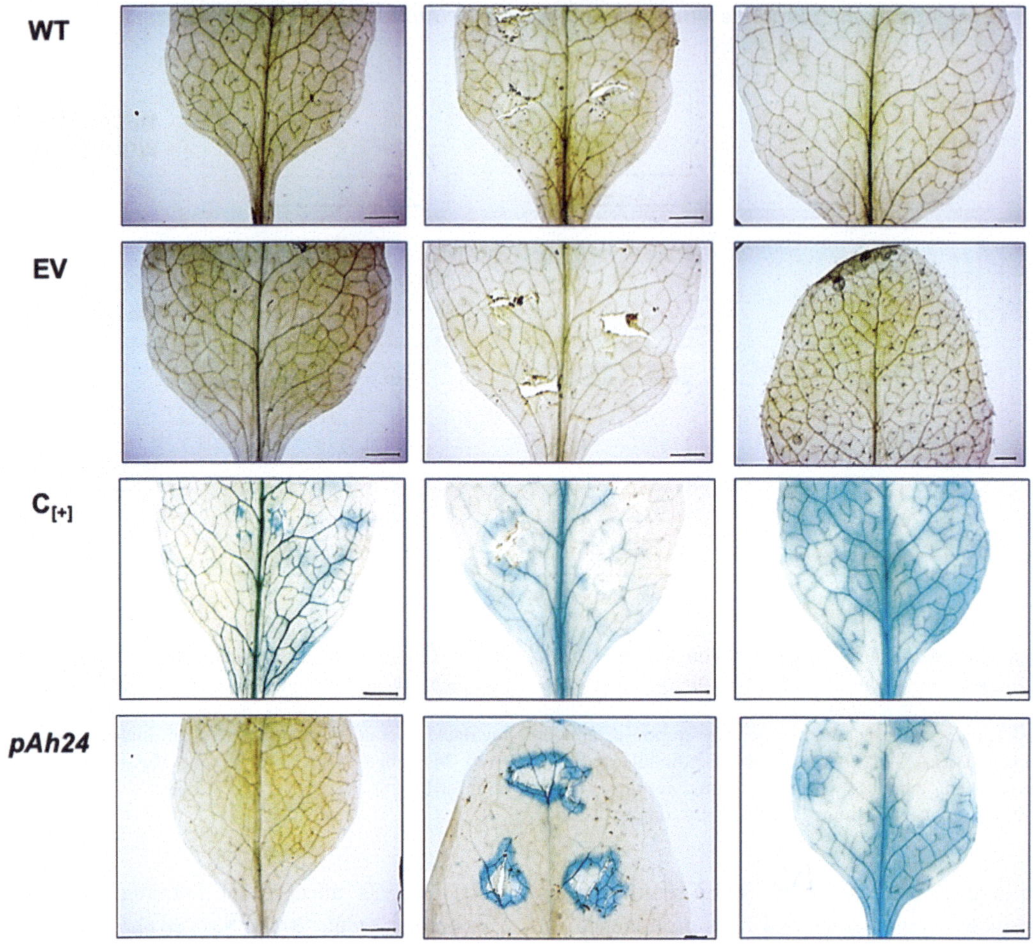

Fig. 7.4 GUS expression of the *pAh24::GUS* recombinant transgene in transgenic *Arabidopsis thaliana* plants. Shown are leaves of untransformed (**WT**) plants, plants transformed with the empty vector (**EV**), transgenic plants transformed with the jasmonic acid *pLOXD* promoter coupled to the GUS reporter gene ($\mathbf{C}_{+}$) and transgenic plants harboring the *pAh24::GUS* recombinant transgene (***pAh24***). Untreated plants (left-side panel) were compared with those that received a foliar application of methyl jasmonate (right side panel) or suffered mechanical wounding (central panel) (original images, modified from Mateo-Hernández 2019)

plants (Fig. 7.7). Further, GO analysis of the microarray data suggested that *Ah24* overexpression in *A. thaliana* transgenic plants triggered mechanisms resembling those designed to re-establish and maintain ion and cellular homeostasis in response to stress (Conde et al. 2011). Particularly, relevant were the enrichment in categories involving stress sensing and signaling mechanisms, plant cell detoxification, transport and compartmentation.

On the other hand, *Ah24* transgenic tobacco plants, whose growth and reproduction did not differ from controls, were more resistant to chewing insect herbivory stress. Increased resistance to insect herbivory in transgenic tobacco plants was in accordance with significantly higher proteinase inhibitor activity levels, in addition to the accumulation of JA, nicotine and several other putative defense-related metabolites. These biochemical traits have been positively associated with the wound/insect herbivory response in several other plants, including other *Nicotiana* species and grain amaranth (Nugroho and Verpoorte 2002; Halitschke and Baldwin

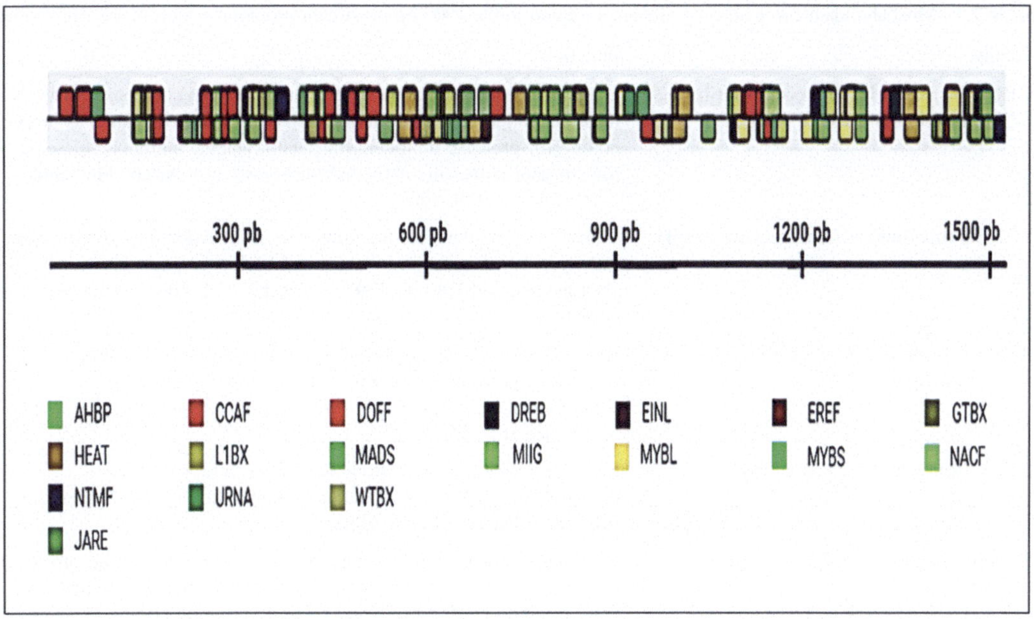

Fig. 7.5 *Cis*-Regulatory elements present in the 1500 bp up-stream promoter region of the *Ah24* gene. Relate to Table 7.1 for a description of the response elements shown in this figure (original images, up-dated from Massange-Sánchez et al. 2015 and adapted from Mateo-Hernández 2019)

2003; Lou and Baldwin 2003; Sánchez-Hernández et al. 2004; Steppuhn et al. 2004; Zavala et al. 2004; Steppuhn and Baldwin 2007; Jassbi et al. 2008; Großkinsky et al. 2012; Mithöfer and Boland 2012; Fürstenberg-Hägg et al. 2013). This body of experimental data supports the proposal suggesting that *Ah24* plays a multifactorial role in *A. hypochondriacus* by contributing to regulate growth, including the profound developmental changes that are required for a rapid recovery post-wounding or defoliation, in addition to responses against biotic and abiotic stress conditions.

7.5 Work in Progress

A sample of other possible TRGs, whose expression was affected by several different stress conditions in *A. hypochondriacus* (Délano-Frier et al. 2011) is shown in Table 7.2. Four of these genes were selected for further study, namely: *Ah2880*, *Ah3240*, *Ah3880* and *Ah16023*. *Ah2880* codes for a 161 aa protein. Its predicted secondary structure includes a predominant 58 aa α-helix, localized practically at the start of the N-terminal end. The rest of the protein is conformed mostly by a random coil interspaced with 5 short (4 aa) to very short (2 aa) helixes (Fig. 7.8a). *Ah3240* codes for a 101 aa protein having a more organized secondary structure consisting of an N-terminal α-helix (13 aa) followed by a slightly longer one starting at a threonine residue at position 44. Two β-strands, of 11 and 4 aa, respectively, are positioned between these two α-helixes. The C-terminal end of this protein is a tandem succession of short α-helixes and β-strands separated by random coils (Fig. 7.8b). *Ah3844* codes for a 113 aa protein. Its secondary structure is organized as a succession of pairs of small β-strands and α-helixes starting from its N-terminal end. Its C-terminal end is dominated by a pair of larger (15 aa each) α-helixes (Fig. 7.8c). Ah16023 was named AhZnF given that the 30 kDa protein product possesses a Zinc-finger domain belonging to the vWA_C3HC4 super-family. Its secondary structure is defined by an N-terminal end dominated by β-strands followed

Table 7.1 Most abundant cis-regulatory elements present in the 1500 bp up-stream promoter region of the *Ah24* gene

Family	Function
GBOX (G and C boxes, recognized by bZIP transcription factors [TFs])	Essential for the regulation of jasmonic acid (JA) responses. They have been identified in wound response defense genes such as those coding for protease inhibitors, in potato, leucine aminopeptidase, in tomato, and putrescine *N*-methyltransferase, in tobacco (Memelink 2009; Jakoby et al. 2002)
TGACG (uncommon JA-responsive motif)	Found to be essential for the induction of the *Nopaline synthase* and *Lipoxygenase 1* genes in tobacco, and barley, respectively (Wasternack and Parthier 1997; Narusaka et al. 2004; Memelink 2009)
AHBP (recognized by homeobox proteins in Arabidopsis)	Regulation of development processes; present in nitric oxide-responsive genes many of which are involved in growth, development, and response to environmental stresses (Shi et al. 2011; Mukherjee et al. 2009)
CCAF (Recognized circadian rhythm regulators)	Involved in the transition from vegetative to reproductive development. In general, regulated by light and Rubisco (Harmer and Kay 2005; McClung et al. 2006)
GT-Box (identified in light-regulated promoters)	Involved in plant growth regulation and response to various stress conditions. Found in the chalcone synthase gene in common and soya bean (Lawton et al. 1991; Terzaghi and Cashmore 1995)
L1-Box (specific motifs for the expression of L1 proteins)	L1 proteins regulate leaf and flower development processes. Present in Proteoderma factor 1, a proline-rich protein, which is exclusively found in the L1 layer of the apical meristem apical, the proteoderm and organ primordia (Abe et al. 2001; Lau et al. 2012)
MADS (Recognized by MADS-box TFs)	MADS-box genes encode a family of transcription factors that control diverse developmental processes in flowering plants ranging from root to flower and fruit development. Regulation of the transition from vegetative to reproductive development and generation of flowering structures and floral meristems (e.g., via Sepallata 3 TF) (Becker and Theißen 2003; Angenent et al. 2009; Erdmann et al. 2010, Smaczniak et al. 2012)
HSE (Heat stress elements)	Heat shock factors (HSFs) are involved in stress sensing and signaling in response to different environmental stresses, such as high temperature and concomitant accumulation of reactive oxygen (ROS) species in plants (Dat et al. 1998). ROS participate in (a)biotic stress responses and signaling pathways
JARE (Jasmonic acid response elements)	*cis*-Regulatory elements that regulate the timing and amplitude of the JA response (Nakata et al. 2013; Sasaki-Sekimoto et al. 2013; Song et al. 2013). They are divided into primary and secondary response elements. The former interact with preexisting transcription factors activated as terminal orchestrators of the JA signal transduction pathway. The latter are bound by transcription factors synthesized de *novo* in response to JA (Pauw and Memelink 2005)
MYB, MYBL (MYB and MYB-like TF-response elements)	The MYB and MYB-like TF family comprises numerous proteins involved in the regulation of biotic and abiotic stress responses, as well as development, differentiation, metabolism and defense (Ambawat et al. 2013)
NACF (NAC transcription factors)	The NAC family proteins participate in several plant biological processes, including the development of root and shoot apical meristems, leaf senescence, organogenesis, phytohormone signaling, fruit ripening, response to (a)biotic stress and secondary cell wall development (Hoang et al. 2020)

(continued)

Table 7.1 (continued)

Family	Function
DREB (Dehydration response element binding)	Dehydration responsive element binding (DREB) factors or CRT element binding factors (CBFs) are members of the AP2/ERF family which is constituted by numerous stress-responsive regulatory genes (Agarwal et al. 2017)
EREF (Ethylene response factors)	Ethylene Response Factors (ERFs), which belong to the transcription factor family APETALA2/ERF are part of the ethylene signaling and response pathway. Ethylene is fundamental for many developmental processes and a key regulator of (a)biotic stress responses in plants. ERFs are believed to regulate abiotic stress responses in plants partly through the integration of ethylene, abscisic acid, jasmonate, and redox signaling (Müller and Munné-Bosch 2015)
MYBS (MYB proteins with a single DNA binding repeat)	The functions of one-repeat (1R) MYBs involve circadian and light regulation, cell differentiation, and telomeric DNA binding (Jin and Martin 1999). They have also been found to control sugar and gibberellin regulation of α-amylase gene expression (Lu et al. 2002)
MIIG (MYB IIG-like binding sites)	Some MYT TFs, such as MYB15, reportedly binds to type IIG Myb recognition sequences that include the 7-bp secondary wall MYB-responsive element (SMRE) [ACC(A/T)A(A/C)(T/C)] that is recognized by SG13-type R2R3-MYBs to drive secondary cell wall biosynthesis (Zhong and Ye 2012). MYB15 TF is also known to control defense-induced cell wall lignification and basal immunity in Arabidopsis (Chezem et al. 2017)
NTMF (NAC transcription factors with transmembrane domain)	Close to 9% of all NAC TFs possess an α-helical transmembrane domain. These domains are found at both the N- and C-terminal ends of these NAC proteins. They are predicted to be anchored to intracellular membranes, mostly to the ER, and to require membrane release, through regulated proteolysis, for activation. Once activated, they regulate diverse plant growth and developmental processes, such as cell division, flowering initiation and stress signaling (Kim et al. 2007; Mohanta et al. 2020)
URNA (U-snARN binding proteins)	Small nuclear ribonucleoproteins (snRNPs) are protein-RNA complexes containing specific snRNP-associated proteins together with small nuclear RNAs (snRNAs). They represent non-coding RNA molecules and are abundant in the nucleus. Their main function is as core components of the spliceosome required for pre-mRNA splicing. snRNP biogenesis is a critical factor for plants, being essential for the determination of a cell's activity through the regulation of gene expression. U1, U2, U4, U5, and U6 snRNPs as core components of the major spliceosome contains, while U11, U12, U4atac, U5, and U6atac snRNPs conform the minor spliceosome (Ohtani 2018)
DOF (DNA-binding One Zinc Finger)	The DOF family of TFs is involved in many critical processes in higher plants, such as responses to light and phytohormones, in addition to seed maturation and germination. DOF TF genes are found in both gymnosperms and angiosperms and are also present in lower plants like mosses and algae (Noguero et al., 2013)
EINL (EIN3/Ethylene insensitive-like transcription factors)	EIN3/EIL1 TF are the crucial regulators of ethylene signaling in plants. They can integrate and process the different signal to modulate the ethylene response in accordance with spatio-temporal and ambient conditions. They are involved in a wide range of ethylene responses that include: de-etiolation, root hair development, salt and cold stress responses, regulated expression of chlorophyll

(continued)

Table 7.1 (continued)

Family	Function
	biosynthesis genes, pigment-binding proteins essential for photosynthesis initiation, immune receptors and apical hook formation regulator. They also influence hormones biosynthesis and signaling (Dolgikh et al. 2019)
WRKY70 (WRKY DNA-binding transcription factor 70)	Transcription factor involved in senescence, biotic and abiotic stress responses. Acts by modulating various phytohormones signaling pathways. Known to be a point of convergence for salicylic acid (SA) and jasmonic acid (JA)-mediated defense signal pathways. Interacts specifically with the W box (5′-(T)TGAC[CT]-3′), a frequently occurring elicitor-responsive cis-acting element. Required for the resistance (R) gene Mi-1-mediated resistance to potato aphids and root-knot nematodes in tomato (Atamian et al. 2012). It is known to negatively modulate cell wall-associated defenses to necrotrophic pathogens in Arabidopsis (Li et al. 2017)

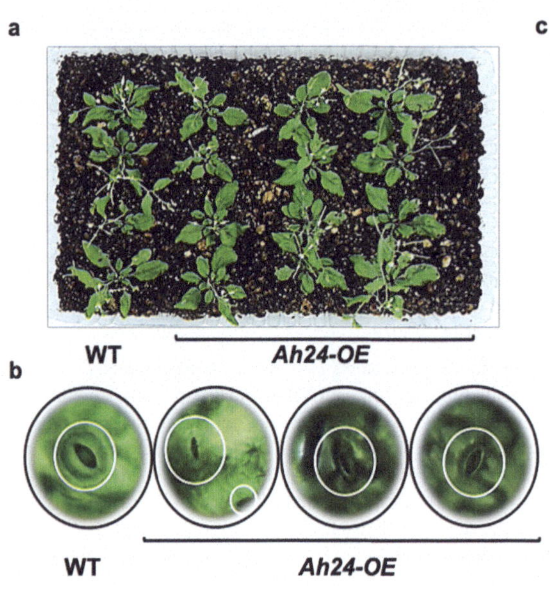

Fig. 7.6 Physiological changes in transgenic *Arabidopsis thaliana* plants overexpressing *Ah24*. **a** Transgenic plants (*Ah24-OE*) showed enhanced vegetative growth and increased leaf number and **b** reduced stomatal opening, compared to untransformed controls (WT). This trait probably contributed to their increased water-deficit stress (WDS) tolerance as shown in **c** where the aspect of WT and *Ah24*-OE plants is shown under well-watered conditions (WW), 8-days after severe WDS and 24 h after recovery (R) produced by re-watering the plants (original images, related to the study by Massange-Sánchez et al. 2015)

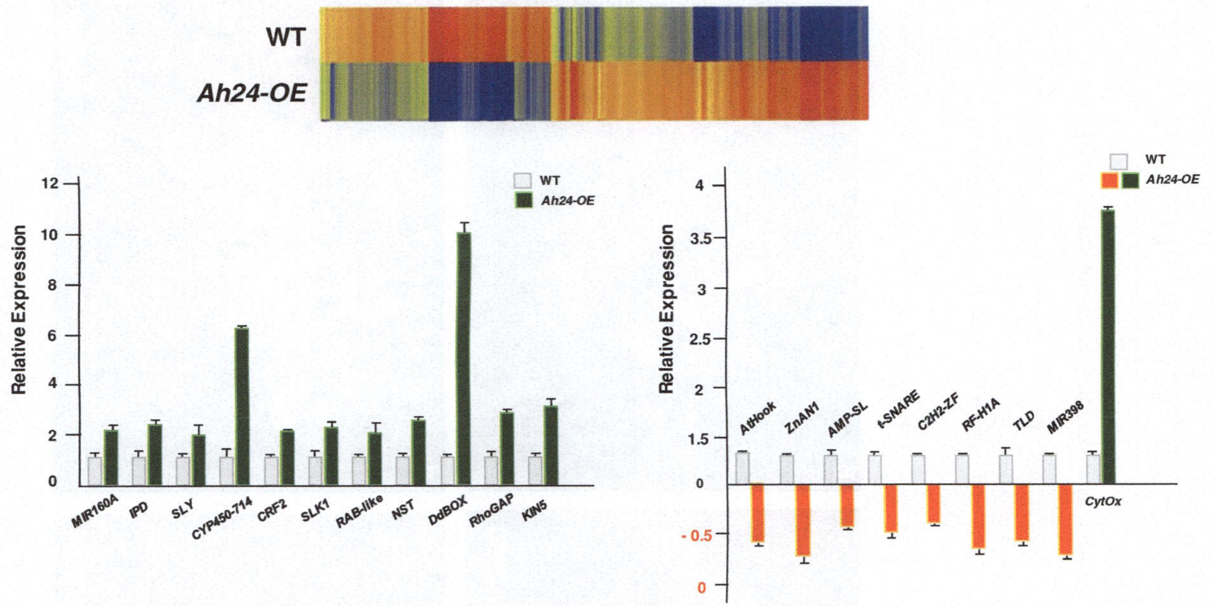

Fig. 7.7 Transcriptomic profile of wild-type and *Ah24* overexpressing *Arabidopsis thaliana* plants by microarray analysis. The heat map shown in **a** represents normalized expression intensity values of 50 differentially regulated genes (fold change $\geq$ 2 and *P*-value < 0.05) in the *Ah24* transgenic line (***Ah24-OE***) as compared with the wild type (**WT**). The intensity of the red- and blue-colored bands indicate over- and under-expressed genes, respectively. The validation, by qRT-PCR, of the microarray data was performed for genes whose expression was either up- **b** or down-regulated **c** in response to the overexpression of the *Ah24* transgene. These were the following: *microRNA 160A* (***MIR160A***); *IMIDAZOLEGLYCEROL-PHOSPHATE DEHYDRATASE* (***IPD***); F-box family *SLEEPY-1* protein (***SLY1***); *CYTOCHROME P450, FAMILY 714, SUBFAMILY A*, POLYPEPTIDE 2 (***CYP450-714***); *CYTOKININ RESPONSE FACTOR 2* (***CRF2***); *SEUSS-like 1* (***SLK1***); *RAB GTPase homolog A1I* (***RABL1***); *NUCLEOTIDE-SUGAR TRANSPORTER* family protein (***NST***); *DEAD BOX RNA HELICASE* family protein (***DdBOX***); *Rho GTPase* activation protein with PH domain (***RhoGAP***); *Kinesin5* (***KIN5***); *AT HOOK MOTIF DNA-BINDING* family protein (***AtHook***); *ZINC FINGER* (*AN1-LIKE*) family protein (***ZnAN1***); *AMP-DEPENDENT SYNTHETASE AND LIGASE* family protein (***AMP-LS***); *SYNTAXIN/T-SNARE* family protein (***SNARE***); *C2H2-LIKE ZINC FINGER* protein (*C2H2-ZF*); *RING-H2 FINGER A1A* (***RH1A***); *TLD-DOMAIN CONTAINING NUCLEOLAR PROTEIN* (***TLD***), and *microRNA 398* (***MIR398***). In **c**, the down-regulated expression of the *MIR398* gene is inversely correlated with increased levels of expression of one of its known targets (i.e., cytochrome C oxidase subunit V, ***CytOx*** [Jones-Rhoades and Bartel (2004)]). *Ah* actin and tubulin were used as internal controls. The bars indicate the relative transcript levels. The mean and SE from three technical replicates are shown (original figure, adapted from data reported in Massange-Sánchez et al. 2015)

Table 7.2 Expression levels of a selected group of genes of unknown function. The genes were detected as induced or repressed in a transcriptomic analysis of leaves of *Amaranthus hypochondriacus* plants subjected to four different (a)biotic stress conditions[a]

Gene	Domain(s)	WDS[b]	Salinity	Herbivory[c]	BI[d]
Ah-6719	DUF642-CBM4_9	**9.5**[e]	**9.8**	**8.8**	**23.9**
Ah-674	Agglutinin superfamily	**7.8**	**5.7**	**10**	**0**
Ah-2880	None	0.7	0.7	**1.6**	*0.4*
Ah-5258	Senescence superfamily	*0.1*[e]	*0.4*	*0.1*	*0.1*
Ah-16023	vWA_C3HC4 superfamily	1	*0.3*	1.3	**ND**[f]
Ah-14119	DUF3407	**2.3**	**2.7**	**2.2**	*0.4*
Ah-3844	None	**1.9**	**3.5**	1.1	**1.6**
Ah-3240	None	0.9	1.2	1.3	**ND**

[a]Data taken from Délano-Frier et al. (2011) and Cabrales-Orona (2017)
[b]WDS = Water-deficit stress
[c]Insect herbivory by chewing insect larvae
[d]BI = Avirulent bacterial infection
[e]Numbers in bold text represent induced (black) or repressed (italics)
[f]ND = Not detected

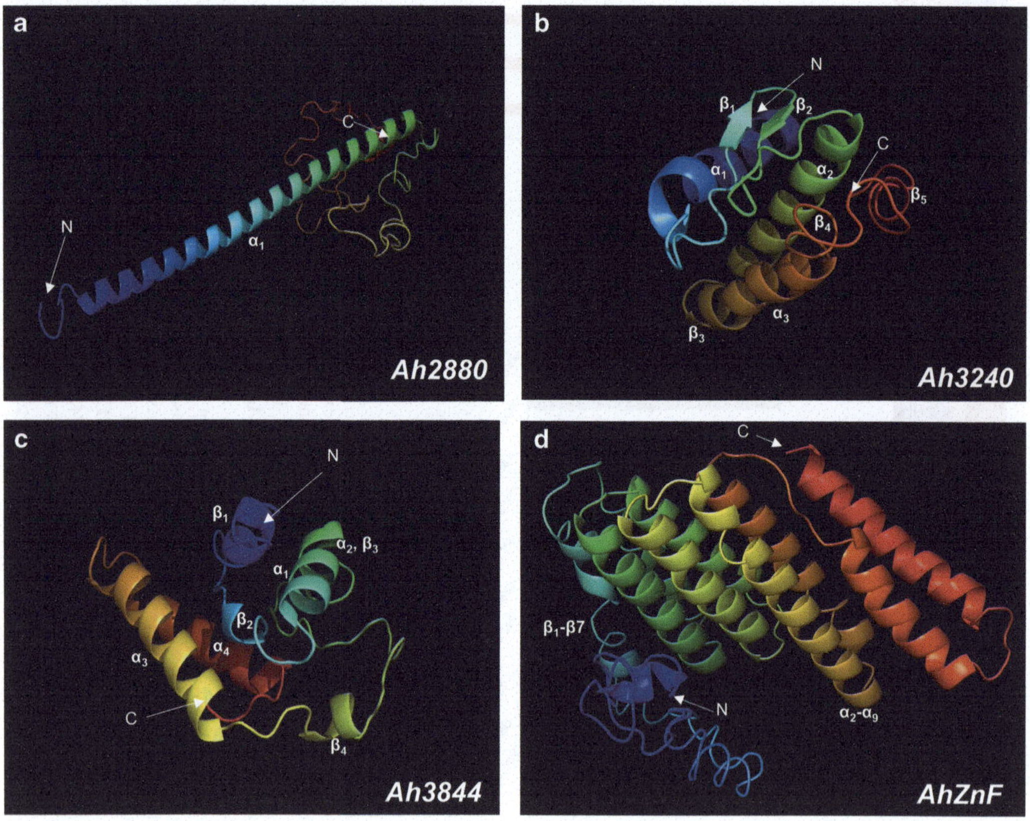

Fig. 7.8 Predicted structural conformation of the **a** Ah2880, **b** Ah3240, **c** Ah3844 and **d** AhZnF proteins. In each protein, the N- and C-terminal ends are shown as well as the main α-helixes and β-strands that conform their structure (shown are original figures generated using https://zhanglab.ccmb.med.umich.edu/I-TASSER)

by a succession of α-helixes towards its C-terminal end (Fig. 7.8d). The overall predicted solvent accessibility of the four proteins is very low, with many buried residues present mostly in organized secondary protein structures. This suggests that their structures are likely to be highly compact (Xu and Zhang 2012, 2013). Apart from the AhZnF protein, they also lacked identifiable domains.

Phylogenetic analysis of these genes confirmed their restricted taxonomic distribution. The most striking result was produced by the Ah3240 protein, which was only found in *A. hypochondriacus*. This finding was similar to the unique *A. thaliana* QQS protein that lacked homology with proteins of closely related species (Li et al. 2009, 2015). The Ah2880 protein was clustered together with homologous proteins present in related species such as *Chenopodium quinoa*, *Beta vulgaris* and *Spinacia oleracea*. This cluster was clearly separated from more distant homologs detected in *Citrus clementina*, *C. sinensis*, *Theobroma cacao*, *Vitis vinifera* and *Gossypium raimondi* (Fig. 7.9a). AhZnFinger was grouped in a cluster that was clearly separated from others that grouped homologs in *Manihot esculenta*, *Ricinus communis*, *Brachypodium stacei*, *B. distachyon*, *Sorghum bicolor* and *Zea mays* (Fig. 7.9b). Ah3880 shared only limited homology with proteins found in *B. vulgaris*, *Sinningia tuberosa* (Lamiales, Gesneriaceae), *Strobilanthes dyeriana* (Lamiales, Acanthaceae) and *Lobelia siphilitica* (Asterales, Campanulaceae).

A deeper transcriptomic analysis of these genes was performed. In addition to the conditions examined in the original transcriptomic study (Delano-Frier et al. 2011), it included additional time points and stress conditions, such as defoliation and infection by virulent and avirulent bacteria and treatments with exogenous abscisic acid (ABA) or BTH. It also incorporated a comparison between the three-grain amaranth species, *A. hypochondriacus*, *A. cruentus* and *A. caudatus* as well as tissue-specific analyses (Cabrales-Orona 2017). The results obtained showed that these genes were expressed in a tissue-, development stage-, stress- and species-specific manner in response to the different stress conditions applied. The results obtained further supported the role of these amaranth TRGs in the regulation of stress responses in plants. In particular, the results derived from the defoliation experiment in *A. cruentus* were suggestive that changes in the expression of these particular genes could contribute to the marked decrease in defoliation tolerance observed in these plants during the flowering development stage (Vargas-Ortiz et al. 2015). The most relevant findings are summarized in Table 7.3.

7.6 Final Comments

TRGs are gradually acquiring increased importance as their participation in several biological processes, including responses to stress, gradually emerge. Grain amaranths appear to have a unique set of these TRGs, a property that could be important to sustain the superior tolerance to diverse stress conditions that characterizes these plants. A deeper understanding of their biological function in grain amaranths, which could require the association with other proteins in order to create regulatory hubs and/ or signaling cross-talk networks, is necessary to unveil their hidden potential. Work is now in progress to get closer to the objective of finding a functional identity for these mystifying proteins.

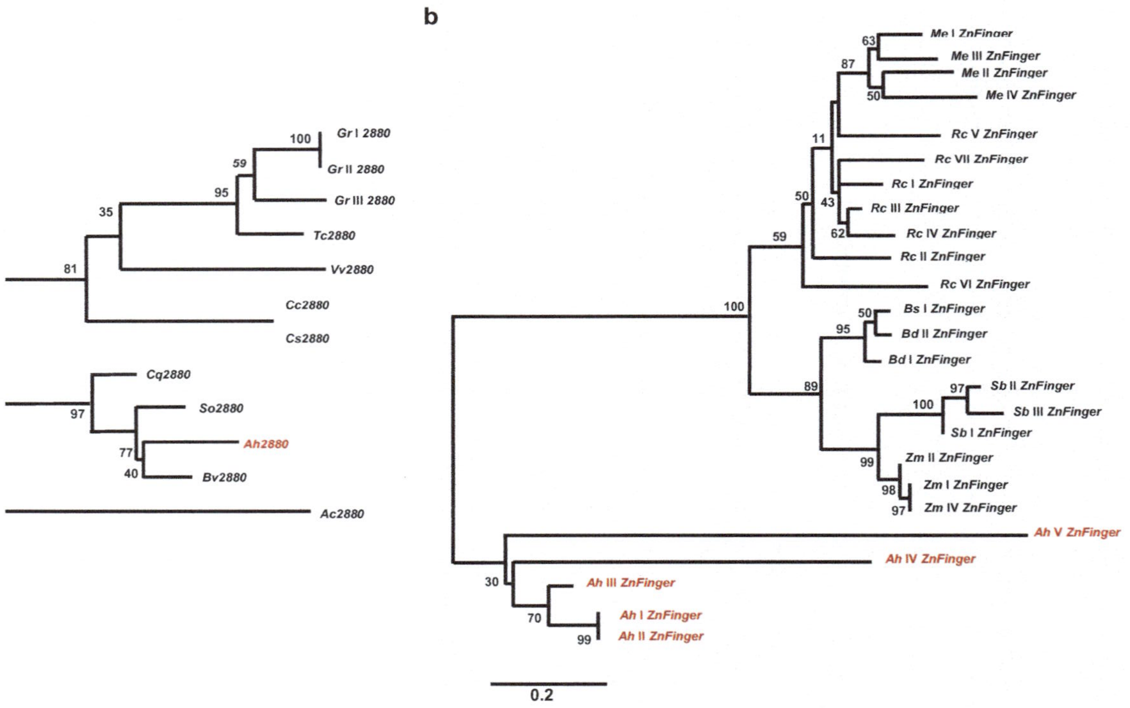

Fig. 7.9 Molecular phylogenetic analysis obtained by the maximum likelihood method shows the relationship of amaranth **a** *Ah2880* and **b** *AhZnFinger* with family members reported in other plant species. Abbreviations: *Ah*, *Amaranthus hypochondriacus*; *Ac*, *Ananas comosus*; *Bd*, *Brachypodium distachyon*; *Bs*, *Brachypodium stacei*; *Bv*, *Beta vulgaris*; *Cq*, *Chenopodium quinoa*; *Cc*, *Citrus clementina*; *Cs*, *Citrus sinensis*; *Gr*, *Gossypium raimondii*; *Me*, *Manihot esculenta*; *Rc*, *Ricinus communis*; *Sb*, *Sorghum bicolor*; *So*, *Spinacia oleracea*; *Tc*, *Theobroma cacao*; *Vt*, *Vitis vinifera*, *and Zm*, *Zea mays* (shown is an original figure generated by applying Neighbor-Join and BioNJ algorithms to a matrix of pairwise distances estimated using the Maximum Composite Likelihood (MCL) approach, and then selecting the topology with superior log likelihood value)

Table 7.3 Most relevant changes in the expression of four grain amaranth[a] TRGs[b] in response to diverse (a)biotic stress conditions or treatment with exogenous activators or stress-related responses

TRG	Response to stress condition or stress-related treatment[c]
AhZnF	**Insect herbivory**[d]: Early (6 h) and intermediate (12 h) local and systemic induction in leaves of all species. Sustained systemic response in roots in all species. Sustained tissue-specific systemic induction in stems of Acau, only. Strongest responses in Acau
	Defoliation in Acru[e]. Sustained induction in all organs (leaf, stems, roots and panicles) of plants defoliated during early (V1) and late (V2) vegetative and panicle emergence (PE) development stages. Induction of this gene sustained during flowering (FL) in leaves but repressed in roots and panicles
	Virulent bacterial infection (*Clavibacter michiganensis* pv. *michiganensis* [*Cmm*])[f]. Mostly repressed in Acau in all organs and time points, particularly in leaves. Repressed local response in damaged leaves of Ahypo. Systemic response repressed in distal leaves and roots of Acru. Local induction only in leaves of Acru and in roots of Ahypo
	Avirulent bacterial infection (*Pseudomonas syringae* pv. *syringae* [*Pss*])[g]. Late (24 and 48 h) and moderate to strong local repression in infected leaves of all species. Mostly systemic repression in distal leaves of Acau and Acru. Pattern contrary to Ahypo. Sustained systemic repression in stems of Acru and Acau and in Ahypo (except at 6 h). Sustained repression in all tissues of Acru, except roots. Strongly repressed in roots of Acau (6-to-24 h)
	Salinity[h]. Repressed expression in leaves of all species. Strongest in Acau. Intermediate (Ahyp; 7 and days) and early (Acru; 4 and 7 days) induction in stems. Strongest sustained induction (4-to-9 days) in stems of Acau. Sustained expression in roots of Ahypo and Acau (strongest in Acau)
	Water deficit stress (WDS) in Acru[i]. Strongly repressed by severe[c] WDS in leaves and roots. Weaker negative effect by moderate[c] WDS. Mostly unaffected during recovery[c]
	ABA[j] treatment in Acru. Early (1–3 h) and intermediate (6–12 h) local induction in treated leaves and slightly retarded (3 to 12 h) systemic induction in untreated distal leaves. Late (24 h) local and systemic repression
	BTH[k] treatment in Acru. Early (1–3 h) and intermediate (6–12 h) local and systemic induction in treated and untreated distal leaves. Late (24 h) local and systemic repression
Ah3844	**Insect herbivory**. Sustained (6-to-24 h) local expression in leaves of Ahypo and Acru only. Sustained systemic response only in leaves of Acru. Late (24 h; Ahypo) or sustained (Acau) repression in distal leaves. Sustained induction (Ahypo) or repression (Acau) in stems. Sustained systemic induction in roots of Ahypo and Acru. Repressed or no systemic responses in all Acau organs
	Defoliation in Acru. Mostly induced in all organs during V1, V2 and PE but sustained repression during FL
	Virulent bacterial infection, *Cmm*. Mostly sustained local induction in leaves and systemic induction in stems of Ahypo and Acru. Sustained systemic induction in stems of Ahypo and Acru (much stronger in Acru) and early to intermediate induction (12–24 h) in roots of Ahypo and Acru. Late systemic repression in leaves of all species (48 h). Strong late local response in leaves (48 h) and sustained repressed systemic response in leaves of Acau
	Avirulent bacterial infection, *Pss*. Sustained (6-to-48 h) local induction in infected leaves of Ahypo. Extensive local induction in infected leaves of Acru (12–48 h) and biphasic in Acau. Sustained systemic expression in distal leaves of Acru only. Sustained systemic induction in stems of Ahypo and Acau; restricted (12 and 24 h) in stems of Acru. Sustained systemic induction in roots of Ahypo contrary to sustained systemic repression in roots of Acru and Acau (strongest in Acau)
	Salinity. Sustained induction in all Ahypo organs. Sustained strong induction in stems and roots of Acau. Sustained strong repression in leaves of Acau. Weak, irregular, response in Acru organs
	WDS in Acru. Strongly repressed by severe WDS in leaves and roots. Weaker negative effect by moderate WDS in leaves. Induced by moderate WDS in roots. Mostly unaffected in the recovery stage
	ABA treatment in Acru. Gradually increasing local and systemic response in leaves, peaking at 12 to 24 h after application

(continued)

Table 7.3 (continued)

TRG	Response to stress condition or stress-related treatment[c]
	BTH treatment in Acru. Gene expression mostly unaffected (local) or repressed (systemic). Temporal systemic induction, 12 h after application
Ah3240	**Insect herbivory**. Sustained local and systemic induction in leaves of Acru only. Sustained systemic induction in roots of Ahypo only. Early (6 h) systemic induction in stems and roots of all species
	Defoliation in Acru. Mostly induced in all organs during V1, V2 and PE but sustained repression during FL, except for strong induction in stems
	Virulent bacterial infection, *Cmm*. Sustained local repression in infected leaves of Ahypo. Sustained induction in roots of Ahypo and Acru (stronger in Ahypo), and biphasic in roots of Acau. Biphasic local induction in infected leaves of Acru. Strong, late (48 h), systemic repression in leaves of Acru and Acau
	Avirulent bacterial infection, *Pss*. Sustained local induction in infected leaves of Ahypo and Acau. Sustained systemic induction in distal leaves of Acru but sustained systemic repression in equivalent leaves of Acau. Sustained systemic induction in roots and stems of Ahypo and in stems of Acru. Biphasic repression in roots of A.cru. Very strong sustained repression in roots of Acau
	Salinity. Early (4d), intermediate (7d) and late (9d) induction in leaves, stems and roots of Ahypo, respectively. Sustained induction in all organs of Acru. Stronger, sustained induction in all organs of Acau
	WDS in Acru. In leaves, repressed by moderate WDS, but induced by severe WDS and recovery. Weak induction in response to moderate WDS in roots. Mostly unaffected by severe WDS and recovery in roots
	ABA treatment in Acru. Gradually increasing response in leaves, peaking at 12 h after application. Systemic response repressed or unchanged
	BTH treatment in Acru. Sustained local induction in treated leaf, peaking at 12 h after application. Transient systemic induction at 12 h
Ah2880	**Insect herbivory**. Local induction in damaged leaves of all species. Systemic induction in distal leaves of Ahypo and Acru; only weak and late systemic response (24 h) in Acau. Sustained induction in roots of all species and in stems of Acau. Repressed or unchanged in stems of the other two species
	Defoliation in Acru. Developmentally-related expression patterns in leaves (increasing with age until PE stage) and stem (decreasing with age until PE stage). Sustained repression in roots. Extensive repression during FL, except for strong induction in stems
	Virulent bacterial infection, Cmm. Strong late (48 h) systemic repression in distal leaves of all species. Late (24–48 h) local induction in infected leaves of Acru and Acau, contrary to late local repression in leaves of Ahypo. Sustained induction in stems of Acau and Acru and biphasic in Ahypo. Temporary induction (12 and 24 h) in roots of all species. Late repression in local leaves, stems and roots (24 and 48 h) of Ahypo
	Avirulent bacterial infection, *Pss*. Sustained local induction in infected leaves of all species. Sustained (Ahypo) to delayed (Acru; 12 to 48 h) systemic induced response in distal leaves; contrasting sustained repression in distal leaves of Acau. Sustained repressed systemic response in stems of Ahypo; no change in Acru and delayed (12–48 h) systemic induction in stems of Acau. Very strong represion of in roots of Acau
	Salinity. Early (4 d) and intermediate (7 d) repression followed by late (9 d) induction in leaves and roots of Ahypo. Early and intermediate induction in stems of Ahypo. Very strong early induction in leaves and strong, late, induction in stems and roots of Acru. Sustained, strong, expression in stems and roots of Acau. Very strong, sustained, repression in leaves of Acau
	WDS in Acru. Repressed by moderate WDS in leaves, but induced by severe WDS. Unaffected by recovery. Sustained repression by WDS and recovery in roots

(continued)

Table 7.3 (continued)

TRG	Response to stress condition or stress-related treatment[c]
	ABA treatment in Acru. Sustained local and systemic repression in leaves
	BTH treatment in Acru. Late (12 to 24 h) local and systemic induction in treated and untreated distal leaves

[a]Ahypo = *Amaranthus hypochondriacus*; Acru = *A. cruentus* and Acau = *A. caudatus*
[b]TRGs = Transcriptionally Restricted Genes
[c]Measured by quantitative RT-PCR
[d]Insect herbivory treatment: Plants were subjected to herbivory by *Trichoplusia ni* leaf chewing larvae. Damaged leaves, distal undamaged leaves, stems and roots were sampled at 6, 12 and 24 h after larval feeding
[e]Defoliation assay in Acru: leaves, stems, roots and panicles were sampled from Acru plants that were gradually and completely defoliated in a 3-day period Defoliation was performed in plants at different development stages: vegetative, early (V1); vegetative, mature (V2); panicle emergence (PE) and flowering (Fl)
[f]Cmm infection: Leaves of plants were infected by injecting aliquots of *Clavibacter michiganensis* pv. *michiganensis* (strain acr42) at OD_{600} = 0.3 in the third and fourth expanded leaf of six-leaf plants. Infected leaves, distal uninfected leaves, stems and roots were sampled separately 6, 12, 24 and 48 h after infection
[g]Pss infection: Leaves of plants were infected by injecting aliquots of *Pseudomonas syringae* pv. *syringae* (strain 3525) at OD_{600} = 0.2 in the third and fourth expanded leaf of six-leaf plants. Infected leaves, distal uninfected leaves, stems and roots were sampled separately 6, 12, 24 and 48 h after infection
[h]Salinity assay: Six-leaf plants were watered daily with 100 ml of a 400 mM NaCl solution for 8 days. Root, stems and leaves were sampled 4, 7 and 9 days after initiating the salt treatment. The final conductivity of the soil was 12.1 dS/m
[i]Water-deficit stress (WDS) treatment: Acru plants having 10 to 12 expanded leaves were watered to soil capacity. Then, water was withheld until the water content of the soil was lowered to 30% (moderate SWDS) or 10% (severe SWDS). A group of plants exposed to SWDS was allowed to recover by re-watering. Sampling of the recovered plants was performed 24 h after re-watering
[j]Refer to Palmeros-Suárez et al. (2015)
[k]Refer to Casarrubias-Castillo et al. (2014)

References

Abe M, Takahashi T, Komeda Y (2001) Identification of a cis-regulatory element for L1 layer-specific gene expression, which is targeted by an L1-specific homeodomain protein. Plant J 26:487–494

Agarwal PK, Gupta K, Lopato S, Agarwal P (2017) Dehydration responsive element binding transcription factors and their applications for the engineering of stress tolerance. J Exp Bot 68:2135–2148

Aguilar-Hernández HS, Santos L, León-Galván F, Barrera-Pacheco A, Espitia-Rangel E, De León-Rodríguez A, Guevara-González RG, Barba de la Rosa AP (2011) Identification of calcium stress induced genes in amaranth leaves through suppression subtractive hybridization. J Plant Physiol 168:2102–2109

Albertin CB, Simakov O, Mitros T, Wang ZY, Pungor JR, Edsinger-Gonzales E, Brenner DSS, Ragsdale CW, Rokhsar DS (2015) The octopus genome and the evolution of cephalopod neural and morphological novelties. Nature 524:220–224

Ambawat S, Sharma P, Yadav NR, Yadav RC (2013) MYB transcription factor genes as regulators for plant responses: an overview. Physiol Mol Biol Plants 19:307–321

Ambrose C, Wasteneys GO (2014) Microtubule initiation from the nuclear surface controls cortical microtubule growth polarity and orientation in *Arabidopsis thaliana*. Plant Cell Physiol 55:1636–1645

Ang D, Georgopoulos C (2012) An ORFan no more: the bacteriophage T4 39.2 gene product, NwgI, modulates GroEL chaperone function. Genetics 190:989–1000

Angenent GC, Immink RGH (2009) Combinatorial action of petunia MADS box genes and their protein products. In: Gerats T, Strommer J (eds) Petunia. Springer, New York, NY, pp 225–245

Arendsee ZW, Li L, Wurtele ES (2014) Coming of age: orphan genes in plants. Trends Plant Sci 19:698–708

Ariizumi T, Lawrence PK, Steber CM (2011) The role of two F-Box proteins, SLEEPY1 and SNEEZY, in *Arabidopsis* gibberellin signaling. Plant Physiol 155:765–775

Atamian HS, Eulgem T, Kaloshian I (2012) SlWRKY70 is required for Mi-1-mediated resistance to aphids and nematodes in tomato. Planta 235:299–309

Baltensperger DD, Weber LE, Nelson LA (1992) Registration of ‘Plainsman’ grain amaranth. Crop Sci 32:1510–1511

Bao F, Azhakanandam S, Franks RG (2010) SEUSS and SEUSS-LIKE transcriptional adaptors regulate floral and embryonic development in *Arabidopsis*. Plant Physiol 152:821–836

Barah P, Winge P, Kusnierczyk A, Tran DH, Bones AM (2013) Molecular signatures in *Arabidopsis thaliana* in response to insect attack and bacterial infection. PLoS ONE 8:e58987

Brenner D, Baltensperger D, Kulakow P, Lehmann J, Myers R, Slabbert M, Sleugh B (2000) Genetic resources and breeding of *Amaranthus*. Plant Breeding Rev 19:227–285

Brockes JP, Gates PB (2014) Mechanisms underlying vertebrate limb regeneration: lessons from the salamander. Biochem Soc Trans 42:625–630

Cabrales-Orona G (2017) Análisis de genes de función desconocida de amaranto de grano que se expresan en múltiples condiciones de estrés. Dissertation. Centro de Investigación y de Estudios Avanzados del IPN, Unidad Irapuato, Guanajuato, México

Carvunis A-R, Rolland T, Wapinski I, Calderwood MA, Yildirim MA, Simonis N, Charloteaux B, Hidalgo CA, Barbette J, Santhanam B, Brar GA, Weissman JS, Regev A, Thierry-Mieg N, Cusick ME, Vidal M (2012) Proto-genes and *de novo* gene birth. Nature 487:370–374

Casarrubias-Castillo K, Martínez-Gallardo NA, Délano-Frier JP (2014) Treatment of *Amaranthus cruentus* with chemical and biological inducers of resistance has contrasting effects on fitness and protection against compatible Gram positive and Gram negative bacterial pathogens. J Plant Physiol 171:927–939

Caselato-Sousa VM, Amaya-Farfán J (2012) State of knowledge on amaranth grain: a comprehensive review. J Food Sci 77:R93–R104

Castrillón-Arbeláez PA, Martínez-Gallardo N, Avilés Arnaut H, Tiessen A, Délano-Frier JP (2012) Metabolic and enzymatic changes associated with carbon mobilization, utilization and replenishment triggered in grain amaranth (*Amaranthus cruentus*) in response to partial defoliation by mechanical injury or insect herbivory. BMC Plant Biol 12:163

Chezem WR, Memon A, Li FS, Weng JK Claya NK (2017) SG2-Type R2R3-MYB transcription factor MYB15 controls defense-induced lignification and basal immunity in Arabidopsis. Plant Cell 29:1907–1926

Clamp M, Fry B, Kamal M, Xie XH, Cuff J, Lin MF, Kellis M, Lindblad-Toh K, Lander ES (2007) Distinguishing protein-coding and noncoding genes in the human genome. Proc Natl Acad Sci USA 104:19428–19433

Clouse JW, Adhikary D, Page JT, Ramaraj T, Deyholos MK, Udall JA, Fairbanks DJ, Jellen EN, Maughan PJ (2016) The amaranth genome: Genome, transcriptome, and physical map assembly. Plant Genome 9:1–14

Conde A, Chaves MM, Gerós H (2011) Membrane transport, sensing and signaling in plant adaptation to environmental stress. Plant Cell Physiol 52:1583–1602

Curaba J, Singh MB, Bhalla PL (2014) miRNAs in the crosstalk between phytohormone signaling pathways. J Exp Bot 65:1425–1438

Cutcliffe JW, Hellmann E, Heyl A, Rashotte AM (2011) CRFs form protein–protein interactions with each other and with members of the cytokinin signalling pathway in *Arabidopsis* via the CRF domain. J Exp Bot 62:4995–5002

da Silva SM, Gates PB, Brockes JP (2002) The newt ortholog of CD59 is implicated in proximodistal identity during amphibian limb regeneration. Dev Cell 3:547–555

Dat JF, Foyer CH, Scott IM (1998) Changes in salicylic acid and antioxidants during induction of thermotolerance in mustard seedlings. Plant Physiol 118:1455–1461

Daubin V, Ochman H (2004) Bacterial genomes as new gene homes: the genealogy of ORFans in *E. coli*. Genome Res 14:1036–1042

Délano-Frier JP, Martínez-Gallardo NA, Martínez-De la Vega O, Salas-Araiza M, Barbosa-Jaramillo ER, Torres A, Vargas P, Borodanenko A (2004) The effect of exogenous jasmonic acid on induced resistance and productivity in amaranth (*Amaranthus hypochondriacus*) is influenced by environmental conditions. J Chem Ecol 30:1001–1034

Délano-Frier JP, Avilés-Arnaut H, Casarrubias-Castillo K, Casique-Arroyo G, Castrillón-Arbeláez PA, Herrera-Estrella L, Massange-Sánchez J, Martínez-Gallardo NA, Parra-Cota FI, Vargas-Ortiz E, Estrada-Hernández MG (2011) Transcriptomic analysis of grain amaranth (*Amaranthus hypochondriacus*) using 454 pyrosequencing: comparison with *A. tuberculatus*, expression profiling in stems and in response to biotic and abiotic stress. BMC Genomics 12:363

Ding Y, Zhou Q, Wang W (2012) Origins of new genes and evolution of their novel functions. Ann Rev Ecol Evol Syst 43:345–363

Dolgikh VA, Pukhovaya EM, Zemlyanskaya EV (2019) Shaping ethylene response: The role of EIN3/EIL1 transcription factors. Front Plant Sci 10:1030

Domazet-Lošo T, Tautz D (2007) A phylostratigraphy approach to uncover the genomic history of major adaptations in metazoan lineages. Trends Genet 23:533–539

Domazet-Lošo T, Tautz D (2008) An ancient evolutionary origin of genes associated with human genetic diseases. Mol Biol Evol 25:2699–2707

Domazet-Lošo T, Tautz D (2010) A phylogenetically based transcriptome age index mirrors ontogenetic divergence patterns. Nature 468:815–818

Donoghue M, Keshavaiah C, Swamidatta S, Spillane C (2011) Evolutionary origins of Brassicaceae specific genes in *Arabidopsis thaliana*. BMC Evol Biol 11:47

Dujon B (1996) The yeast genome project: what did we learn? Trends Genet 12:263–270

Elhaik E, Sabath N, Graur D (2006) The "inverse relationship between evolutionary rate and age of mammalian genes" is an artifact of increased genetic distance with rate of evolution and time of divergence. Mol Biol Evol 23:1–3

Elzinga DA, Jander G (2013) The role of protein effectors in plant–aphid interactions. Curr Opin Plant Biol 16:451–456

Erdmann R, Gramzow L, Melzer R, Theißen G, Becker A (2010) *GORDITA* (AGL63) is a young paralog of the

Arabidopsis thaliana Bsister MADS box gene *ABS* (*TT16*) that has undergone neo-functionalization. Plant J 63:914–924

Fan P, Nie L, Jiang P, Feng J, Lv S, Chen X, Bao H, Guo J, Tai F, Wang J, Jia W, Li Y (2013) Transcriptome analysis of *Salicornia europaea* under saline conditions revealed the adaptive primary metabolic pathways as early events to facilitate salt adaptation. PLoS ONE 8:e80595

Fisher RA (1930) The genetical theory of natural selection. Oxford University Press, Oxford

Fischer D, Eisenberg D (1999) Finding families for genomic ORFans. Bioinformatics 15:759–762

Flowers TJ, Colmer TD (2008) Salinity tolerance in halophytes. New Phytol 179:945–963

Foote AD, Liu Y, Thomas GWC, Vinař T, Alföldi J, Deng J, Dugan S, van Elk CE, Hunter ME, Joshi V, Khan Z, Kovar C, Lee SL, Lindblad-Toh K, Mancia AL, Nielsen R, Qin X, Qu J, Raney BJ, Vijay N, Wolf JBW, Hahn MW, Muzny DM, Worley KC, Gilbert MTP, Gibbs RA (2015) Convergent evolution of the genomes of marine mammals. Nature Genet 47:272–275

Fraune S, Augustin R, Anton-Erxleben F, Wittlieb J, Gelhaus C, Klimovich VB, Samoilovich MP, Bosch TCG (2010) In an early branching metazoan, bacterial colonization of the embryo is controlled by maternal antimicrobial peptides. Proc Natl Acad Sci USA 107:18067–18072

Fröbisch NB, Shubin NH (2011) Salamander limb development: Integrating genes, morphology, and fossils. Develop Dynam 240:1087–1099

Fürstenberg-Hägg J, Zagrobelny M, Bak S (2013) Plant defense against insect herbivores. Int J Mol Sci 14:10242–10297

Giarola V, Krey S, Frerichs A, Bartels D (2015) Taxonomically restricted genes of *Craterostigma plantagineum* are modulated in their expression during dehydration and rehydration. Planta 241:193–208

Ghiurcuta CG, Moret BM (2014) Evaluating synteny for improved comparative studies. Bioinformatics 30:i9-18

González-Rodríguez T, Cisneros-Hernández I, Acosta Bayona J, Ramírez-Chavez E, Martínez-Gallardo N, Mellado-Mojica E, López-Pérez MG, Molina-Torres J, Délano-Frier J (2019) Identification of factors linked to higher water-deficit stress tolerance in *Amaranthus hypochondriacus* compared to other grain amaranths and *A. hybridus*, their shared ancestor. Plants 8:239

Großkinsky DK, van der Graaff E, Roitsch T (2012) Phytoalexin transgenics in crop protection-Fairy tale with a happy end? Plant Sci 195:54–70

Guo Z, Zhang Z, Yang X, Yin K, Chen Y, Zhang Z, Shin K, Zhu Q, Wang ZY, Wang W (2020) *PSBR1*, encoding a mitochondrial protein, is regulated by brassinosteroid in moso bamboo (*Phyllostachys edulis*). Plant Mol Biol 103:63–74

Halitschke R, Baldwin IT (2003) Antisense LOX expression increases herbivore performance by decreasing defense responses and inhibiting growth-related transcriptional reorganization in *Nicotiana attenuata*. Plant J 36:794–807

Harmer SL, Kay SA (2005) Positive and negative factors confer phase-specific circadian regulation of transcription in Arabidopsis. Plant Cell 17:1926–1940

Heinen TJAJ, Staubach F, Häming D, Tautz D (2009) Emergence of a new gene from an intergenic region. Curr Biol 19:1527–1531

Hernández-Mateo AC (2019). Estudio funcional de la región promotora del gen *Ah24* de amaranto de grano mediante su fusión transcripcional en *Arabidopsis thaliana*. Dissertation. Tecnológico de Tuxtla Gutiérrez, Tuxtla Gutiérrez, Chiapas, México

Hoang XLT, Nguyen YNH, Thao NP, Tran LSP (2020) NAC transcription factors in drought and salinity tolerance. In: Hasanuzzaman M, Tanveer M (eds) Salt and drought stress tolerance in plants. Signaling and communication in plants. Springer, Cham, pp 351–366

Huerta-Ocampo JA, Briones-Cerecero EP, Mendoza-Hernandez G, De Leon-Rodriguez A, Barba de la Rosa AP (2009) Proteomic analysis of amaranth (*Amaranthus hypochondriacus* L.) leaves under drought stress. Int J Plant Sci 170:990–998

Huerta-Ocampo JA, Leon-Galvan MF, Ortega-Cruz LB, Barrera-Pacheco A, De Leon-Rodriguez A, Mendoza-Hernández G, Barba de la Rosa AP (2011) Water stress induces up-regulation of DOF1 and MIF1 transcription factors and down-regulation of proteins involved in secondary metabolism in amaranth roots (*Amaranthus hypochondriacus* L.). Plant Biol 13:472–482

Huerta-Ocampo JA, Barrera-Pacheco A, Mendoza-Hernández CS, Espitia-Rangel E, Mock HP, Barba de la Rosa AP (2014) Salt stress-induced alterations in the root proteome of *Amaranthus cruentus* L. J. Proteome Res 13:3607–3627

Jacob F (1977) Evolution and tinkering. Science 196:1161–1166

Jakoby M, Weisshaar B, Dröge-Laser W, Vicente-Carbajosa J, Tiedemann J, Kroj T, Parcy F, bZIP Research Group (2002) bZIP transcription factors in *Arabidopsis*. Trends Plant Sci 7:106–111

Jassbi AR, Gase K, Hettenhausen C, Schmidt A, Baldwin IT (2008) Silencing geranylgeranyl diphosphate synthase in *Nicotiana attenuata* dramatically impairs resistance to tobacco hornworm. Plant Physiol 146:974–986

Jackson DJ, McDougall C, Woodcroft B, Moase P, Rose RA, Kube M, Reinhardt R, Rokhsar DS, Montagnani C, Joubert C, Piquemal D, Degnan BM (2010) Parallel evolution of nacre building gene sets in molluscs. Mol Biol Evol 27:591–608

Jasper WC, Linksvayer TA, Atallah J, Friedman D, Chiu JC, Johnson BR (2015) Large-scale coding sequence change underlies the evolution of post developmental novelty in honey bees. Mol Biol Evol 32:334–346

Jeon J, Choi J, Lee GW, Park SY, Huh A, Dean RA, Lee YH (2015) Genome-wide profiling of DNA methylation provides insights into epigenetic regulation of fungal development in a plant pathogenic fungus *Magnaporthe oryzae*. Sci Rep 5:8567

Jin H, Martin C (1999) Multi-functionality and diversity within the plant MYB-gene family. Plant Mol Biol 41:577–585

Johnson BL, Henderson TL (2002) Water use patterns of grain amaranth in the northern Great Plains. Agron J 94:1437–1443

Johnson BR, Tsutsui ND (2011) Taxonomically restricted genes are associated with the evolution of sociality in the honey bee. BMC Genomics 12:164

Johnson BR (2018) Taxonomically restricted genes are fundamental to biology and evolution. Front Genet 9:407

Jones DC, Zheng W, Huang S, Du C, Zhao X, Yennamalli RM, Sen TZ, Nettleton D, Wurtele ES, Li L (2016) A clade-specific *Arabidopsis* gene connects primary metabolism and senescence. Front Plant Sci 7:983

Jones-Rhoades MW, Bartel DP (2004) Computational identification of plant microRNAs and their targets, including a stress-induced miRNA. Mol Cell 14:787–799

Kachout SS, Mansoura AB, Leclerc JC, Mechergui R, Rejeb MN, Ouerghi Z (2010) Effects of heavy metals on antioxidant activities of: *Atriplex hortensis* and *A. rosea*. Electron J Environ Agric Food Chem 9:444–457

Kalinina EB, Keith BK, Kern AJ, Dyer WE (2012) Salt- and osmotic stress-induced choline monooxygenase expression in *Kochia scoparia* is ABA-independent. Biol Plant 56:699–704

Kellis M, Wold B, Snyder MP, Bernstein BE, Kundaje A, Marinov GK, Ward LD, Birney E, Crawford GE, Dekker J, Dunham I, Elnitski LL, Farnham PJ, Feingold EA, Gerstein M, Giddings MC, Gilbert DM, Gingeras TR, Green ED, Guigo R, Hubbard T, Kent J, Lieb JD, Myers RM, Pazin MJ, Ren B, Stamatoyannopoulos JA, Weng Z, White KP, Hardison RC (2014) Defining functional DNA elements in the human genome. Proc Natl Acad Sci USA 111:6131–6138

Khalturin K, Anton-Erxleben F, Sassmann S, Wittlieb J, Hemmrich G, Bosch TCG (2008) A novel gene family controls species-specific morphological traits in hydra. PLoS Biol 6:e278

Khalturin K, Hemmrich G, Fraune S, Augustin R, Bosch T (2009) More than just orphans: are taxonomically-restricted genes important in evolution? Trends Genet 25:404–413

Kim S-Y, Kim S-G, Kim Y-S, Seo PJ, Bae M, Yoon H-K, Par C-M (2007) Exploring membrane-associated NAC transcription factors in *Arabidopsis*: Implications for membrane biology in genome regulation. Nucleic Acids Res 35:203–213

Knowles DG, McLysaght A (2009) Recent *de novo* origin of human protein coding genes. Genome Res 19:1752–1759

Lal A, Edwards E (1996) Analysis of inhibition of photosynthesis under water stress in the C4 species *Amaranthus cruentus* and *Zea mays*: Electron transport, CO_2 fixation and carboxylation capacity. Aust J Plant Physiol 23:403–412

Laloum T, De Mita S, Gamas P, Baudin M, Niebel A (2013) CCAATbox binding transcription factors in plants: Y so many? Trends Plant Sci 18:157–166

Lancíková V, Tomka M, Žiarovská J, Gažo J, Hricová A (2020) Morphological responses and gene expression of grain amaranth (*Amaranthus* spp.) growing under Cd. Plants 9:572.

Lau S, Slane D, Herud O, Kong J, Jürgens G (2012) Early embryogenesis in flowering plants: Setting up the basic body pattern. Annu Rev Plant Biol 63:483–506

Lawton MA, Dean S, Dron M, Kooter JM, Kragh KM, Harrison MJ, Yu L, Tanguay L, Dixon RA, Lamb CJ (1991) Silencer region of a chalcone synthase promoter contains multiple binding sites for a factor, SBF-1, closely related to GT-1. Plant Mol Biol 16:235–249

Levine MT, Jones CD, Kern AD, Lindfors HA, Begun DJ (2006) Novel genes derived from noncoding DNA in *Drosophila melanogaster* are frequently X-linked and exhibit testis-biased expression. Proc Natl Acad Sci USA 103:9935–9939

Li L, Foster CM, Gan Q, Nettleton D, James MG, Myers AM, Wurtele ES (2009) Identification of the novel protein QQS as a component of the starch metabolic network in *Arabidopsis* leaves. Plant J 58:485–498

Li J, Zhong R, Palva ET (2017) WRKY70 and its homolog WRKY54 negatively modulate the cell wall-associated defenses to necrotrophic pathogens in Arabidopsis. PLoS ONE 12:e0183731

Li D, Dong Y, Jiang Y, Cai J, Wang W (2010a) A de novo originated gene depresses budding yeast mating pathway and is repressed by the protein encoded by its antisense strand. Cell Res 20:408–420

Li CY, Zhang Y, Wang Z, Zhang Y, Cao C, Zhang PW, Lu SJ, Li XM, Yu Q, Zheng X, Du Q, Uhl GR, Liu QR, Wei L (2010b) A human-specific de novo protein-coding gene associated with human brain functions. PLoS Comput Biol 6:e1000734

Li L, Wurtele ES (2015) The QQS orphan gene of *Arabidopsis* modulates carbon allocation in soybean. Plant Biotechnol J 13:177–187

Li L, Zheng W, Zhu Y, Ye H, Tang B, Arendsee ZW, Jones D, Li R, Ortiz D, Zhao X, Du C, Nettleton D, Scott MP, Salas-Fernandez MG, Yin Y, Wurtele ES (2015) QQS orphan gene regulates carbon and nitrogen partitioning across species via NF-YC interactions. Proc Natl Acad Sci USA 112:14734–14739

Li ZW, Chen X, Wu Q, Hagmann J, Han TS, Zou YP, Ge S, Guo YL (2016) On the origin of *de novo* genes in *Arabidopsis thaliana* populations. Genome Biol Evol 8:2190–2202

Lipman D, Souvorov A, Koonin E, Panchenko A, Tatusova T (2002) The relationship of protein conservation and sequence length. BMC Evol Biol 2:20

Liu Z, Persson S, Sánchez-Rodríguez C (2015) At the border: the plasma membrane–cell wall continuum. J Exp Bot 66:1553–1563

Lou Y, Baldwin IT (2003) *Manduca sexta* recognition and resistance among allopolyploid *Nicotiana* host plants. Proc Natl Acad Sci USA 100:14581–14586

Lu C, Ho TD, Ho S, Yu S (2002) Three novel MYB proteins with one DNA binding repeat mediate sugar and hormone regulation of α-amylase gene expression. Plant Cell 14:1963–1980

Luhua S, Hegie A, Suzuki N, Shulaev E, Luo X, Cenariu D, Ma V, Kao S, Lim J, Gunay MB, Oosumi T, Lee SC, Harper J, Cushman J, Gollery M, Girke T, Bailey-Serres J, Stevenson RA, Zhu JK, Mittler R (2013) Linking genes of unknown function with abiotic stress responses by high-throughput phenotype screening. Physiol Plant 148:322–333

Ma J, Zhang M, Xiao X, You J, Wang J, Wang T, Wang T, Yao Y, Tian C (2013) Global transcriptome profiling of *Salicornia europaea* L. shoots under NaCl treatment. PLoS ONE 8:e65877

Maekawa S, Inada N, Yasuda S, Fukao Y, Fujiwara M, Sato T, Yamaguchi J (2014) The carbon/nitrogen regulator ARABIDOPSIS TOXICOS EN LEVADURA31 controls papilla formation in response to powdery mildew fungi penetration by interacting with SYNTAXIN OF PLANTS121 in Arabidopsis. Plant Physiol 164:879–887

Mandal SM, Porto WF, Dey P, Maiti MK, Ghosh AK, Franco OL (2013) The attack of the phytopathogens and the trumpet solo: Identification of a novel plant antifungal peptide with distinct fold and disulfide bond pattern. Biochimie 95:1939–1948

Marshall E, Costa LM, Gutierrez-Marcos J (2011) Cysteine-rich peptides (CRPs) mediate diverse aspects of cell–cell communication in plant reproduction and development. J Exp Bot 62:1677–1686

Massange-Sánchez JA (2015) Caracterización funcional de genes de *Amaranthus hypochondriacus* L. que se expresan en respuesta a estrés biótico y abiótico. Dissertation, Centro de Investigación y de Estudios Avanzados del IPN, Unidad Irapuato.

Massange-Sanchez JA, Palmeros-Suarez PA, Martinez-Gallardo NA, Castrillon-Arbelaez PA, Avilés-Arnaut H, Alatorre-Cobos F, Tiessen A, Délano-Frier JP (2015) The novel and taxonomically restricted *Ah24* gene from grain amaranth (*Amaranthus hypochondriacus*) has a dual role in development and defense. Front Plant Sci 6:602

Mateo-Hernández AC (2019) Estudio funcional de la región promotora del gen *Ah24* de amaranto de grano mediante su fusión transcripcional en *Arabidopsis thaliana*. Dissertation. Tecnológico de Tuxtla Gutiérrez, Chiapas, México

Matsushita A, Furumoto T, Ishida S, Takahashi Y (2007) AGF1, an AT-Hook protein, is necessary for the negative feedback of *AtGA3ox1* encoding GA 3-Oxidase1. Plant Physiol 143:1152–1162

Mayer MG, Rödelsperger C, Witte H, Riebesell M, Sommer RJ (2015) The orphan gene *dauerless* regulates dauer development and intraspecific competition in nematodes by copy number variation. PLOS Genet 11:e1005146

McClung CR (2006) Plant circadian rhythms. Plant Cell 18:792–803

McLysaght A, Hurst LD (2016) Open questions in the study of *de novo* genes: what, how and why. Nat Rev Genet 17:567–578

Memelink J (2009) Regulation of gene expression by jasmonate hormones. Phytochemistry 70:1560–1570

Mithöfer A, Boland W (2012) Plant defense against herbivores: chemical aspects. Annu Rev Plant Biol 63:431–450

Miller TE, Wing JS, Huete AR (1984) The agricultural potential of selected C4 plants in arid environments. J Arid Environ 7:275–286

Mira A, Martín-Cuadrado AB, D'Auria G, Rodríguez-Valera F (2010) The bacterial pan-genome: a new paradigm in microbiology. Int Microbiol 13:45–57

Mohanta TK, Yadav D, Khan A, Hashem A, Tabassum B, Khan AL, Abd-Allah EF, Al-Harrasi A (2020) Genomics, molecular and evolutionary perspective of NAC transcription factors. PLoS ONE 15:e0231425

Mukherjee K, Brocchieri L, Bürglin TR (2009) A comprehensive classification and evolutionary analysis of plant homeobox genes. Mol Biol Evol 26: 2775–2794

Müller M, Munné-Bosch S (2015) Focus on ethylene ethylene response factors: A key regulatory hub in hormone and stress signaling. Plant Physiol 169:32–41

Nakata M, Mitsuda N, Herde M, Koo AJ, Moreno JE, Suzuki K, Howe GA, Ohme-Takagi M (2013) A bHLHtype transcription factor, ABA-INDUCIBLE BHLH-TYPE TRANSCRIPTION FACTOR/JA-ASSOCIATED MYC2-LIKE1, acts as a repressor to negatively regulate jasmonate signaling in Arabidopsis. Plant Cell 25:1641–1656

Nardini M, Gnesutta N, Donati G, Gatta R, Forni C, Fossati A, Vonrhein C, Moras D, Romier C, Bolognesi M, Mantovani R (2013) Sequence-specific transcription factor NF-Y displays histone-like DNA binding and H2B-like ubiquitination. Cell 152:132–143

Narra HP, Cordes MHJ, Ochman H (2008) Structural features and the persistence of acquired proteins. Proteomics 8:1–10

Narusaka Y, Narusaka M, Park P, Kubo Y, Hirayama T, Seki M, Shiraishi T, Ishida J, Nakashima M, Enju A, Sakurai T, Satou M, Kobayashi M, Shinozaki K (2004) *RCH1*, a locus in *Arabidopsis* that confers resistance to the hemibiotrophic fungal pathogen *Colletotrichum higginsianum*. Mol Plant-Microbe Interact 17:749–762

Navarro-Meléndez AL (2009) Análisis Proteómico del Amaranto, Sometido a Herbivoría (*Spodoptera exigua*) y/o Evocadores Relacionados con la Resistencia a Insectos, MeJA. Dissertation. Centro de investigación y de Estudios Avanzados del IPN, Unidad Irapuato

Nelson P, Buggs R (2016) Next-generation apomorphy: the ubiquity of taxonomically restricted genes. In:

Hughes J, Cotton JA (eds) Olson PD. Next generation systematics, Cambridge University Press, pp 237–264
Neme R, Tautz D (2013) Phylogenetic patterns of emergence of new genes support a model of frequent *de novo* evolution. BMC Genomics 14:117
Newman AM, Cooper JB (2011) Global analysis of proline-rich tandem repeat proteins reveals broad phylogenetic diversity in plant secretomes. PLoS ONE 6:e23167
Noguero M, Atif RM, Ochatt S, Thompson RD (2013) The role of the DNA-binding One Zinc Finger (DOF) transcription factor family in plants. Plant Sci 209:32–45
Nugroho LH, Verpoorte R (2002) Secondary metabolism in tobacco. Plant Cell Tiss Org 68:105–125
Oh DH, Dassanayake M, Bohnert HJ, Cheeseman JM (2012) Life at the extreme: lessons from the genome. Genome Biol 13:241
Ohtani M (2018) Plant snRNP biogenesis: A perspective from the nucleolus and Cajal bodies. Front Plant Sci 8:218
Omamt EN, Hammes PS, Robbertse PJ (2006) Differences in salinity tolerance for growth and water-use efficiency in some amaranth (*Amaranthus* spp.) genotypes. NZ J Crop Hort Sci 34:11–22
Palmeros-Suárez PA, Massange-Sánchez JA, Martínez-Gallardo NA, Montero-Vargas JM, Gómez-Leyva JF, Délano-Frier JP (2015) The overexpression of an *Amaranthus hypochondriacus NF-YC* gene modifies growth and confers water deficit stress resistance in *Arabidopsis*. Plant Sci 240:25–40
Pauw B, Memelink J (2005) Jasmonate-responsive gene expression. J Plant Growth Regul 23:200–210
Perochon A, Jianguang J, Kahla A, Arunachalam C, Scofield SR, Bowden S, Wallington E, Doohan FM (2015) *TaFROG* encodes a *Pooideae* orphan protein that interacts with SnRK1 and enhances resistance to the mycotoxigenic fungus *Fusarium graminearum*. Plant Physiol 169:2895–2906
Perochon A, Kahla A, Vrani M, Jia J, Malla KB Craze M, Wallington E, Doohan FM (2019) A wheat NAC interacts with an orphan protein and enhances resistance to *Fusarium* head blight disease. Plant Biotechnol J 17:1892–1904
Qi M, Zheng W, Zhao X, Hohenstein JD, Kandel Y, O'Conner S, Wang Y, Du C, Nettleton D, MacIntosh GC, Tylka GL, Wurtele ES, Whitham SA, Li L (2019) QQS orphan gene and its interactor NF-YC4 reduce susceptibility to pathogens and pests. Plant Biotechnol J 17:252–263
Rasko DA, Rosovitz MJ, Myers GS, Mongodin EF, Fricke WF, Gajer P, Crabtree J, Sebaihia M, Thomson NR, Chaudhuri R, Henderson IR, Sperandio V, Ravel J (2008) The pangenome structure of *Escherichia coli*: comparative genomic analysis of *E. coli* commensal and pathogenic isolates. J Bacteriol 190:6881–6893
Sánchez-Hernandez C, Martinez-Gallardo N, Guerrero-Rangel A, Valdes-Rodriguez S, Delano-Frier J (2004) Trypsin and alpha-amylase inhibitors are differentially induced in leaves of amaranth (*Amaranthus hypochondriacus*) in response to biotic and abiotic stress. Physiol Plant 122:254–264
Saripella GV, Sonnhammer EL, Forslund K (2016) Benchmarking the next generation of homology inference tools. Bioinformatics 32:2636–2641
Sasaki-Sekimoto Y, Jikumaru Y, Obayashi T, Saito H, Masuda S, Kamiya Y, Ohta H, Shirasu K (2013) Basic helix-loop-helix transcription factors JASMONATE-ASSOCIATED MYC2-LIKE1 (JAM1), JAM2, and JAM3 are negative regulators of jasmonate responses in Arabidopsis. Plant Physiol 163:291–304
Schommer C, Palatnik JF, Aggarwal P, Chételat A, Cubas P, Farmer EE, Nath U, Weigel D (2008) Control of jasmonate biosynthesis and senescence by miR319 targets. PLoS Biol 6:e230
Schroder F, Lisso J, Müssig C (2011) EXORDIUM-LIKE1 promotes growth during low carbon availability in Arabidopsis. Plant Physiol 156:1620–1630
Schwechheimer C, Willige BC, Zourelidou M, Dohmann EMN (2009) Examining protein stability and its relevance for plant growth and development. Methods Mol Biol 479:147–171
Seo PJ, Hong SY, Kim SG, Park CM (2011a) Competitive inhibition of transcription factors by small interfering peptides. Trends Plant Sci 16:541–549
Seo PJ, Kim MJ, Ryu JY, Jeong EY, Park CM (2011b) Two splice variants of the IDD14 transcription factor competitively form nonfunctional heterodimers which may regulate starch metabolism. Nat Commun 2:303
Seo PJ, Park MJ, Park CM (2013) Alternative splicing of transcription factors in plant responses to low temperature stress: mechanisms and functions. Planta 237:1415–1424
Shi CL, Stenvik GE, Vie AK, Bones AM, Pautot V, Proveniers M, Aalen RB, Butenko MA (2011) Arabidopsis class I KNOTTED-like homeobox proteins act downstream in the IDA-HAE/HSL2 floral abscission signaling pathway. Plant Cell 23:2553–2567
Shigenobu S, Stern DL (2013) Aphids evolved novel secreted proteins for symbiosis with bacterial endosymbiont. Proc Biol Sci 280:20121952
Siepel A (2009) Darwinian alchemy: human genes from noncoding DNA. Genome Res 19:1693–1695
Siew N, Fischer D (2003) Unraveling the ORFan puzzle. Comp Funct Genom 4:432–441
Siré C, Moreno AB, Garcia-Chapa M, López-Moya JJ, San Segundo B (2009) Diurnal oscillation in the accumulation of Arabidopsis microRNAs, miR167, miR168, miR171 and miR398. FEBS Lett 583:1039–1044
Skovgaard M, Jensen LJ, Brunak S, Ussery D, Krogh A (2001) On the total number of genes and their length distribution in complete microbial genomes. Trends Genet 17:425–428
Smaczniak C, Immink RGH, Muiño JM, Blanvillain R, Busscher M, Busscher-Lange J, Dinh QD, Liu S, Westphal AH, Boeren S, Parcy F, Xu L, Carles CC, Angenent GC, Kaufmann K (2012) Characterization of MADS-domain transcription factor complexes in

Arabidopsis flower development. Proc Natl Acad Sci USA 109:1560–1565

Sommer RJ, Ogawa A (2011) Hormone signalling and phenotypic plasticity in nematode development and evolution. Curr Biol 21:R758–R766

Song S, Qi T, Fan M, Zhang X, Gao H, Huang H, Wu D, Guo H, Xie D (2013) The bHLH subgroup IIId factors negatively regulate jasmonate-mediated plant defense and development. PLoS Genet 9:e1003653

Stein M (2006) Arabidopsis PEN3/PDR8, an ATP binding cassette transporter, contributes to non-host resistance to inappropriate pathogens that enter by direct penetration. Plant Cell 18:731–746

Steppuhn A, Gase K, Kroch B, Halitschke R, Baldwin IT (2004) Nicotine's defensive function in nature. PLoS Biol 2:e217

Steppuhn A, Baldwin IT (2007) Resistance management in a native plant: nicotine prevents herbivores from compensating for plant protease inhibitors. Ecol Lett 10:499–511

Stergiopoulos I, de Wit PJGM (2009) Fungal effector proteins. Annu Rev Phytopathol 47:233–263

Stoppin V, Vantard M, Schmit AC, Lambert AM (1994) Isolated plant nuclei nucleate microtubule assembly-the nuclear surface in higher-plants has centrosome-like activity. Plant Cell 6:1099–1106

Tautz D, Domazet-Lošo T (2011) The evolutionary origin of orphan genes. Nat Rev Genet 12:692–702

Terzaghi WB, Cashmore AR (1995) Light-regulated transcription. Annu Rev Plant Physiol Plant Mol Biol 46:445–474

Tettelin H, Masignani V, Cieslewicz MJ, Donati C, Medini D, Ward NL, Angiuoli SV, Crabtree J, Jones AL, Durkin AS, Deboy RT, Davidsen TM, Mora M, Scarselli M, Margarit y Ros I, Peterson JD, Hauser CR, Sundaram JP, Nelson WC, Madupu R, Brinkac LM, Dodson RJ, Rosovitz MJ, Sullivan SA, Daugherty SC, Haft DH, Selengut J, Gwinn ML, Zhou L, Zafar N, Khouri H, Radune D, Dimitrov G, Watkins K, O'Connor KJB, Smith S, Utterback TR, White O, Rubens CE, Grandi G, Madoff LC, Kasper DL, Telford JL, Wessels MR, Rappuoli R, Fraser CM (2005) Genome analysis of multiple pathogenic isolates of *Streptococcus agalactiae*: Implications for the microbial "pan-genome". Proc Natl Acad Sci USA 102:13950–13955

Tomalova I, Iachia C, Mulet K, Castagnone-Sereno P (2012) The *map-1* gene family in root-knot nematodes, *Meloidogyne* spp.: a set of taxonomically restricted genes specific to clonal species. PLoS ONE 7:e38656

Tompa P, Schad E, Tantos A, Kalmar L (2015) Intrinsically disordered proteins: emerging interaction specialists. Curr Opin Struct Biol 35:49–59

Touchon M, Hoede C, Tenaillon O, Barbe V, Baeriswyl S, Bidet P, Bingen E, Bonacorsi S, Bouchier C, Bouvet O, Calteau A, Chiapello H, Clermont O, Cruveiller S, Danchin A, Diard M, Dossat C, El Karoui M, Frapy E, Garry L, Ghigo JM, Gilles AM, Johnson J, Le Bouguénec C, Lescat M, Mangenot S, Martinez-Jéhanne V, Matic I, Nassif X, Oztas S, Petit MA, Pichon C, Rouy Z, Ruf CS, Schneider D, Tourret J, Vacherie B, Vallenet D, Médigue C, Rocha EPC, Denamur E (2009) Organised genome dynamics in the *Escherichia coli* species results in highly diverse adaptive paths. PLoS Genet 5: e1000344

Turner M, Nizampatnam NR, Baron M, Coppin S, Damodaran S, Adhikari S, Arunachalam SP, Yu O, Subramanian S (2013) Ectopic expression of miR160 results in auxin hypersensitivity, cytokinin hyposensitivity, and inhibition of symbiotic nodule development in soybean. Plant Physiol 162:2042–2055

Vakirlis N, Hebert AS, Opulente DA, Achaz G, Hittinger CT, Fischer G, Coon JJ, Lafontaine I (2018) A molecular portrait of *de novo* genes in yeasts. Mol Biol Evol 35:631–645

van Oss SB, Carvunis A-R (2019) *De novo* gene birth. PLoS Genet 15:e1008160

Vargas-Ortiz E, Espitia-Rangel E, Tiessen A, Délano-Frier JP (2013) Grain amaranths are defoliation tolerant crop species capable of utilizing stem and root carbohydrate reserves to sustain vegetative and reproductive growth after leaf loss. PLoS ONE 8: e67879

Vargas-Ortiz E, Délano-Frier JP, Tiessen A (2015) The tolerance of grain amaranth (*Amaranthus cruentus* L.) to defoliation during vegetative growth is compromised during flowering. Plant Physiol Biochem 91:36–40

Wasternack C, Parthier B (1997) Jasmonate-signalled plant gene expression. Trends Plant Sci 8:302–307

Werner MS, Sieriebriennikov B, Prabh N, Loschko T, Lanz C, Sommer RJ (2018) Young genes have distinct gene structure, epigenetic profiles, and transcriptional regulation. Genome Res 28:1675–1687

Wilson GA, Bertrand N, Patel Y, Hughes JB, Feil EJ, Field D (2005) Orphans as taxonomically restricted and ecologically important genes. Microbiology 151:2499–2501

Wilson GA, Feil EJ, Lilley AK, Field D (2007) Large-scale comparative genomic ranking of taxonomically restricted genes (TRGs) in bacterial and archaeal genomes. PLoS ONE 2:e324

Wilson BA, Foy SG, Neme R, Masel J (2017) Young genes are highly disordered as predicted by the preadaptation hypothesis of *de novo* gene birth. Nat Ecol Evol 1:0146

Wolf YI, Novichkov PS, Karev GP, Koonin EV, Lipman DJ (2009) The universal distribution of evolutionary rates of genes and distinct characteristics of eukaryotic genes of different apparent ages. Proc Natl Acad Sci USA 106:7273–7280

Wright S (1931) Evolution in Mendelian populations. Genetics 16:97–159

Wright PE, Dyson HJ (2015) Intrinsically disordered proteins in cellular signalling and regulation. Nat Rev Mol Cell Biol 16:18

Wu DD, Irwin DM, Zhang YP (2011) *De novo* origin of human protein-coding genes. PLoS Genet 7:e1002379

Xiao W, Liu H, Li Y, Li X, Xu C, Long M, Wang S (2009) A rice gene of de novo origin negatively regulates pathogen-induced defense response. PLoS ONE 4:e4603

Xu D, Zhang Y (2012) Ab initio protein structure assembly using continuous structure fragments and optimized knowledge-based force field. Proteins 80:1715–1735

Xu D, Zhang Y (2013) Toward optimal fragment generations for ab initio protein structure assembly. Proteins 81:229–239

Yadeta KA, Valkenburg D-J, Hanemian M, Marco Y, Thomma BPHJ (2014) The Brassicaceae-specific EWR1 gene provides resistance to vascular wilt pathogens. PLoS ONE 9:e88230

Yang Z, Huang J (2011) De novo origin of new genes with introns in *Plasmodium vivax*. FEBS Lett 585:641–644

Zavala JA, Patankar AG, Gase K, Hui D, Baldwin IT (2004) Manipulation of endogenous trypsin proteinase inhibitor production in *Nicotiana attenuata* demonstrates their function as antiherbivore defenses. Plant Physiol 134:1181–1190

Zhang Z, Lenk A, Andersson MX, Gjetting T, Pedersen C, Nielsen ME, Newman MA, Hou BH, Somerville SC, Thordal-Christensen H (2008) A lesion-mimic syntaxin double mutant in *Arabidopsis* reveals novel complexity of pathogen defense signaling. Mol Plant 1:510–527

Zhang Y, Zhang B, Yan D, Dong W, Yang W, Li Q, Zeng L, Wang J, Wang L, Hicks LM, He Z (2011) Two *Arabidopsis* cytochrome P450 monooxygenases, CYP714A1 and CYP714A2, function redundantly in plant development through gibberellin deactivation. Plant J 67: 342–353

Zhao L, Saelao P, Jones CD, Begun DJ (2014) Origin and spread of de novo genes in *Drosophila melanogaster* populations. Science 343:769–772

Zhao CY, Escalante LN, Chen H, Benatti TR, Qu JX, Chellapilla S, Waterhouse RM, Wheeler D, Andersson MN, Bao R, Batterton M, Behura SK, Blankenburg KP, Caragea D, Carolan JC, Coyle M, El-Bouhssini M, Francisco L, Friedrich M, Gill N, Grace T, Grimmelikhuijzen CJP, Han Y, Hauser F, Herndon N, Holder M, Ioannidis P, Jackson LR, Javaid M, Jhangiani SN, Johnson AJ, Kalra D, Korchina V, Kovar CL, Lara F, Lee SL, Liu X, Löfstedt C, Mata R, Mathew T, Muzny DM, Nagar S, Nazareth LV, Okwuonu G, Ongeri F, Perales L, Peterson BF, Pu LL, Robertson HM, Schemerhorn BJ, Scherer SE, Shreve JT, Simmons DN, Subramanyam S, Thornton RL, Xue K, Weissenberger GM, Williams CE, Worley KC, Zhu D, Zhu Y, Harris MO, Shukle RH, Werren JH, Zdobnov EM, Chen MS, Brown SJ, Stuart JJ, Richards S (2015) A massive expansion of effector genes underlies gall-formation in the wheat pest *Mayetiola destructor*. Curr Biol 25:613–620

Zhong R, Ye ZH (2012) MYB46 and MYB83 bind to the SMRE sites and directly activate a suite of transcription factors and secondary wall biosynthetic genes. Plant Cell Physiol 53:368–380

Zhou J, Wang B, Li Y, Wang Y, Zhu L (2007) Responses of chrysanthemum cells to mechanical stimulation require intact microtubules and plasma membrane-cell wall adhesion. J Plant Growth Regul 26:55–68

Zhou Q, Zhang G, Zhang Y, Xu S, Zhao R, Zhan Z et al (2008) On the origin of new genes in *Drosophila*. Genome Res 18:1446–1455

Zhu Y, Nomura T, Xu Y, Zhang Y, Peng Y, Mao B, Hanada A, Zhou H, Wang R, Li P, Zhu X, Mander LN, Kamiya Y, Yamaguchi S, He Z (2006) *ELONGATED UPPERMOST INTERNODE* encodes a cytochrome P450 monooxygenase that epoxidizes gibberellins in a novel deactivation reaction in rice. Plant Cell 18:442–456

Zhu C, Ding Y, Liu H (2011) MiR398 and plant stress responses. Physiol. Plant 143:1–9

How to Overcome Recalcitrance? Novel Strategies and Recent Advances in the Genetic Transformation of Grain Amaranth

8

Andrea Paola Castellanos-Arévalo,
José Luis Cabrera-Ponce, Cecilia Nava-Sandoval,
and John P. Délano-Frier

Abstract

Genetic transformation of grain amaranth has been a challenging enterprise due to its notorious recalcitrance to *Agrobacterium*-mediated transformation processes. This chapter will describe the relatively small number of studies, most of them involving leafy amaranths, that have been successful in overcoming this natural limitation. It will also describe the use of the up to now ineffectual alternative genetic transformation methods. A focus on recent findings that promise to be the breakthrough required to permit the straightforward genetic transformation of grain amaranths will close this chapter in a positive note. This development will represent a powerful experimental tool to achieve the full agronomic potential of this environmentally resilient crop.

A. P. Castellanos-Arévalo · J. L. Cabrera-Ponce ·
J. P. Délano-Frier (✉)
Departamento de Biotecnología y Bioquímica, Centro de Investigación y de Estudios Avanzados del I.P.N. (Cinvestav), Unidad Irapuato. Km 9.6 del Libramiento Norte Carretera Irapuato-León. Apartado Postal 629, C.P. 36824 Irapuato, Gto., México
e-mail: john.delano@cinvestav.mx

C. Nava-Sandoval
Escuela Nacional de Ciencias Biológicas del Instituto Politécnico Nacional, Unidad Profesional Lázaro Cárdenas, Prolongación de Carpio y Plan de Ayala s/n, Col. Santo Tómas, C.P. 11340 Alcaldía Miguel Hidalgo, CDMX, México

8.1 How It All Started: *Agrobacterium*-Mediated Transformation

Genetic transformation of plants can be performed directly or indirectly, the latter almost exclusively mediated via *Agrobacterium*-mediated procedures (Zupan et al. 2000; Meyers et al. 2010). These procedures take advantage of the infection strategy used by *Agrobacterium tumefaciens* and *A. rhizogenes* to infect their host via the integration of virulent DNA (or T-DNA) into the plant nuclear genome (Ream 2009). The delivery of this infective component from bacterial cells into host plant cells occurs through a bacterial type IV secretion system (Tzfira and Citovsky 2002; Tzfira et al. 2004; Gelvin 2012). This T-DNA is defined by conserved 25-base pair imperfect repeats at the ends of the T-region called border sequences and by a virulence (*vir*) region composed of at least seven major *vir* loci that encode components of the bacterial protein machinery needed for T-DNA processing and transfer into the plant genome (Hooykaas and Beijersbergen 1994). Several plant factors are involved in the *Agrobacterium*-mediated plant transformation process. These include various proteins that favor the initial attachment of *Agrobacterium* to the plant cell surface followed by the transport, integration and/or expression of its T-DNA. Plant hormones and specific plant secondary metabolites also play a prominent role

D. Adhikary et al. (eds.), *The Amaranth Genome*, Compendium of Plant Genomes,
https://doi.org/10.1007/978-3-030-72365-1_8

in this process (Hwang et al. 2017). However, some metabolites, such as HDMBOA and o-imidoquinone, are to known negatively affect *A. tumefaciens* infection of maize and other plants and may contribute to their recalcitrance to *Agrobacterium*-mediated transformation (Zhang et al. 2000; Maresh et al. 2006; Lin et al. 2008; Gohlke and Deeken 2014; Subramoni et al. 2014).

Additionally, other direct or *in planta* gene transfer methods have been developed (see below) mostly to solve the most pressing problem of the former method, which is the recalcitrance to transformation, or regeneration, inherent to some plant species or mutant lines. Other drawbacks are the strict requirement of sterile conditions for plant regeneration, which is time-consuming and requires considerable technical skills. The efficiency of transformation also depends on the efficacy of the selection procedure and the frequency of shoot regeneration and (pollen or somatic) embryogenesis (Sobhanian et al. 2012). Several other tissue culture parameters need to be optimized for each target plant in regular tissue culture-based *Agrobacterium*-mediated gene transformations, many times in a trial-and-error basis. These include the combination and concentration of plant growth regulators, bacterial strains and cell densities, nature of inducing phenolic compounds, plant genotypes, explant type, light, temperature, use of antibiotics, wounding or not of the target tissue and a suitable selection method of transgenic cells (Niazian et al. 2017). Vacuum infiltration, sonication and additives, such as the Silwet L-77 surfactant, are frequently employed to increase transformation efficiency.

Thus, simpler and quicker in planta *Agrobacterium*-mediated gene transformation methods that are independent of tissue culture procedures that avoid somaclonal variation have been tested in some plants and crops. The most common direct transformation methods include vacuum or ultrasonic infiltration, direct injection of *Agrobacterium* cultures into plant tissues, the pollen-tube pathway method, shoot apex transformations, floral dip or floral spray and silicon carbide mediated transformation (Rao et al. 2009). The main limitation of these methods is that DNA integration into the plant is often random and largely unpredictable (Table 8.1) (Hadi et al. 1996).

8.2 Variants of the Classical *Agrobacterium* Transformation Method: Are They Better?

The meristem-based regeneration or shoot apex method for *Agrobacterium*-mediated gene transformation offers the advantages of causing a low incidence of tissue culture-induced genetic changes. Besides, it is a simple and direct plant transformation method (Gould et al. 1991), having been used successfully in *Agrobacterium*-mediated transformation of cotton, rice, cucumber and broccoli (Niazian et al. 2017). The sonication-assisted *Agrobacterium*-mediated transformation has been reported to enhance the transformation of monocots and certain dicot tissues that are recalcitrant to *Agrobacterium* infection. During sonication, plant tissues are subjected to brief periods of ultrasound in the presence of *Agrobacterium* to enhance the transfer of exogenous, naked or plasmid DNA by overcoming host specificity barriers and helping the bacteria reach the proper cells in the target tissue (Trick and Finer 1997; Joersbo and Brunstedt 1990, 1992; Zhang et al. 1991). Important factors involve the addition of antioxidants to the co-culture medium and wounding of the target plant tissue to enhance transformation rates (Trick and Finer 1997). Other wounding methods, apart from the simple wounds made during the normal course of explant preparation, include particle gun-mediated micro-wounding (Bidney et al. 1992), and the use of syringe infiltration (Chee et al. 1989).

In other alternative transformation procedures, *Agrobacterium* is used to directly infect the explants without the need of in *in vitro* culture and regeneration of plant cells or tissues (Mayavan et al. 2013, 2015; Kalbande and Patil 2016). The main methods included in this category are vacuum infiltration, *Agrobacterium*

Table 8.1 Summary of direct and indirect methods employed for the genetic transformation of plants (Adapted from Rivera et al. 2012)

Method	General procedure	Advantages	Disadvantages	Transformation efficiency
Agrobacterium tumefaciens/A. rhizogenes	A modified bacterial phytopathogen introduces a plasmid DNA amenable to plants that express the transgene in a constitutive or inducible way	The integration of the transgene into the plant genome is simple and precise, with defined borders. Stable and durable inheritance of the transgene through various generations. Different types of cells can be used. Reproducible and efficient protocols have been developed for Dicot plants and some Monocots	Several factors that are difficult to control affect the transformation efficiency and subsequent regeneration of the transformed plant. The process is time-consuming. It introduces expression vectors that contain elements that might have unforeseen deleterious effects on the environment	High
Electroporation	Electric impulses induce membrane permeability and supply the energy for the transport of ions and macromolecules across the transient membrane pores	It can be used to transform plant protoplasts from different cell types. It is simple, rapid and affordable	Protocols are laborious; they frequently require the formation of protoplasts. It is highly dependent on the electrophysiological characteristics of the plant	Low
Biobalistics	It employs minuscule particles covered with genetic material that can traverse the plant cell when accelerated	The procedure is easy to implement. There is no need to degrade the plant cell wall. It is independent of the physiological properties of the plant cells. Simultaneous transformation events with multiple transgenes are possible	It is expensive. It requires a constant supply of reagents and other consumables. Damage to the DNA is likely. Multiple foreign DNA insertions into the target genome con occur, leading to gene silencing	Low
Vacuum infiltration	Exposure to vacuum reduces the number of air spaces between cells, thereby facilitating the infiltration of *Agrobacterium* or other bacterial cells	Simple and quick. It has the potential to produce large number of transgenic plants	Free entry of bacteria into the plant cells could have unintended deleterious effects on the transformed plants	Medium

(continued)

Table 8.1 (continued)

Method	General procedure	Advantages	Disadvantages	Transformation efficiency
Ultrasound and shock waves	Acoustic cavitation alters the cell's membrane permeability, facilitating the absorption of foreign DNA	Simple and safe to use	Expensive. Can cause damage to plants, leading to morphological alterations. Standardized protocols for plants are rare	Medium
Transformation mediated by silicon carbide whiskers	The silicon carbide whiskers are vortexed together in a suspension containing plant tissues and the foreign DNA. The abrasion produced permits foreign DNA to enter the cells	Simple, rapid and un-expensive. Can be used without limitations in several plant species	Abrasive damage caused to plant cells may negatively affect their capacity to regenerate. Precautions must be taken to avoid inhalation of silicon carbide fibers by the operator	Very low
Micro-injection	Direct introduction of foreign DNA into the plant cell via an injection pipette	It allows the introduction of large DNA plasmids and chromosomes into plant cells	Expensive, tedious and time consuming	Extremely high
Macro-injection	The inheritable materials are injected into the plant using a hypodermic syringe	Stability and reproducibility	Only a section of the plant is transformed	Very low
Laser micro-rays	The laser micro-rays perforate the plant cell wall, allowing entry of the foreign DNA	It is very precise	Costly and laborious	High
Electrophoresis	Plant embryos are placed between two pipettes connected to electrodes. The foreign DNA flows through the cells from the cathode to the anode	Simple and inexpensive	The viability of the embryos is seriously compromised	It has only been used for the transformation of orchids (Griesbach and Hammond 1994)

injection, pollen tube-mediated gene transfer (Zhou et al. 1983; Luo and Wu 1988; Liu et al. 2009; Bibi et al. 2013; Wang et al. 2013; Ali et al. 2015; Hwang et al. 2017), floral dip and floral spray methods. (Bechtold and Pelletier 1998; Clough and Bent 1998; Richardson et al. 1998; Ye et al. 1999; Bent 2000, 2006; Chung et al. 2000; Desfeux et al. 2000; Mysore et al. 2000; Trujillo et al. 2004; Tague and Mantis 2006; Baranski 2008; Lin et al. 2009; Ali et al. 2015; Hussain Shah et al. 2015; Kalbande and Patil 2016; Hwang et al. 2017; Niazian et al. 2017). Cell suspensions have been also injected into epidermal onion cells to successfully transform onion plants (Xu et al. 2014). Important disadvantages of this method are its limitation to few plant cell types, the difficulty of regenerating viable plants and the production of chimeric transgenic plants, especially when the shoot tip is the site of *Agrobacterium* inoculation (Kalbande and Patil 2016; Hwang et al. 2017). Nevertheless, a recently reported and promising gene-editing method employed a simplified tissue-culture-free procedure that reprogrammed genome-edited somatic cells into meristems through the co-expression of developmental regulators and gene editing components. This method permitted the direct regeneration of genome-edited plants from somatic cells (Maher et al. 2020).

8.3 Increasing Transformation Efficiency in Recalcitrant Crops

Transgenesis and genome editing are limited by recalcitrance to tissue culture and transformation (Shrawat and Lörz 2006; Hiei et al. 2014). Contributing factors include the complex set of conditions that control the susceptibility of cells to *Agrobacterium*, a situation that results in highly variable delivery efficiencies between ecotypes and species (Lacroix and Citovsky 2019). This circumstance limits the efficiency of *Agrobacterium*-mediated transformation to a narrow range of genotypes within a species (Nam et al. 1997). In addition, regenerated plants can undergo changes to their genome and epigenome (Phillips et al. 1994; Kaeppler et al. 2000).

The optimization of cell and tissue culture has mostly involved the manipulation of plant developmental programs *in vitro* by the exogenous application of plant growth regulators (PGRs), principally auxins and cytokinins. Stress is now understood to play an important role in this process. The implementation of this sequence of key steps nowadays needs to be determined empirically for each species and frequently requires genotype-related adjustments (Altpeter et al. 2016). As mentioned already, several other tissue culture protocol parameters in addition to PGRs and stress, must be fine-tuned to establish a successful gene transformation procedure. Callus formation and the subsequent re-differentiation of organs and somatic embryos from these undifferentiated tissues requires both genetic and epigenetic mechanisms that are strictly modulated by hormonal signaling in combination with epigenetic regulation via DNA methylation and histone modification. The rather complex regulatory process involves a whole gamut of regulatory protein and transcription factors (Altpeter et al. 2016). A relevant benefit that will be derived from reaching reliable solutions to the complications associated with genetic transformation in several recalcitrant plant species is that contemporary protocols that permit precise genome editing methods, such as the clustered regularly interspaced short palindromic repeats (CRISPR)-associated protein (CRISPR/Cas) system, require a suitable delivery method to efficiently transform plants (Kumar and Jain 2015).

8.4 The Long and Winding Road Leading to the Best Procedure to Transform Grain Amaranth

Experimental evidence gathered in the last two decades, during which only a relatively reduced number of efforts to genetically transform amaranth plants have been described, confirmed the widely accepted notion that *Agrobacterium*-mediated procedures are successful in a narrow range of genotypes within a species (Lacroix and Citovsky 2019). As described subsequently, leafy amaranth species appear to be more amenable to genetic transformation than grain amaranths, and within the latter, *Amaranthus hypochondriacus* seems to be the most promising species for transformation, either by *A. tumefaciens* or *A. rhizogenes* (Jofre-Garfias et al. 1997; Pal et al. 2103a, b; Castellanos-Arévalo et al. 2020).

Somatic embryogenesis (SE) is one of the most widely applied plant tissue culture technique for plant regeneration, micro-propagation and genetic improvement by stable genetic transformation and genome editing. The importance of SE for genetic transformation is due mainly to the unicellular origin of a somatic embryo. In amaranth species, SE induction, proliferation and plant regeneration is poorly understood. The development of *in vitro* methods for shoot formation from callus derived from hypocotyl explants of *A. hypochondriacus*, *A. cruentus*, *A. tricolor*, and *A. paniculatus* established the basis from which the *Agrobacterium*-mediated genetic transformation of amaranth plants could be attempted (Flores et al. 1982; Flores and Teutonico 1982; Bagga et al. 1987). The first of these reports, by Flores et al. (1982), described how the use of high concentrations of the synthetic auxin 2,4-dichlorophenoxyacetic acid (2, 4-D; from 1-to-10 mg/ L), induced compact embryogenic-like callus, but that further attempts to convert these structures to plantlets were unsuccessful. Only shoots could be regenerated when using cytokinins (i.e., zeatin) and auxins (i.e., naphthalene acetic acid, NAA). Later, Flores et al. (1982) tried to correlate the plant tissue culture competence of amaranth callus and cell suspension cultures with known amaranth nutritional stress factors, such as saponins, phenolic acids and oxalates. Part of the reasoning was that levels of calcium and nitrates could be dependent on oxalate concentration. Following this hypothesis, they were able to decrease nitrate concentration by using modified B5 media. However, they were unable to demonstrate that such modification had a positive influence on the growth and development of amaranth.

In parallel, and defying the belief that amaranth plants were resistant to infection by *Agrobacterium* (De Cleene and De Ley 1976), pioneer efforts by Jofre-Garfias and co-workers (1997) found that two oncogenic strains of *A. tumefaciens* were capable of infecting hypocotyl and cotyledon explants from *A. hypochondriacus* seedlings. These subsequently produced the expected tumors when cultivated on Murashige and Skoog (MS) media without growth regulators. Having demonstrated that a particular grain amaranth genotype (*A. hypochondriacus* cv. Azteca) was apt for genetic transformation by *A. tumefaciens*, this group proceeded to establish the first genetic transformation method for amaranth using mature embryos of *A. hypochondriacus* as the basis of an *in vitro* culture system capable of regenerating genetically transformed plants (Jofre-Garfias et al. 1997). Regenerated plants were produced from embryo-derived callus cultured on MS subsequently to the infection of mature embryo explants with a disarmed *A. tumefaciens* strain harboring genes conferring antibiotic resistance and β-glucuronidase/GUS (*uid*A) reporter activity. The regeneration procedure involved the use of media supplemented with a cocktail of phytohormones and coconut liquid endosperm. Transgenic plants, detected by Southern blot hybridization, also carried a pea chlorophyll a/b-binding protein promoter, which was used to test the transgenic amaranth plants and their progeny for tissue-specific and light-inducible expression. Nevertheless, the ontogeny of embryogenic-like callus generation was not well understood since globular structures were unable to produce the latter stages of SE until maturation and plant regeneration (Jofre-Garfias et al. 1997, 2001). Curiously, it was observed that plant regeneration was achieved in certain periods of the year in central México, usually in the dry season, between March–May, in which ambient relative humidity ranged from 5 to 10%. However, all efforts to simulate these circumstances for plant regeneration were unsuccessful. This group was also able to induce tumors in *A. hypochondriacus* explants using *A. tumefaciens* wild type strains, pTiC58 (nopaline type) and A281 (octopine type) in MS medium lacking growth regulators. Tumors were propagated and nopaline type tumors were eventually differentiated and regenerated plants in MS medium without growth regulators.

Later, Swain et al. (2010) described a procedure for *A. rhizogenes*-mediated genetic transformation of leafy *A. tricolor* via the co-cultivation of different explants with two *A. rhizogenes* strains. This study reported the optimization of key factors able to influence amaranth´s root genetic transformation, including *Agrobacterium* strain, explant type, infection strategy and bacterial cell density, among others. It recorded a better transformation efficiency: (i) using explants from garden-grown plants instead of those generated from *in vitro* cultures; (ii) using stem internodes instead of leaves, and (iii) immersion of pre-pricked explants in the bacterial suspensions instead of direct bacterial injection. The maximum frequency (ca. 97%) of hairy roots emergence was obtained when internode explants were infected with a selected strain followed by co-cultivation on growth regulator-free MS medium for five days. The transformation was confirmed by positive agropine and mannopine synthesis in all selected transformed rhizoclones. Additionally, shoot regeneration was stimulated in zeatin-complemented media. The same research group (Pal et al. 2103a), subsequently described an optimized *A. tumefaciens*-mediated genetic transformation protocol for *A. tricolor* using epicotyl explants. Two bacterial strains carrying a binary plasmid designed for constitutive genetic expression and harboring the *neomycin phosphotransferase II* selection gene and the *uid*A reporter gene, were compared to select the one having higher transforming efficiency. As previously reported, several critical factors influencing genetic transformation were optimized. This achievement was followed shortly after by the transformation of *A. spinosus*, a vegetable amaranth with nutraceutical properties, via the combined use of different wild type *A. rhizogenes* strains and plant explants. Although the results were variable, an experimental protocol was established to generate transgenic plants able to accumulate opines (Pal et al. 2103b). In parallel, Munusamy et al. (2013) developed a floral dip transformation method similar to a modified method employed routinely in *Arabidopsis thaliana* (Martinez-Trujillo et al. 2004) using several constructs carrying *hygromycin phosphotransferase* (*hph*) as a selectable marker gene. These were introduced individually into an *A. tumefaciens* strain. The resulting bacterial suspensions were applied drop-wise to flowering *Amaranthus* (species not mentioned) panicles to generate T1 seeds. Germination of T1 seeds in selective media resulted in the identification of 12 putative transformants, which were subsequently confirmed by PCR amplification of an *hph* gene fragment of ca.750 bp which was further verified by sequencing. This study indicated that *Amaranthus* inflorescences can be easily transformed and that this event can be efficiently transmitted to the progeny. Kuluev et al. (2017) generated transgenic *A. retroflexus* plants able to express the *A. thaliana ARGOS-LIKE (ARL)* gene under the control of the constitutive dahlia mosaic virus promoter via *A. tumefaciens*-mediated transformation. These workers rationalized that the overexpression of *ARL*, coding for a transmembrane endoplasmic reticulum protein presumably involved in signal cross-talk between phytohormones and transcription factors (Hu et al. 2006; Feng et al. 2011), would lead to greater vegetative growth in *A. retroflexus* and improve its use as a forage crop. The authors reported a 1.4% transformation effectiveness, resulting in genetically transformed amaranth plants showing notable increases in stem and leaf length, and in fresh weight. Recent reports have further described the successful transformation of *A. caudatus* L. and *A. caudatus* × *A. paniculatus* hybrids using either floral dip or vacuum infiltration methods (Yaroshko et al. 2018, 2020; Yaroshko and Kuchuk 2019).

Other contemporary studies concentrated on the exploration of alternative methods to optimize the genetic transformation of grain amaranth. Initially, *in vitro*, *ex vitro* and *in planta* transformation strategies were tested to transform different grain amaranth genotypes. The use of two different strains of *A. tumefaciens* (GV2260 and AGL1) and four strains of *A. rhizogenes* (A4 RS, A4 PC, AR1500 and K599) was tested in two cultivars of *A. hypochondriacus* (Nutrisol and Revancha) and in *A. cruentus* (cv. Dorada). These cultivars are part of a small group of grain amaranths recommended for cultivation in Mexico (Espitia-Rangel et al. 2012; García-Canales 2014). The first drawback encountered was the inability to replicate the results using the methodology reported by Jofre-Garfias et al. (1997). *In vitro* generation of callus, which mostly proceeded to form embryos in the globular stage, were generated from explants obtained from aseptically grown amaranth plantlets (Fig. 8.1a). This transformation occurred after ca. 6 weeks under the conditions described by these workers, and favored the formation of roots in the presence of gibberellins (Fig. 8.1b). Also, care was needed to avoid excessive accumulation of oxidized phenolic compounds in explants (Fig. 8.1c). However, less than 10% of the callus formed were *uidA*-positive (Fig. 8.1d) and none of them formed somatic embryos, even after a five-month period of constant sub-cultivation with segmentation events every two weeks.

The second, occurred while testing *in planta* transformation by direct infection of apical meristems with *A. tumefaciens* strain AGL1, as described by Pal et al. (2103a, b) but with slight modifications. However, the emergence of multiple adventitious shoots was only observed in uninfected control plants after a two-month period that included sub-cultivation onto fresh media every two weeks (Fig. 8.2).

Similar to Jofre-Garfias et al. (1997), shoot regeneration from apical meristems of grain amaranth hypocotyl explants infected with *A. tumefaciens* did not occur when placed on herbicide-containing selective media. Possible explanations of this outcome were that *A.*

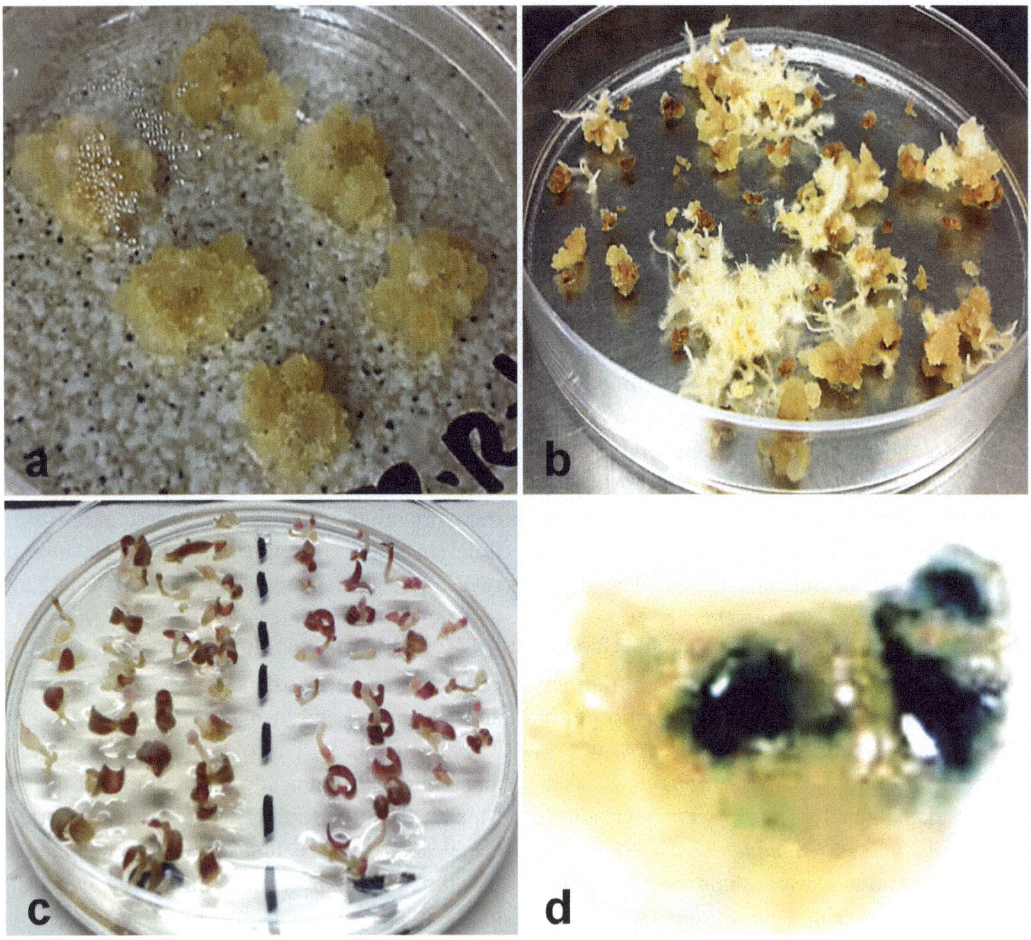

Fig. 8.1 Tissue culture of *A. hypochondriacus* explants. **a** Callus were generated from cotyledon explants in MS media complemented with 1 μg/ml 1-naphthaleneacetic acid, 6-benzylaminopurine (BAP) and gibberellins; **b** root emergence was activated in these callus at some time after culture; **c** the presence of BAP promoted the oxidation of the cultured plantlets as shown in the left side of the image, and **d** transformation efficiency of callus, monitored by GUS staining was low, ca. 20% (original figure, adapted from García-Canales 2014)

Fig. 8.2 Emergence of fully developed plantlets from decapitated shoot apical meristems (SAM) of *A. hypochondriacus* cotyledon explants. Multiple shoots developed from wounded SAMs of cotyledon explants in MS complemented with 1-naphthaleneacetic acid, 6-benzylaminopurine and gibberellins. This response was inhibited in the presence of *A. tumefaciens* (original figure, adapted from García-Canales 2014)

tumefaciens was incapable of infecting meristematic cells or that the infection rate was too small to allow the proper regeneration of transgenic shoots. The third, while testing direct infection of mechanically damaged or scarified seeds by the same *A. tumefaciens* strain. Seeds, pre-hydrated for 4 h, were subjected to the following treatments: (i) longitudinal incision of the seed testa to expose the embryo; (ii) scarification of the seed surface with sterile sandpaper for short periods (5–15 s), and (iii) scarification by sonication for longer periods (10–30 min). A fourth pre-treatment involved direct submergence in a 30% HCl solution for 90 to 300 s. All treated seeds were subsequently infected by submersion in bacterial suspensions having different titers ($O.D._{600}$ = 0.6–2.0) with/without different concentrations of Silwett L-77 (0.1–0.8%) for 72 h in the dark at room temperature. All infected seeds were then transferred to MS media spiked with glufosinate ammonium herbicide for selection of the transformed seedlings. These experimental strategies were not successful. Chemical scarification was lethal and although they facilitated germination, physical scarification methods damaged the seeds to such an extent that they negatively affected the vitality of the emerging seedlings and in many cases produced anomalous seedling morphologies when infected with surfactant-containing bacterial suspensions and allowed to germinate in non-selective media (Fig. 8.3). More important was that no seedlings survived cultivation on selective media. The negative results obtained contrasted with the successful genetic transformation from seeds reported in tomato (Yasmeen et al. 2009), *M. truncatula* (Trieu et al. 2000), peanuts (Rohini and Rao 2000), *A. thaliana* (Bent 2000),

Fig. 8.3 The infection of scarified seeds of *A. hypochondriacus* with *A. tumefaciens* causes the germination of deformed seedlings in MS media without herbicide selection. No germination was produced in selective MS media (original figure, adapted from García-Canales 2014)

Fig. 8.4 The "Floral-Dip" transformation procedure was tested in panicles of *Amaranthus cruentus* at three different development stages. In **a** the anthers are visible in the youngest panicle; **b** approximately 5-days later, mature anthers can be observed at the base of the panicle, and in **c** after 10 days, 50% of the anthers have reached maturity. At his stage, pollen granules have started to senesce and most ovules have been fertilized (original figure, adapted from García-Canales 2014)

sugar cane (Mayavan et al. 2013), wheat (Supartana et al. 2006), maize (Chumakov et al. 2006) and rice (Supartana et al. 2005). And the fourth, after several attempts to achieve transformation by the floral-dip method were tested.

The two *A. tumefaciens* strains selected were used to infect individual dichasia of amaranth panicles at the beginning of anthesis, the period in which at least some florets are open for sexual activity. Other experiments involved two additional re-infections applied at five-day intervals after the initial one (Fig. 8.4). Two different surfactant concentrations (0.1 and 0.2%) and three bacterial cell densities (OD_{600} = 0.6, 1.0 and 1.6) were also tested parting from the "classic" conditions used for transformation in *A. thaliana* (Clough and Bent 1998). An added modification involved carefully drenching the whole panicle in the bacterial suspension. Infected plants were maintained in a greenhouse under ambient conditions to ensure that sufficient light was available to allow the full development of the panicle. Six-to eight hundred seeds per panicle were harvested at maturity and were set to germinate on herbicide-containing media. Transformation events were followed by monitoring green fluorescence emission from the green fluorescent marker protein in seeds or by detection of the bialaphos resistance (*bar*) selection gene. The frequency of bacterial infection events did not affect seed yield nor did the surfactant concentration used. However, and independently of the concentration employed, the use of a surfactant negatively affected the morphology and integrity of the panicle and flowers that acquired deformed shapes or suffered chemical-related burns, respectively (Fig. 8.5a, b).

Altered panicle morphology, increased time for seed maturation and lower seed number were also observed in response to bacterial infection of the panicle irrespective of the method employed (Fig. 8.5c–e). Theoretically, the high seed number in grain amaranth panicles should have favorably complemented the low transformation efficiency of this method and have produced a sufficiently high number of transformation events. However, no seeds were able to germinate in selective MS media, whereas a putatively transformed GFP-positive seedling was found by further analysis to be a false positive. One possible factor that may have negatively affected the floral dip method in grain amaranth was its indeterminate flower maturation period. This is characterized by a maturation process that progresses from the bottom to the top of the inflorescence during a period of several days, thereby resulting in a flowering habit that

Fig. 8.5 Negative effects produced during the "Floral-Dip" transformation in panicles of *Amaranthus cruentus*. Panels **a**, **b** show that the surfactant employed caused necrosis at the panicle primary rachis and the anthers, respectively. Surfactant concentrations of 0.1 and 0.2% also caused panicle morphological deformations, as shown by the generation of tertiary rachis in "glomerular" panicles, as shown in panel **c**. In contrast to control plants (panel **d**), the infection with high titers of *A. tumefaciens* (i.e., OD_{600} = 1.6) in 0.2% surfactant led to the emergence of deformed panicles (panel **e**), none of which yielded herbicide-resistant seeds (original figure, adapted from García-Canales 2014)

produces flowers maturing at different times in a single plant (Brenner et al. 2000).

A possible alternative method to palliate the lack of success obtained using the previous genetic transformation strategies was the use of underwater shock waves (UWSWs). This method has been widely practicable for the genetic transformation of bacteria and fungi, many of which were previously known to be recalcitrant to transformation, and of animal and human cells. This flexibility suggested that it could also work for the transformation of amaranth (Lauer et al. 1997; Jagadeesh et al. 2004; Magaña-Ortíz et al. 2013; Escobar-Tovar et al. 2015; Divya Prakash et al. 2011). UWSWs were originally devised for the controlled and non-invasive disintegration of kidney stones (Magaña-Ortiz 2010) and their use for the transformation of amaranth seeds was explored by Nava-Sandoval (2018). For this effect, a generator of UWSWs employing the principle shown in Fig. 8.6a was employed. This equipment generates UWSWs using two electrodes or foci (F) arranged underwater in such a way that the generation of an electric spark (15–30 kV) in F1 simultaneously evaporates and expands the surrounding fluid that gives rise to the UWSWs that are concentrated by an ellipsoidal reflector in F2 (Esteban-Fuertes et al. 1994; Loske 2017). Machines that use this principle generate, within microseconds, both negative and positive pressure impulses (Fig. 8.6b). The positive pressure peak can reach 150 MPa and have a duration of 0.5–3 µs. As a result, the surrounding micro-bubbles are forcefully compressed and acquire a temporary kinetic energy potential that is quickly dissipated, causing a pressure drop to ca. −20 MPa during which the water bubbles undergo more than a 100-fold expansion in size, occurring in 50–100 µs, as the kinetic energy of the UWSWs is transformed to potential energy in its path. The bubbles continue their expansion and reach a stable state that lasts between 200 and 450 µs and then collapse violently, generating secondary UWSWs and water micro-jets that can reach speeds of up to 400 m/s (Ohl 2002; Klaseboer et al. 2007). The considerable speed of these micro-jets is responsible of causing a transient increase in the permeability of the cell membranes that permits any macromolecules present in their surroundings, included naked plasmid DNA, to be absorbed (Gambihler et al. 1994; Lauer et al. 1997; Ohl and Ikink 2003; Arora et al. 2005; Zarnitsyn et al. 2008). However, the mechanisms by means of which these high-speed micro-jets change the cell permeability remain mostly unknown (Loske 2017).

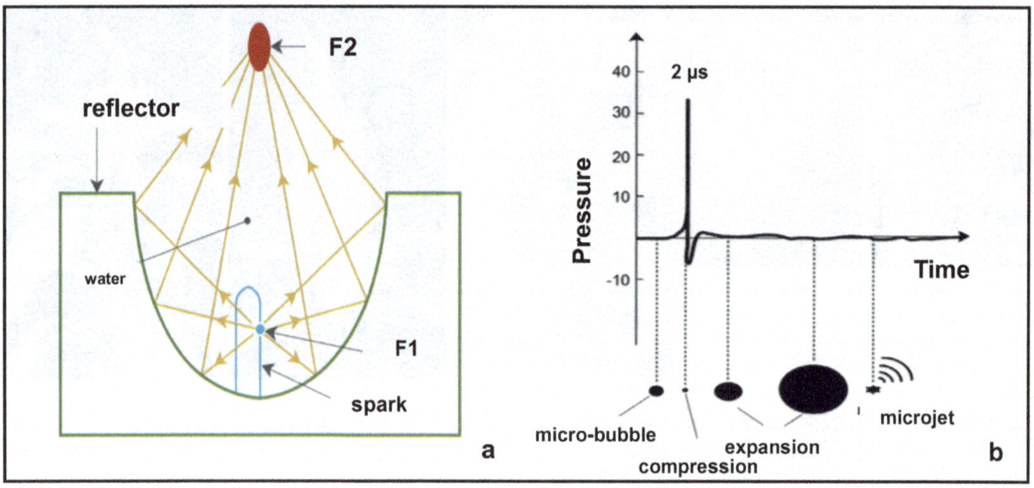

Fig. 8.6 **a** Simplified scheme of an electro-hydraulic generator of underwater shock waves. The shock wave generator is localized in foci 1 (F1) embedded within an ellipsoidal reflector that concentrates the reflected shock-waves in a second foci (F2) where the sample to be treated is placed. **b** Changes produced in the water milieu in response to the pressure profile of an underwater shock wave (original figure, modified from Loske et al. 2017)

The only recent study reporting the use of this particular technique for the transformation of plants involved the use of *Nicotiana benthamiana* seeds that were genetically transformed with the pCambia plant transformation vector after exposure to UWSWs. Remarkably, the successful transformation events were found to be inherited by the F1 generation with a 100% efficiency (D. Guzmán, unpublished data). This led to an attempt to transform amaranth seeds by sonication with two sucrose synthase genes (i.e., *AhSuS2* and *AhSUS3*), believed to be important for tolerance to water-deficit stress (González-Rodríguez et al. 2019) and severe defoliation (Castrillón-Arbeláez et al. 2012) in *A. hypochondriacus* and *A. cruentus*, respectively (Nava-Sandoval 2018). The open reading frames of these genes were inserted in the pB7WG2D expression vector downstream of the CaMV35S constitutive promoter. This plasmid is designed to confer herbicide resistance via the *bar* gene originally cloned from *Streptomyces hygroscopicus*. It also harbors the *GFP* marker gene (Karimi et al. 2002). Prior to transformation, the optimal herbicide concentrations for the selection of transformed seeds were 5 mg/ml (*Finale* Ultra), 9 mg/ml (Basta 20S) and 10 mg/mL (glufosinate, ammonium salt/phosphinothricin). The optimal quantity of UWSWs (100 and 200 at 20 kV and 0.5 Hz), DNA concentration (50 and 100 ng/µl) and seed scarification treatment (5 min in 1 N H_2SO_4) was defined after several trials. This led to the generation of 16 plants transformed with the empty vector only, at ca. 1.25% efficiency. Twenty and 11 plants transformed with the *AhSUS2* and *AhSUS*3 genes, respectively, resulted positive either by their capacity to grow in the selective media or by detection of GFP fluorescence emission (Fig. 8.7). However, quantitative gene expression analysis of the transformed *A. hypochondriacus* plants showed a significantly lower transcript abundance of both *AhSUS* genes (Fig. 8.8) compared to controls transformed with the empty vector. These plants also showed stunted vegetative and abnormal reproductive growth phenotypes (Fig. 8.9). This was in agreement with the negative effect on the overall growth that the silencing of a *SUS* gene had in carrot (Tang and Sturm 1999), which was contrary to several studies that showed that *SUS* gene overexpression in diverse plant models had a positive effect on biomass production, plant height, accelerated early development and augmented foliar surface, among others (Coleman

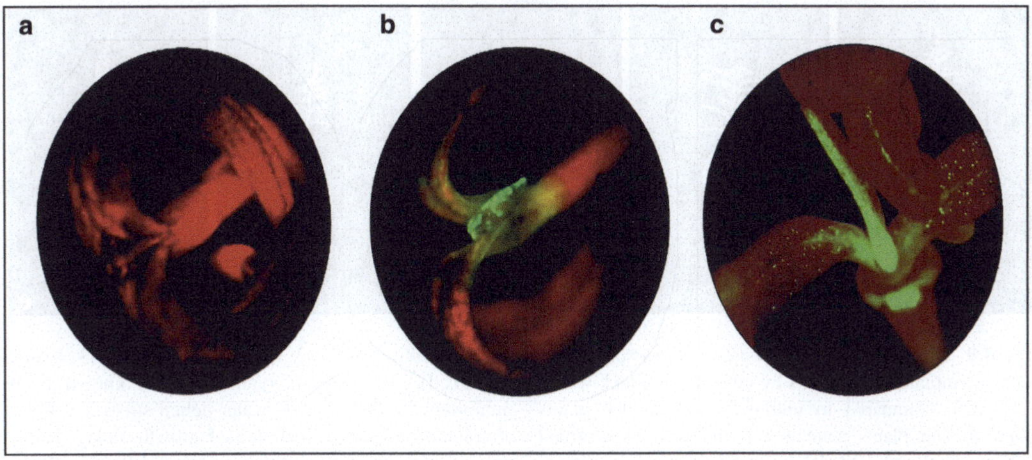

Fig. 8.7 Transformation by underwater shock waves (UWSWs). Monitoring GFP emission in seedlings germinated from *Amaranthus hypochondriacus* seeds transformed with a the pB7WG2D plant expression vector harboring the GFP marker gene. Compared to seedlings germinated from untreated control seeds (panels **a**), seedlings that germinated from USSWs-transformed seeds (panels **b** and **c**) were GFP positive. However, GFP emission was highly localized in mosaic-like sectors (original figure, adapted from Nava-Sandoval 2018)

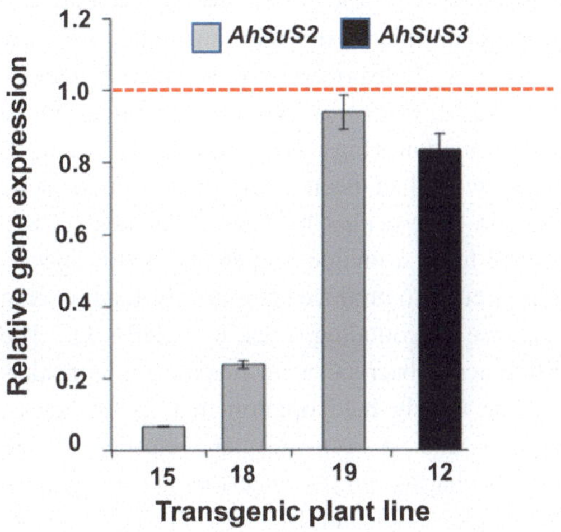

Fig. 8.8 *AhSuS2* and *AhSuS3* expression levels in *Amaranthus hypochondriacus* plants germinated from seeds by underwater shock waves. Transcript abundance of these two genes, particularly of *AhSuS2*, was lower than in control plants, which was indicative of a possible silencing of these genes (original figure, adapted from Nava-Sandoval 2018)

et al. 2006; Xu et al. 2012; Wei et al. 2015; Goren et al. 2017). In this respect, a factor that has been frequently associated with gene silencing is an excessive number of DNA copies inserted in the genome (Stam et al. 1997).

This effect could offer a plausible explanation for the transcriptional silencing of the *AhSUS2 and AhSUS3* genes recorded in these *A. hypochondriacus* plants, which probably occurred as a consequence of having several copies of foreign DNA copies inserted in their plant genomes. A similar phenomenon was observed in a percentage of *Mycosphaerella fijiensis* fungal cells that were transformed by a very similar sonication procedure (Escobar-Tovar et al. 2015). Furthermore, and contrary to the results obtained in *N. benthaminana*, the transformation event was not inherited into the next generation. All seeds generated from the selected transformed plants produced plants unable to grow on selective media and lacking the transformation marker genes. This failure could be attributed to the formation of chimeric or mosaic plants, in which only some cells of the seed embryo were transformed during the sonication events, greatly reducing the possibility of transferring the transgene into the seeds. This possibility is avoided in most sonication transformation procedures by the use of individual cells, such as prokaryotic unicellular organisms (Jagadeesh et al. 2004; Divya-Prakash et al. 2011) mammalian suspension

Fig. 8.9 Phenotypes of *Amaranthus hypochondriacus* plants germinated from seeds by underwater shock waves (UWSWs). Compared to untransformed control plants (panel **a**) and plants generated from seeds transformed with an empty pB7WG2D transformation vector (panel **e**), USSWs-transformed *A. hypochondriacus* plants overexpressing *AhSus2* (panel **b**) or *AhSuS3* (panels **c** and **d**), two sucrose synthase genes known to be important for this species' water-deficit stress and severe defoliation tolerance, had stunted growth, delayed flowering and reduced seed number (original figure, adapted from Nava-Sandoval 2018)

cell cultures or fungal conidia (Lauer et al. 1997; Magaña-Ortíz et al. 2013; Escobar-Tovar et al. 2015).

8.5 A Light at the End of the Tunnel? Combined Use of *A. rhizogenes* and *A. tumefaciens* for the Generation of Transformed *A. hypochondriacus* Plants

As mentioned above, a previous study by Swain et al. (2010), reported the successful generation of transgenic hairy roots from *A. tricolor* explants infected with *A. rhizogenes*, which were subsequently employed to produce whole transformed plants by the precise application of hormonal changes in the culture media. Also, García-Canales (2014) reported the abundant ectopic emergence of hairy roots from *A. cruentus* hypocotyl explants injected with *A. rhizogenes* bacterial cell suspensions. These results supported the possibility that the recalcitrance to genetic transformation characteristic of grain amaranths could be by-passed by first generating transgenic roots and then concentrating on triggering the morfo-physiological changes required to generate fully developed transformed plants. Thus, an experimental strategy was implemented to establish an optimized protocol for root genetic transformation of the three grain amaranth species together with *A. hybridus* (Castellanos-Arévalo et al. 2020). Consistent with the species-specific nature of recalcitrance, root transformation of grain amaranths was only possible in *A. hypochondriacus* since *A. caudatus* and *A. cruentus* remained recalcitrant to transformation (Figs. 8.10 and 8.11). Interestingly, what had been interpreted as transgenic roots in *A. cruentus* by García-Canales (2014), proved to be a unique propensity of this species in particular to produce ectopic wild-type roots in response to wounding (Figs. 8.10 and 8.11). The differences observed were unexpected considering the widely held opinion that *A. hypochondriacus* and *A. cruentus* are closely related species sharing similar ancestors and geographical origins (Wu and Blair 2017). Their subsequent diversification was probably set in motion by the need to adapt to different altitudes, which may explain the variability observed between these two grain amaranth species. Presumably, this factor could have determined whether genomic regions important for successful *Agrobacterium*-mediated transformation (see below) were present in *A. hypochondriacus* but not in *A. cruentus*. This remains to be determined experimentally However, the wide diversity of transcriptomic and metabolic responses observed between the three grain amaranths species when subjected to diverse stress conditions support the

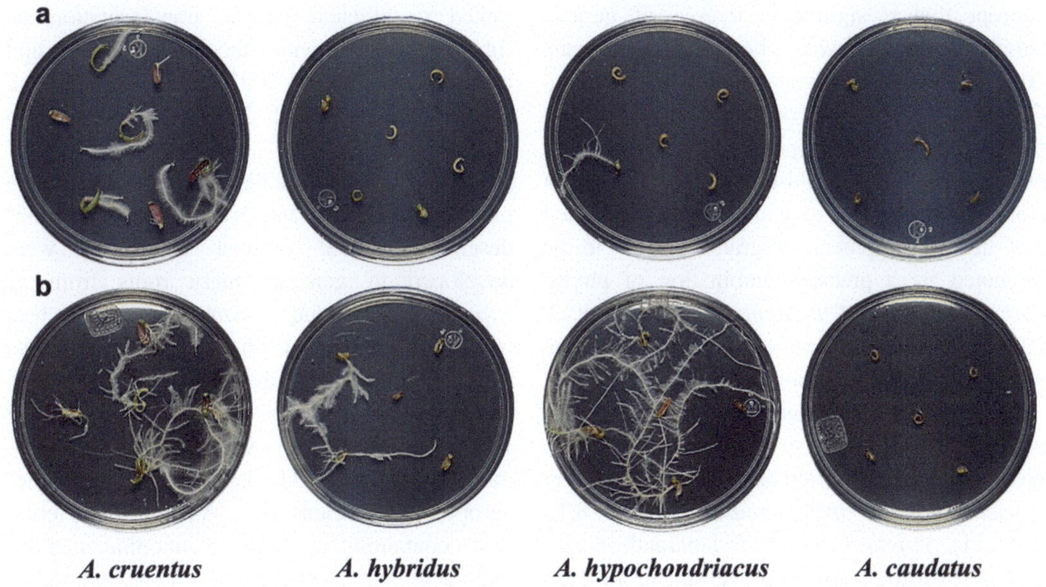

Fig. 8.10 Hairy root formation in grain amaranth cotyledon explants infected with *A. rhizogenes*. The capacity to generate hairy roots was tested in cotyledon explants of *Amaranthus cruentus, A. hybridus, A. hypochondriacus* and *A. caudatus* infected with the BVG strain of *Agrobacterium rhizogenes* and cultivated on non-selective media. Compared to uninfected controls (**a**), hairy roots were more abundant in **b** cotyledon explants of *A. cruentus* and *A. hypochondriacus,* whereas *A. caudatus* remained recalcitrant (original figure, generated as part of the study described in Castellanos-Arévalo et al. 2020)

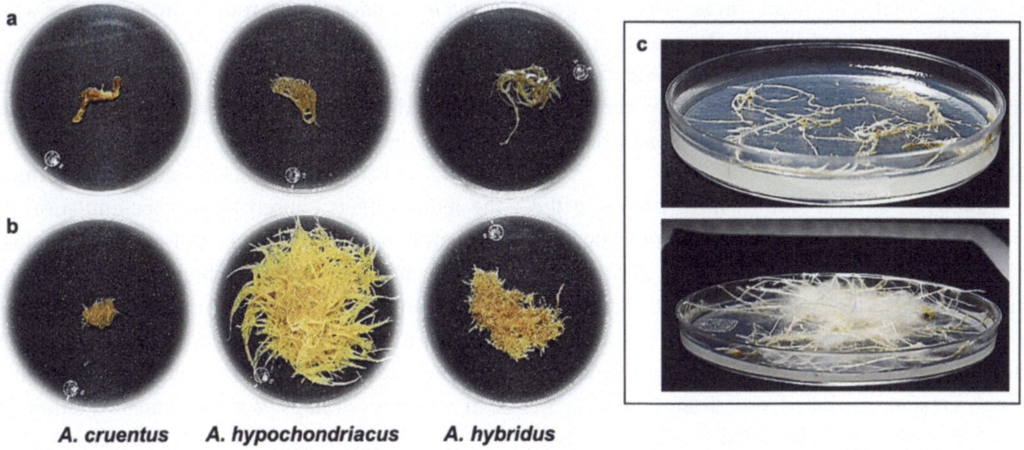

Fig. 8.11 Hairy root cultivation in selective media. Roots obtained from **a** control un-infected and **b** *Agrobacterium rhizogenes*-infected explants of *Amaranthus cruentus, A. hypochondriacus* and *A. hybridus* were cultured in selective MS media complemented with cefotaxime and Basta 20S. In **c** transgenic hairy roots of *A. hypochondriacus* show the typical hairy, prolifically branched, densely knotted and agravitropic phenotype that characterizes *A. rhizogenes*-transformed roots (original figure, generated as part of the study described in Castellanos-Arévalo et al. 2020)

concept that a significant degree of genetic variability exists between them (Cabrales-Orona 2017; González-Rodríguez et al. 2019; refer to Chap. 7 [by Cabrales-Orona and Délano Frier] in this book). In this respect, the generation of prolific adventitious wild-type roots observed in wounded *A. cruentus* explants possibly indicated that this development program, known to be promoted by a precise combination of phytohormone levels in *A. thaliana* (Verstraeten et al. 2014), is much more actively induced by wounding in this grain amaranth species.

Another important finding of this study was that *A. hybridus* was also amenable to *A. rhizogenes* transformation, although at a much lower efficiency than *A. hypochondriacus* (Figs. 8.10 and 8.11). *A. hybridus* is a leafy amaranth species considered to be a common ancestor of all three grain amaranths (Stetter and Schmid 2017). In this respect, the marked differences observed between *A. caudatus* and *A. hypochondriacus* could be explained by their closer relationship to *A. hybridus* than to each other (Kietlinski et al. 2014). Likewise, it could be speculated that the difference observed between *A. hybridus* and *A. caudatus* could have presumably arisen from the possibility that the latter arose from a geographically separated South American lineage that was different from the Central American *A. hybridus* lineage from which the two other grain amaranth species originated (Stetter and Schmid 2017).

Moreover, several studies have proposed the role of *A. quitensis* in *A. caudatus* domestication (Trucco and Tranel 2011; Kietlinski et al. 2014; Adhikary and Pratt 2015). It could, therefore, be proposed that genetic differences were a determinant factor influencing the outcome of *Agrobacterium*-mediated root transformation of grain amaranths, as advocated by studies showing: (i) the ecotype-dependent efficiency of T-DNA integration into the host genome of *A. thaliana* (Nam et al. 1997); (ii) the impossibility of using similar transformation protocols for different varieties of the same species, as shown in chrysanthemum (Sánchez-Velázquez et al. 2016) and barley (Hisano and Sato 2016), and (iii) the contribution of certain QTLs or gene loci linked to efficient genetic transformation by *Agrobacterium* of wild cabbage, potato, rice and barley (Cogan et al. 2002; El-Kharbotly et al. 1995; Nishimura et al. 2005; Hisano and Sato 2016).

As done in the establishment of the previous protocols for leafy and grain amaranth species described above, optimal conditions were developed to generate hairy roots from *A. hypochondriacus* and *A. hybridus*. Thus, the best results were obtained when cotyledon explants were infected with the BVG strain, which belongs to the of *A. rhizogenes* 15,834 agropine strains (ATCC 15,834) which harbor the agropine type pRi 15,834 plasmid. This and other agropine-type strains (i.e., A4 and HRI) elicit roots containing agropine, mannopine, mannopinic acid, and agropinic acid (White and Nester 1980). Optimal OD_{600} bacterial cell densities were also species-dependent, being 0.4 and 0.8 for *A. hypochondriacus* and *A. hybridus*, respectively. Hairy roots of both amaranth species were validated by the amplification of appropriate marker genes and, when pertinent, by monitoring green fluorescent protein emission or β-glucuronidase activity (Figs. 8.12 and 8.13). *A. hypochondriacus* hairy roots were generated with either constitutive or ethanol-inducible expression vectors (Fig. 8.12). The use of the latter was visualized as an alternative experimental tool to control the expression of transgenes at determining developmental stages and/or to reduce the lethal effects caused by the accumulation of biotechnologically useful, but toxic proteins, such as amaranth lectins, which are presumed to be effective for the control of soil-dwelling pests or the treatment of chronic human ailments (Lannoo et al. 2014). The bottleneck of this experimental strategy was the transformation of *A. hypochondriacus* hairy roots into shoots and, eventually, into fully developed plantlets.

First, embryogenic callus was generated from *A. hypochondriacus* rhizoclones using the methodology initially described by Jofre-Garfias et al. (1997). The subsequent induction of somatic embryo maturation and germination required the examination of dozens of different media. Maturation was found to be best using

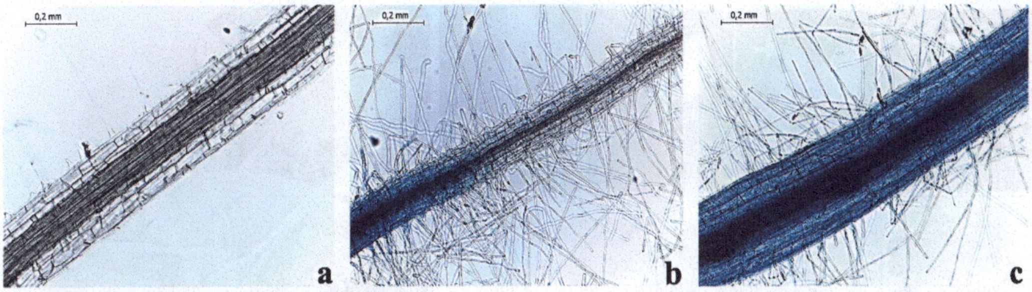

Fig. 8.12 Inducible transgenic hairy roots. β-glucuronidase (GUS) activity was monitored in *A. hypochondriacus* hairy roots transformed with the pJL67 ethanol (EtOH)-inducible expression vector **a** without EtOH and **b**, **c** 16 h after the addition of 0.5% EtOH (original figure, generated as part of the study described in Castellanos-Arévalo et al. 2020)

Fig. 8.13 Callus formation from hairy roots. **a** Abundant callus emerged shortly after placing hairy roots on MS media and remained undifferentiated in the absence of growth regulators. Compared to those kept under **b** natural light, hairy root-derived callus strongly emitted GFP fluorescence **c** when excited by UV light (original figure generated as part of the study described in Castellanos-Arévalo et al. 2020)

SE-M9 media constituted by MS medium supplemented with 0.2 mg/L 6-benzyladenine, 0.1 mg/L kinetin, 1% glucose, 9.0 g/L gelrite (−1.47 MPa), and pH 5.8 as described previously (Valencia-Lozano et al. 2019) (Fig. 8.14). In addition to the activation of cytokinin signaling and osmotic stress responses, all subsequent morphological modifying steps were strictly dependent on the red light. To activate the germination and pigmentation of mature somatic embryos, these were sub-cultured in tandem media. First, in MS medium supplemented with 5 g/L gelrite, 50 g/L of sucrose, 0.2 mg/L 2-isopentenyladenine (2iP), 0.1 mg/L kinetin, pH 5.8, to complete the axial axis previously activated by cytokinin signaling and, second, in a high salt content medium (400 mM NaCl) supplemented with 100 mg/L $CaCl_2$, and 3.0 g/L gelrite. The incubation under red filters was needed to activate phytochrome PhyB- and

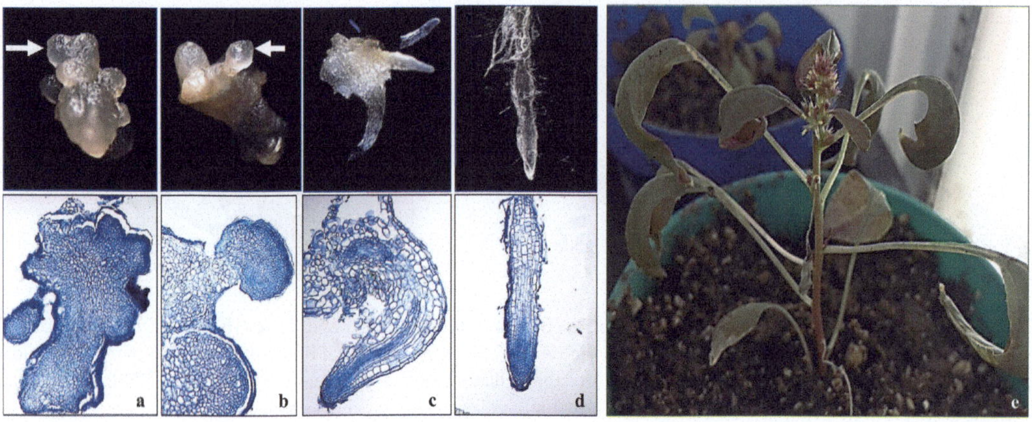

Fig. 8.14 Induction of somatic embryo maturation and germination from amaranth hairy roots. Microscopic images of embryos, in the **a** heart and **b** torpedo stages generated from *Amaranthus hypochondriacus* rhizoclones. **c** Somatic embryo (SE) structures in the torpedo stage with an established vascular system generated in response to cytokinins and osmotic stress. **d** View of SE germination, showing a well-developed root apical meristem **e** Fully regenerated plant produced ca. 8 weeks after pre-treatment of the individualized germinating SEs in media acidified with phosphoric acid (original figure generated as part of the study described in Castellanos-Arévalo et al. 2020)

PhyA-related signaling. These changes were suggested, in part, by previously reported transcriptomic data obtained from transformed *A. thaliana* plants overexpressing the *AhDof* transcription factor (Massange-Sánchez et al. 2016). The final plant regeneration step from the germinated somatic embryos required their individualization, by cutting, from their massive induction in hairy root cultures and subsequent incubation in Petri dishes containing MS media supplemented with 5 g/L gelrite, 50 g/L sucrose, 0.2 mg/L 2iP and acidified to pH 4.0 with concentrated phosphoric acid. After a 3-day culture period in this media, they were transferred to MS media supplemented with 5 g/L gelrite, 50 g/L of sucrose, 0.2 mg/ L 2iP and 100 mg/L $CaCl_2$. Again, this process was strictly red-light dependent. The morphological changes leading to well-defined plantlets having at least two expanded leaves were produced in 2-to-3 days. These later developed into fully formed plants (Fig. 8.14e). No plantlets were produced in the absence of the phosphoric acid pre-treatment nor in the presence of NaCl in the plant regeneration media.

The fact that cytokinin signaling pathway together with osmolyte-stress were key factors to induce somatic embryo (SE) maturation was in agreement with the critical role that cytokinin signaling is known to play during root and stem cell niche establishment leading to the RAM system initiation in SE (Su et al. 2015; Wang and Chong 2016, Valencia-Lozano et al. 2019). A dynamic cross-talk of this signaling pathway with abscisic acid, ethylene, light, stress and glucose signaling is also fundamental. These interactions, incorporated in the context of previous data showing that the overexpression of an amaranth *AhDof* transcription factor (TF) increases salt-stress tolerance in transgenic *A. thaliana* plants (Massange-Sánchez et al. 2016), and with the established connection of this TF with embryogenic development programs in *A. thaliana*, via calcium and red light, validated the hypothesis that these factors are critical for the induction of SE germination in amaranth. Moreover, the fundamental role played by phosphoric acid and calcium supplementation in the activation of the morphological changes required for plant regeneration suggests that these are highly dependent on calcium-related kinases and downstream MAP kinase signaling pathways that may require the activation of

CDPK1, MAPK3, MAPK2, MPK6, protein phosphatase 2C, AHK1, AHK2 and others.

In summary, Castellanos-Arévalo et al. (2020) first induced similar embryogenic-like structures as those generated by Flores et al. (1982) and Jofre-Garfias et al. (1997). They later applied ontogeny analysis and cytokinin signaling pathway to induce SE maturation and conversion to plantlets. Ontogeny analysis revealed that the globular compact structures resembling somatic embryos were, in fact, pro-embryogenic structures having the shoot apical meristem (SAM) but no root apical meristem, thereby indicating that bipolar structures were not yet formed. This phenomenon explained the lack of success in several efforts to regenerate plants from these structures. Modifications of the maturation medium were directed to induce de novo meristem by means of osmotic stress and the presence of cytokinins. Calcium content was also increased, supporting the observations of Flores et al. (1982) concerning the negative effect on calcium levels due to naturally high content of oxalates. Also red light was a critical factor for the activation of germination and plantlet conversion.

The work of Castellanos-Arévalo et al. (2020) indicated that *A. hypochondriacus* is the best option for the efficient generation of transgenic grain amaranth hairy roots. Grain amaranth hairy roots, or ideally whole transgenic plants generated from them, have the potential to be used to study poorly understood physiological processes that negatively affect the agronomic performance of this marginal crop, such as small seed size, unsynchronized plant maturation and flowering, seed shattering, relatively low seed yields, among several others. They could also be used as a "factory" for the biosynthesis of valuable secondary metabolites or proteins, including several unknown function genes known to codify for transcription factors and proteins coded by transcriptionally restricted genes that are strongly induced by (a)biotic stresses in *A. hypochondriacus* (Délano-Frier et al. 2011; see Chap. 7 in this book). Their expression in transgenic *A. hypochondriacus* hairy roots or plants, could lead to a better understanding of the role they play in grain amaranths. Other candidates include several lectins and seed proteins believed to have a potential therapeutic effect against chronic human pathologies, such as diabetes and cancer. Additionally, the use of inducible vectors offers the added advantage of triggering the expression of the above genes in a controlled manner to optimize production and/or to avoid possible toxic effects. Finally, the development of experimental conditions able to accelerate the transition to somatic embryo maturation and germination, and then to plant regeneration, opens the possibility of developing faster and highly efficient genetic transformation of all grain amaranths.

8.6 Genetic Improvement of Amaranth Through γ-Irradiation

Besides classical breeding and biotechnological approaches, the genetic improvement of other undervalued crops such as amaranth has also been attempted by irradiation-induced mutagenesis. This approach has the potential to produce small genetic changes that can significantly affect critical agronomic traits. These can subsequently be transferred to well-adapted local cultivars. An effort to improve certain agronomic traits of grain amaranths, such as seed yield, adaptation to marginal cultivation areas in high mountainous regions, reduction of post-harvest losses by seed shattering and lowering of anti-nutritional factors in seeds was attempted with limited success (Gajdosova et al. 2004). Thus, seeds of two-grain amaranth genotypes, namely *A. cruentus* Fichaand the K-433 hybrid were chosen for further experimentation due to their superior seed quality and yield. The seeds were γ-irradiated with 175 Gy and M1–M5 generations were established in a 5-year period. The main achievement of this study was a tendency towards increased seed weight, measured as the weight of a thousand seeds, in both genotypes. In addition, increased water-deficit stress (WDS) tolerance, ability to grow under artificial illumination and photoperiod insensitivity were achieved in leafy *A. tricolor* (Slabbert et al. 2004). In this case, seeds were γ-irradiated and M1 seedlings were planted in the field or in greenhouses and screened for early WDS tolerance. WDS tolerance was subsequently screened in M2–M5 plants. Putative *A. tricolor* mutants showed strong WDS avoidance (i.e., leaf shedding followed by regrowth upon watering) and tolerance during severe WDS. They recovered fast after re-watering and re-growth occurred within a few days. High levels of γ-ray dosimetry (300–600 Gy) did not significantly affect germination but negatively influenced the survival of seedlings after germination, whereas surviving seedlings presented stunted growth and chlorosis. Note was made that further research is needed to determine the physiological/ biochemical/ molecular alterations associated with the mutation-induced changes observed (Slabbert et al. 2004). Other attractive agronomic-trait targets identified, mostly for grain amaranths, were the increment of seed size and seedling vigor, reduced plant height, and synchronized flowering and maturation of plants (Brenner et al. 2000).

8.7 Conclusions

The development of efficient and reproducible methods for the genetic transformation of grain amaranth has been elusive. However, recent experimental advances suggest that the utilization of an adequate combination of cytokinins, salt-stress, calcium ions, phosphoric acid-based acidification and red light may allow the establishment of faster and more efficient genetic transformation methods of recalcitrant grain amaranths, including *A. cruentus* and *A. caudatus*. The use of this experimental tool will be key to study poorly understood physiological processes that have hindered the utilization of this still marginalized crop, such as small seed size, unsynchronized plant and flower maturation, lodging and seed shattering, among several others. It could also be successfully employed to gain a better understanding of the role played by many orphan genes discovered in grain amaranth, some of which were found to be physiologically active when expressed in *A. thaliana* and tobacco. Additionally, genetic transformation, mostly of amaranth roots, offers the possibility of using them as a "bio-factories" for the biosynthesis of attractive secondary metabolites or proteins. Suitable candidates include a number of lectins and some seed proteins, or related peptides, believed to have a potential therapeutic effect against chronic human pathologies. Inducible systems will offer the added advantage of triggering the expression of the above genes in a controlled and optimized manner.

References

Adhikary D, Pratt DB (2015) Morphologic and taxonomic analysis of the weedy and cultivated *Amaranthus hybridus* species complex. Syst Bot 40:604–610

Altpeter F, Springer NM, Bartley LE, Blechl AE, Brutnell TP, Citovsky V, Conrad LJ, Gelvin SB, Jackson DP, Kausch AP, Lemaux PG, Medford JI, Orozco-Cárdenas ML, Tricoli DM, Van Eck J, Voytas DF, Walbot V, Wang K, Zhang ZJ, Stewart CN Jr (2016) Advancing crop transformation in the era of genome editing. Plant Cell 28:1510–1520

Ali A, Bang SW, Chung SM, Staub JE (2015) Plant transformation via pollen tube-mediated gene transfer. Plant Mol. Biol Rep 33:742–747

Arora M, Ohl JL, CD, (2005) Cavitation cluster dynamics in shock-wave lithotripsy: Part 1. Free field. Ultrasound Med Biol 31:827–839

Bagga S, Venkateswari K, Sopory SK (1987) In vitro regeneration of plants from hypocotyl segments of *Amaranthus paniculatus*. Plant Cell Rep 6:183–184

Baranski R (2008) Genetic transformation of carrot (*Daucus carota*) and other Apiaceae species. Transgenic Plant J 2:18–38

Bechtold N, Pelletier G (1998) *In-planta Agrobacterium*-mediated transformation of adult *Arabidopsis thaliana* plants by vacuum infiltration. Methods Mol Biol 82:259–326

Bent AF (2000) *Arabidopsis* in planta transformation. Uses, mechanisms, and prospects for transformation of other species. Plant Physiol 124:1540–1547

Bent AF (2006) *Arabidopsis thaliana* floral dip transformation method. Methods Mol Biol 343:87–104

Bibi N, Kai F, Shuna Y, Mi N, Mosaddek IM, Waqas M, Xuede W (2013) An efficient and highly reproducible approach for the selection of upland transgenic cotton produced by pollen tube pathway method. Aust J Crop Sci 7:1714–1722

Bidney D, Scelonge C, Martich J, Burrus M, Sims L, Huffman G (1992) Microprojectile bombardment of plant tissues increases transformation frequency by *Agrobacterium tumefaciens*. Plant Mol Biol 18:301–313

Brenner DM, Baltensperger DD, Kulakow PA, Lehmann JW, Myers RL, Slabbert MM, Sleugh BB (2000) Genetic resources and breeding of *Amaranthus*. In: Janick J (ed) Plant breeding reviews, vol 19. John Wiley & Sons, pp 227–285

Cabrales-Orona G (2017) Análisis de genes de función desconocida de amaranto de grano que se expresan en múltiples condiciones de estrés. Dissertation. Centro de Investigación y de Estudios Avanzados del IPN, Unidad Irapuato, Guanajuato, México.

Castellanos-Arévalo AP, Estrada-Luna A, Cabrera-Ponce JL, Valencia-Lozano E, Herrera-Ubaldo H, de Folter S, Blanco-Labra A, Delano-Frier JP (2020) *Agrobacterium rhizogenes*-mediated transformation of grain (*Amaranthus hypochondriacus*) and leafy (*A. hybridus*) amaranths. Plant Cell Rep 39:1143–1160

Castrillón-Arbeláez PA, Martínez-Gallardo N, Avilés HA, Tiessen A, Délano-Frier JP, (2012) Metabolic and enzymatic changes associated with carbon mobilization, utilization and replenishment triggered in grain amaranth (*Amaranthus cruentus*) in response to partial defoliation by mechanical injury or insect herbivory. BMC Plant Biol 12:163

Chee PP, Fober KA, Slightom JL (1989) Transformation of soybean (*Glycine max*) by infecting germinating seeds with *Agrobacterium tumefaciens*. Plant Physiol 91:1212–1218

Chumakov MI, Rozhok NA, Velikov VA, Tyrnov VS, Volokhina IV (2006) *Agrobacterium*-mediated *in planta* transformation of maize via pistil filaments. Russ J Genet 42:893–897

Chung MH, Chen MK, Pan SE (2000) Floral spray transformation can efficiently generate *Arabidopsis* transgenic plants. Transgenic Res 9:471–476

Clough SJ, Bent AF (1998) Floral dip: a simplified method for *Agrobacterium*-mediated transformation of *Arabidopsis thaliana*. Plant J 16:735–743

Cogan NOI, Lynn JR, King GJ, Kearsey MJ, Newbury HJ, Puddephat IJ (2002) Identification of genetic factors controlling the efficiency of *Agrobacterium rhizogenes*-mediated transformation in *Brassica oleracea* by QTL analysis. Theor Appl Genet 105:568–576

Coleman HD, Ellis DD, Gilbert M, Mansfield SD (2006) Up-regulation of sucrose synthase and UDP-glucose pyrophosphorylase impacts plant growth and metabolism. Plant Biotechnol J 4:87–101

De Cleene M, De Ley J (1976) The host range of crown gall. Bot Rev 42:389–466

Délano-Frier JP, Avilés-Arnaut H, Casarrubias-Castillo K, Casique-Arroyo G, Castrillón-Arbeláeez PA, Herrera-Estrella L, Massange-Sánchez J, Martínez-Gallardo NA, Parra-Cota FI, Vargas-Ortiz E, Estrada-Hernández MG (2011) Transcriptomic analysis of grain amaranth (*Amaranthus hypochondriacus*) using 454 pyrosequencing: comparison with *A. tuberculatus*, expression profiling in stems and in response to biotic and abiotic stress. BMC Genomics 12:363

Desfeux C, Clough SJ, Bent AF (2000) Female reproductive tissues are the primary target of *Agrobacterium*-mediated transformation by the *Arabidopsis* floral-dip method. Plant Physiol 123:895–904

Divya-Prakash G, Anish RV, Jagadeesh G, Chakravortty D (2011) Bacterial transformation using micro-shock waves. Anal Biochem 419:292–301

El-Kharbotly A, Jacobsen E, Stiekema WJ, Pereira A (1995) Genetic localisation of transformation competence in diploid potato. Theor Appl Genet 91:557–562

Escobar-Tovar L, Magaña-Ortíz D, Fernández F, Guzmán-Quesada M, Sandoval-Fernández JA, Ortíz-Vázquez E, Loske AM, Gómez-Lim MA (2015) Efficient transformation of *Mycosphaerella fijiensis* by underwater shock waves. J Microbiol Methods 119:98–105

Espitia Rangel E, Escobedo López D, Aguilar Delgado M (2012) Estrategia y metodología para el mejoramiento genético del amaranto. In: Espitia Rangel E (ed) Amaranto: Ciencia y Tecnología, Libro Científico No. 2. INIFAP/SINAREFI. México, pp 227–247

Esteban-Fuertes M, Hernández-Ajubita H, Mohamed-Mohamed Z, Ramírez-Fernández J, Resel L, Samblas-García R, Senovilla-Pérez J (1994) Estado actual de la litotricia extracorpórea mediante ondas de choque. Clin Urol Complut 3:529–561

Feng G, Qin Z, Yan J, Zhang X, Hu Y (2011) Arabidopsis *ORGAN SIZE RELATED1* regulates organ growth and final organ size in orchestration with *ARGOS* and *ARL*. New Phytol 191:635–646

Flores HE, Thier A, Galston AW (1982) In vitro culture of grain and vegetable amaranths (*Amaranthus* spp.). Am J Bot 69:1049–1054

Flores HE, Teutonico RA (1982) Amaranth (*Amaranthus* spp.): potential grain and vegetable crops. In: Bajyj YPS (ed) Biotechnology in agriculture and forestry, vol 2: Crops. Springer, Berlin Heidelberg, New York, pp 568–578

Gajdosova A, Libiakova G, Huska J (2004) Improvement of selected *Amaranthus* cultivars by means of mutation techniques and biotechnological approaches. Genetic improvement of under-utilized and neglected crops in low income food deficit countries through irradiation and related techniques: plant breeding and genetics section. International Atomic Energy Agency, Vienna, Austria, pp 25–36

Gambihler S, Delius M, Ellwart JW (1994) Permeabilization of the plasma membrane of L1210 mouse leukemia cells using lithotripter shock waves. J. Membrane Biol 141:267–275

Gelvin SB (2012) Traversing the Cell: *Agrobacterium* T-DNA's journey to the host genome. Front Plant Sci 3:52

Gohlke J, Deeken R (2014) Plant responses to *Agrobacterium* tumefaciens and crown gall development. Front Plant Sci 5:155

González-Rodríguez T, Cisneros-Hernández I, Acosta Bayona J, Ramírez-Chavez E, Martínez-Gallardo N,

Mellado-Mojica E, López-Pérez MG, Molina-Torres J, Délano-Frier J (2019) Identification of factors linked to higher water-deficit stress tolerance in *Amaranthus hypochondriacus* compared to other grain amaranths and *A. hybridus*, their shared ancestor. Plants 8:239

Gould J, Banister S, Hasegawa O, Fahima M, Smith RH (1991) Regeneration of *Gossypium hirsutum* and *G. barbadense* from shoot apex tissues for transformation. Plant Cell Rep 10:12–16

García Canales Y (2014) Transformación genética de *Amaranthus* spp. mediada por *Agrobacterium* spp. Dissertation. Centro de Investigación y de Estudios Avanzados del IPN, Unidad Irapuato, Guanajuato, México

Goren S, Lugassi N, Stein O, Yeselson Y, Schaffer AA, Granot D-S, D, (2017) Suppression of sucrose synthase affects auxin signaling and leaf morphology in tomato. PLoS ONE 12:e0182334

Griesbach RJ, Hammond J (1994) An improved method for transforming plants through electrophoresis. Plant Sci 102:81–89

Hadi MZ, McMullen MD, Finer JJ (1996) Transformation of 12 different plasmids into soybean via particle bombardment. Plant Cell Rep 15:500–505

Hiei Y, Ishida Y, Komari T (2014) Progress of cereal transformation technology mediated by *Agrobacterium tumefaciens*. Front Plant Sci 5:628

Hisano H, Sato K (2016) Genomic regions responsible for amenability to *Agrobacterium*-mediated transformation in barley. Sci Rep 6:37505

Hooykaas PJJ, Beijersbergen AGM (1994) The virulence system of *Agrobacterium tumefaciens*. Ann Rev Phytopathol 32:157–181

Hu Y, Poh H, Chua N (2006) The *Arabidopsis ARGOS-LIKE* gene regulates cell expansion during organ growth. Plant J 47:1–9

Hussain Shah S, Ali S, Jan SA, Din JU, Ali GM (2015) Piercing and incubation method of *in planta* transformation producing stable transgenic plants by overexpressing *DREB1A* gene in tomato (*Solanum lycopersicum* Mill.). Plant Cell Tiss Org 120:1139–1157

Hwang HH, Yu M, Lai EM (2017) *Agrobacterium*-mediated plant transformation: biology and applications. Arabidopsis Book. 15(e0186):31

Joersbo M, Brunstedt J (1990) Direct gene transfer to plant protoplasts by mild sonication. Plant Cell Rep 9:207–210

Joersbo M, Brunstedt J (1992) Sonication: A new method for gene transfer to plants. Physiol Plant 85: 230–234. Jagadeesh G, Nataraja KN, Udayakumar M (2004) Shock waves can enhance bacterial transformation with plasmid DNA. Curr Sci India 87:734–735

Jofre-Garfias AE, Villegas-Sepúlveda N, Cabrera-Ponce JL, Adame-Alvarez RM, Herrera-Estrella L, Simpson J (1997) *Agrobacterium*-mediated transformation of *Amaranthus hypochondriacus*: light- and tissue-specific expression of a pea chlorophyll a/b-binding protein promoter. Plant Cell Rep 16:847–852

Jofre-Garfias A.E., Cabrera-Ponce J.L., Adame-Alvarez R.M., Herrera-Estrella L., Simpson J. (2001) Transgenic amaranth (*Amaranthus hypochondriacus*). In: Bajaj YPS (ed) Transgenic crops II. biotechnology in agriculture and forestry, vol 47. Springer, Berlin, Heidelberg, pp 13–25

Kaeppler SM, Kaeppler HF, Rhee Y (2000) Epigenetic aspects of somaclonal variation in plants. In: Matzke MA, Matzke AJM (eds) Plant gene silencing. Springer, Netherlands, pp 59–68

Kalbande BB, Patil AS (2016) Plant tissue culture independent *Agrobacterium tumefaciens* mediated *in-planta* transformation strategy for upland cotton (*Gossypium hirsutum*). J Genet Eng Biotechnol 14:9–18

Karimi M, Inzé D, Depicker A (2002) GATEWAY vectors for *Agrobacterium*-mediated plant transformation. Trends Plant Sci 7:193–195

Kietlinski KD, Jimenez F, Jellen EN, Maughan PJ, Smith SM, Pratt DB (2014) Relationships between the weedy *Amaranthus hybridus* (Amaranthaceae) and the grain amaranths. Crop Sci 54:220

Klaseboer E, Fong SW, Turangan CK, Khoo BC, Szeri AJ, Sankin CML, MN, Zhong P, (2007) Interaction of lithotripter shockwaves with single inertial cavitation bubbles. J Fluid Mech 593:37–56

Kuluev BR, Mikhaylova EV, Taipova RM, Chemeris AV (2017) Changes in phenotype of transgenic amaranth *Amaranthus retroflexus* L., overexpressing *ARGOS-LIKE* gene. Russ J Genet 53:67–75

Kumar V, Jain M (2015) The CRISPR-Cas system for plant genome editing: advances and opportunities. J Exp Bot 66:47–57

Lacroix B, Citovsky V (2019) Pathways of DNA transfer to plants from *Agrobacterium tu*mefaciens and related bacterial species. Annu Rev Phytopathol 57:231–251

Lannoo N, Van Damme EJM, Albenne C, Jamet E (2014) Plant glycobiology-a diverse world of lectins, glycoproteins, glycolipids and glycans. Front Plant Sci 5:604

Lauer U, Bürgelt E, Squire Z, Messmer K, Hofschneider PH, Gregor M, Delius M (1997) Shock wave permeabilization as a new gene transfer method. Gene Ther 4:710–715

Lin YH, Gao R, Binns AN, Lynn DG (2008) Capturing the VirA/VirG TCS of *Agrobacterium tumefaciens*. Adv Exp Med Biol 631:161–177

Lin J, Zhou B, Yang Y, Mei J, Zhao X, Guo X, Huang X, Tang D, Liu X (2009) Piercing and vacuum infiltration of the mature embryo: a simplified method for *Agrobacterium*-mediated transformation of *indica* rice. Plant Cell Rep 28:1065–1074

Liu J, Su Q, An L, Yang A (2009) Transfer of a minimal linear marker-free and vector-free smGFP cassette into soybean via ovary-drip transformation. Biotechnol Lett 31:295–303

Loske AM (2017) Medical and biomedical applications of shock waves. Shock wave high pressure phenomena, Springer International Publishing, México

Luo ZX, Wu RA (1988) Simple method for the transformation of rice via the pollen-tube pathway. Plant Mol Biol Rep 6:165–174

Magaña-Ortíz D (2010) Desarrollo de un nuevo método de transformación genética en el reino fungi.

Dissertation. Centro de Investigación y de Estudios Avanzados del IPN, Unidad Irapuato, Guanajuato, México

Magaña-Ortíz D, Coconi-Linares N, Ortiz-Vazquez E, Fernández F, Loske AM, Gómez-Lim MA (2013) A novel and highly efficient method for genetic transformation of fungi employing shock waves. Fungal Genet Biol 56:9–16

Maher MF, Nasti RA, Vollbrecht M, Starker CG, Clark MD, Voytas DF (2020) Plant gene editing through de novo induction of meristems. Nat Biotechnol 38:84–89

Maresh J, Zhang J, Lynn DG (2006) The innate immunity of maize and the dynamic chemical strategies regulating two-component signal transduction in *Agrobacterium tumefaciens*. ACS Chem Biol 3:165–175

Martinez-Trujillo M, Limones-Briones V, Cabrera-Ponce JL, Herrera-Estrella L (2004) Improving transformation efficiency of *Arabidopsis thaliana* by modifying the floral dip method. Plant Mol Biol Rep 22:63–70

Massange-Sánchez JA, Palmeros-Suárez PA, Espitia-Rangel E, Rodríguez-Arévalo I, Sánchez-Segura L, Martínez-Gallardo NA, Alatorre-Cobos F, Tiessen A, Délano-Frier JP (2016) Overexpression of grain amaranth (*Amaranthus hypochondriacus*) AhERF or AhDOF transcription factors in *Arabidopsis thaliana* increases water deficit- and salt-stress tolerance, respectively, via contrasting stress-amelioration mechanisms. PLoS ONE 11:e0164280

Mayavan S, Subramanyam K, Arun M, Rajesh M, Kapil Dev G, Sivanandhan G, Jaganath B, Manickavasagam M, Selvaraj N, Ganapathi A (2013) *Agrobacterium tumefaciens*-mediated *in planta* seed transformation strategy in sugarcane. Plant Cell Rep 32:1557–1574

Mayavan S, Subramanyam K, Jaganath B, Sathish D, Manickavasagam M, Ganapathi A (2015) *Agrobacterium*-mediated *in planta* genetic transformation of sugarcane setts. Plant Cell Rep 34:1835–1848

Meyers B, Zaltsman A, Lacroix B, Kozlovsky SV, Krichevsky A (2010) Nuclear and plastid genetic engineering of plants: Comparison of opportunities and challenges. Biotechnol Adv 28:747–756

Munusamy U, Abdullah SNA, Aziz MA, Khazaai H (2013) Female reproductive system of *Amaranthus* as the target for *Agrobacterium*-mediated transformation. Adv Biosci Biotechnol 4:188–192

Mysore KS, Kumar CT, Gelvin SB (2000) *Arabidopsis* ecotypes and mutants that are recalcitrant to *Agrobacterium* root transformation are susceptible to germ-line transformation. Plant J 21:9–16

Nam J, Matthysse AG, Gelvin SB (1997) Differences in susceptibility of Arabidopsis ecotypes to crown gall disease may result from a deficiency in T-DNA integration. Plant Cell 9:317–333

Nava-Sandoval C (2018) Transformación genética de amaranto por ondas de choque. Dissertation. Centro de Investigación y de Estudios Avanzados del IPN, Unidad Irapuato, Guanajuato, México.

Niazian M, Sadat Noori SA, Galuszka P, Mortazavian SMM (2017) Tissue culture-based *Agrobacterium*-mediated and *in planta* transformation methods. Czech J Genet Plant Breed 53:133–143

Nishimura A, Ashikari M, Lin S, Takashi T, Angeles ER, Yamamoto T, Matsuoka M (2005) Isolation of a rice regeneration quantitative trait loci gene and its application to transformation systems. Proc Natl Acad Sci USA 102:11940–11944

Ohl CD (2002) Cavitation inception following shock wave passage. Phys Fluids 14:3512–3521

Ohl CD, Ikink R (2003) Shock-wave-induced jetting of micron-size bubble. Phys Rev Lett 90:214502–214505

Pal A, Swain SS, Das AB, Mukherjee AK, Chand PK (2013a) Stable germ line transformation a leafy vegetable crop amaranth (*Amaranthus tricolor* L.) mediated by *Agrobacterium tumefaciens*. In Vitro Cell Dev Biol-Plant 49:114–128

Pal A, Swain SS, Mukherjee AK, Chand PK (2013b) *Agrobacterium* pRi TL-DNA *rol*B and TR-DNA opine genes transferred to the spiny amaranth (*Amaranthus spinosus* L.), a nutraceutical crop. Food Technol Biotechnol. 51:26–35

Phillips RL, Kaeppler SM, Olhoft P (1994) Genetic instability of plant tissue cultures: breakdown of normal controls. Proc Natl Acad Sci USA 91:5222–5226

Rao AQ, Bakhsh A, Kiani S, Shahzad K, Shahid AA, Husnain T, Riazuddin S (2009) The myth of plant transformation. Biotechnol Adv 27:753–763

Ream W (2009) *Agrobacterium tumefaciens* and *A. rhizogenes* use different proteins to transport bacterial DNA into the plant cell nucleus. Microb Biotechnol 2:416–427

Richardson K, Fowler S, Pullen C, Skelton C, Morris B, Putterill J (1998) T-DNA tag-ging of a flowering-time gene and improved gene transfer by *in planta* transformation of *Arabidopsis*. Funct Plant Biol 25:125–130

Rivera AL, Gómez-Lim M, Fernández F, Loske AM (2012) Physical methods for genetic plant transformation. Phys Life Rev 9:308–345

Rohini VK, Rao KS (2000) Transformation of peanut (*Arachis hypogaea* L.): a non-tissue culture based approach for generating transgenic plants. Plant Sci 150:41–49

Sánchez-Velázquez JU, López Puc G, Ramos-Díaz AL, Cano-Sosa JS, Rodríguez Buenfil IM, Rómulo García-Velasco R, Uc Varguez A (2016) Main factors affecting the genetic transformation of chrysanthemum var. Micromargara. Plant Omics J 9:121–125

Shrawat AK, Lörz H (2006) Agrobacterium-mediated transformation of cereals: a promising approach crossing barriers. Plant Biotechnol J 4:575–603

Slabbert MM, de Ronde K, Caetano T, Spreeth M, van den Heever E (2004) Development and evaluation of mutant germplasm of *Amaranthus*. Genetic improvement of under-utilized and neglected crops in low income food deficit countries through irradiation and related techniques: plant breeding and genetics section. International Atomic Energy Agency, Vienna, Austria, pp 13–23

Sobhanian N, Habashy AA, Farshad Far E, Tohidfar M (2012) Optimizing regeneration and reporter gene (*gus*) transformation of alfalfa (*Medicago sativa*). Ann Biol Res 3:2419–2427

Stam M, Mol JNM, Kooter JM (1997) The silence of genes in transgenic plants. Ann Bot 79:3–12

Stetter MG, Schmid KJ (2017) Analysis of phylogenetic relationships and genome size evolution of the *Amaranthus* genus using GBS indicates the ancestors of an ancient crop. Mol Phylogenet Evol 109:80–92

Su YH, Liu YB, Bai B, Zhang XS (2015) Establishment of embryonic shoot-root axis is involved in auxin and cytokinin response during Arabidopsis somatic embryogenesis. Front Plant Sci 5:792

Subramoni S, Nathoo N, Klimov E, Yuan ZC (2014) *Agrobacterium tumefaciens* responses to plant-derived signaling molecules. Front Plant Sci 5:322

Supartana P, Shimizu T, Shioiri H, Nogawa M, Nozue M, Kojima M (2005) Development of simple and efficient in planta transformation method for rice (*Oryza zativa* L.) using *Agrobacterium tumefaciens*. J Biosci Bioeng 100:391–397

Supartana P, Shimizu T, Nogawa M, Shioiri H, Nakajima T, Haramoto N, Nozue M, Kojima M (2006) Development of simple and efficient *in planta* transformation method for wheat (*Triticum aestivum* L.) using *Agrobacterium tumefaciens*. J Biosci Bioeng 102:162–170

Swain SS, Sahu L, Banik DP, Chand PK (2010) *Agrobacterium* × plant factors influencing transformation of Joseph's coat (*Amaranthus tricolor* L.). Sci Hortic 125:461–468

Tague B, Mantis J (2006) *In planta Agrobacterium*-mediated transformation by vacuum infiltration. In: Salinas J, Sanchez-Serrano JJ (eds) Arabidopsis protocols. Humana Press, New York, pp 215–223

Tang GQ, Sturm A (1999) Antisense repression of sucrose synthase in carrot (*Daucus carota* L.) affects growth rather than sucrose partitioning. Plant Mol Biol 4:465–479

Trick HN, Finer JJ (1997) SAAT: sonication-assisted *Agrobacterium*-mediated transformation. Transgenic Res 6:329–336

Trieu AT, Burleigh SH, Kardailsky IV, Maldonado-Mendoza IE, Versaw WK, Blaylock LA, Shin H, Chiou TJ, Katagi H, Dewbre GR, Weigel D, Harrison MJ (2000) Transformation of *Medicago truncatula* via infiltration of seedlings or flowering plants with *Agrobacterium*. Plant J 22:531–541

Trucco F, Tranel PJ (2011) Amaranthus. In: Kole C (ed) Wild crop relatives: genomic and breeding resources. Springer, Berlin, Heidelberg, pp 11–21

Trujillo M, Briones V, Ponce JL, Estrella L (2004) Improving transformation efficiency of *Arabidopsis thaliana* by modifying the floral dip method. Plant Mol Biol Rep 22:63–70

Tzfira T, Citovsky V (2002) Partners-in-infection: host proteins involved in the transformation of plant cells by *Agrobacterium*. Trends Cell Biol 12:121–129

Tzfira T, Li J, Lacroix B, Citovsky V (2004) *Agrobacterium* T-DNA integration: molecules and models. Trends Genet 20:375–383

Valencia-Lozano E, Cabrera-Ponce JL, Gómez-Lim MA, Ibarra JE (2019) Development of an efficient protocol to obtain transgenic coffee, *Coffea arabica* L., expressing the Cry10Aa toxin of *Bacillus thuringiensis*. Int J Mol Sci 20:5334

Verstraeten I, Schotte S, Geelen D (2014) Hypocotyl adventitious root organogenesis differs from lateral root development. Front Plant Sci 5:495

Wang M, Zhang B, Wang Q (2013) Cotton transformation via pollen tube pathway. In: Zhang B (ed) Transgenic cotton: methods and protocols. Humana Press, New York, Methods in Molecular Biology, pp 71–77

Wang L, Chong K (2016) The essential role of cytokinin signaling in root apical meristem formation during somatic embryogenesis. Front Plant Sci 6:1196

Wei Z, Qu Z, Zhang L, Zhao S, Bi Z, Ji X, Wang X, Wei H (2015) Overexpression of poplar xylem sucrose synthase in tobacco leads to a thickened cell wall and increased height. PLoS ONE 10:e0120669

White FF, Nester EW (1980) Hairy root: plasmid encodes virulence traits in *Agrobacterium rhizogenes*. J Bacteriol 141:1134–1141

Wu X, Blair MW (2017) Diversity in grain amaranths and relatives distinguished by genotyping by sequencing (GBS). Front Plant Sci 8:1960

Xu SM, Brill E, Llewellyn DJ, Furbank RT, Ruan YL (2012) Overexpression of a potato sucrose synthase gene in cotton accelerates leaf expansion, reduces seed abortion, and enhances fiber production. Mol Plant 5:430–441

Xu K, Huang X, Wu M, Wang Y, Chang Y, Liu K, Zhang J, Zhang Y, Zhang F, Yi L, Li T, Wang R, Tan G, Li C (2014) A rapid, highly efficient and economical method of *Agrobacterium*-mediated *in planta* transient transformation in living onion epidermis. PLoS ONE 9:e83556

Yaroshko O, Vasylenko M, Gajdošová A, Morgun B, Khrystan O, Velykozhon L, Kuchuk M (2018) "Floral-dip" transformation of *Amaranthus caudatus* L. and hybrids *A. caudatus* ×*A. paniculatus* L. Biologija 64:321–330

Yaroshko OM, Kuchuk MV (2019) *Agrobacterium*-caused transformation of cultivars *Amaranthus caudatus* L. and hybrids of *A. caudatus* L. × *A. paniculatus* L. Int J Sec Metabolite 5:312–318

Yaroshko OM, Morgun BV, Velykozhon LG, Gajdośova A, Andrushenko OL, Kuchuk MV (2020) PCR analyses of first generation plants of Amaranthus caudatus L. after "floral-dip" genetic transformation. Fiziol Rast Genet 52:128–139

Yasmeen A, Mirza B, Inayatullah S, Safdar N, Jamil M, Ali S, Choudhry MF (2009) *In planta* transformation of tomato. Plant Mol Biol Rep 27:20–28

Ye GN, Stone D, Pang SZ, Creely W, Gonzalez K, Hinchee M (1999) *Arabidopsis* ovule is the target for *Agrobacterium* in planta vacuum infiltration transformation. Plant J 19:249–257

Zarnitsyn V, Rostad CA, Prausnitz MR (2008) Modeling transmembrane transport through cell membrane wounds created by acoustic cavitation. Biophys J 95:4124–4138

Zhang L, Cheng L, Xu N, Zhao N, Li C, Yuan J, Jia S (1991) Efficient transformation of tobacco by ultrasonication. Nat Biotechnol 9:996–997

Zhang J, Boone L, Kocz R, Zhang C, Binns AN, Lynn DG (2000) At the maize/*Agrobacterium* interface: natural factors limiting host transformation. Chem Biol 7:611–621

Zhou GY, Weng J, Zeng Y, Huang J, Qian S, Liu G (1983) Introduction of exogenous DNA into cotton embryos. Method Enzymol 101:433–481

Zupan J, Muth TR, Draper O, Zambryski PC (2000) The transfer of DNA from *Agrobacterium tumefaciens* into plants: a feast of fundamental insights. Plant J 23:11–28

Polyploidy and Hybridization for Amaranth Crop Improvement

9

Rita Andini, Muhammad Ikhsan Sulaiman, Muhammad Kadapi, and Ryo Ohsawa

Abstract

Amaranthus spp. or known as amaranth is a genus belonging to Amaranthaceae that includes 60–75 species. It is distinguished as the grain, vegetable, and weed or weedy types. Within the genus, a high genetic variation exists and this is perceived as a result of outcrossing and plant hybridization that promotes gene exchange and introgression between wild relatives. Outcrossing rate in amaranth is ranging from 5 to 30% and this unique characteristic would have enabled the hybridization in amaranth, distinctively among the weedy ones. This chapter also highlights the phylogenetic relationships among the three major grain amaranths and the three postulates referred to as "*A. hybridus* species complex". Other arguments had also highlighted some fertility barriers. Polyploidy accounted as an important evolutionary process in many important crop species that provide food, fodder, and fiber has also occurred in amaranth. Natural polyploids are generated in response to changes in the environment. Ploidy variations usually cause alterations in morphology as this is performed in *A. dubius* attributed with larger plants with many side branches, shorter, and thicker stems. It is believed as the only allotetraploid form with $2n = 4x = 32$, although current analysis based on flow cytometer did not confirm the result.

9.1 Introduction

Crop wild relatives including under-utilized crops, such as grain- and leaf-producing *Amaranthus* species, are receiving greater attention from researchers other than conventional modern crop nowadays. This is mostly because many of them are robust against abiotic- and biotic stresses, can be grown with minimal external inputs, and are adapted to specific marginal soil and climatic conditions.

Crops are regularly exposed to altered growing conditions, and it is necessary to create and develop novel genotypes with specific agronomic traits, *e.g.*, increasing adaptability against abiotic and biotic stresses. In fact, many weedy related

R. Andini (✉) · M. I. Sulaiman
Faculty of Agriculture, Universitas Syiah Kuala (USK), Kecamatan Syiah Kuala, Tgk. Hasan Krueng Kalee No. 3, Darussalam, Banda Aceh, Provinsi Aceh 23111, Indonesia
e-mail: rita.andini@unsyiah.ac.id

M. Kadapi
Faculty of Agriculture, Agrotechnology Department, Universitas Padjajaran (UNPAD), Jl. Raya Bandung-Sumedang KM 21Kecamatan JatinangorJawa Barat, 45363 Hegarmanah, Kabutapen Sumedang, Indonesia

R. Ohsawa
Institute of Life and Environmental Sciences, University of Tsukuba, 1-1-1 Tennodai, Tsukuba, Ibaraki Ken, Japan

D. Adhikary et al. (eds.), *The Amaranth Genome*, Compendium of Plant Genomes,
https://doi.org/10.1007/978-3-030-72365-1_9

species with exotic genes have been incorporated into formal crop improvement programs for more than a century; especially for increasing, nutritional values, enhancing pest and disease resistance and improving tolerance of stressful abiotic conditions (drought, high-salinity, and high temperature) (Jarvis et al. 2008). Therefore, they deserve to be more deeply researched as potential candidates for reducing the adverse effects of climate change for humans, which are predominantly affecting the tropical regions.

On the other hand, plant hybridization that promotes gene exchange has been commonly used in many important feed and food crops such as several Brassicaceae species and *Amaranthus* spp., or amaranth. The latter, once the staple crop of the ancient pre-Columbian civilizations in the Central- and South America, was strictly banned for many centuries in their center of origin due to their close relatedness with the indigenous and cultural ceremonies. Since the seventeenth century, it has played a great role on three pertinent continents (Africa, America, and Asia) and can be found in different morphological types (or *morphotypes*). The crops are of the grain, vegetable, or weedy types with a minor use as ornamentals in temperate regions. Due to its extraordinary nutritional content, easy cultivation, mostly managed by the subsistence farmers in developing countries, high adaptability against biotic or abiotic stresses, amaranth is deserved to be promoted as the future crop against the climate change challenges. A high genetic variability has been reported, however, *A. hybridus* is commonly perceived as the putative progenitor species among grain amaranths. Moreover, the versatile genetic variability, especially in regard to the *A. hybridus* complex, has led to the misapplication of nomenclature, thereby creating taxonomic confusion, species delimitation, and taxonomic disputes such as by the case of *A. quitensis* H.B.K, which is actually the synonym of *A. hybridus* (Das 2012).

Amaranth is perceived to possess the capacity for interspecific hybridization between the weedy and cultivated types, the first one acting as the "reservoir" of unique genes. Thus, gene exchange may occur in Nature, while the utilization of this natural phenomenon has increased the crop's ability to challenge pests, disease, and/ or adverse environmental effects, including those increasingly associated with global climate change.

9.2 Potential of Amaranth as a Future Crop for the Management of Climate's Change Impact

9.2.1 Climate Change

A strong reliance on a handful of major crops bears inherent agronomic, ecological, nutritional, and economic risks and is deliberately unsustainable in the long run, especially in view of global climate change, where a significant average temperature raise between 3.7 and 7.8 °C by 2100 has been predicted. Many scientists have perceived that climate change will increase air-temperature, which will have a major impact on both biotic and abiotic stresses in agricultural production systems. In addition to that (Ebert 2014) predicted a worst scenario: climate change will threaten yield and crop sustainability at the global level. On the other hand, the world Food and Agriculture Organization (UN-FAO) has set a somehow ambitious target, namely "zero-hunger" by 2030 that involves securing future global food needs of more than 800 million tones. This objective remains despite facing five big challenges such as (i) protracted conflict; (ii) growing inequality; (iii) rapid urbanization; (iv) global economic challenges; and (v) drought and natural disasters.

The *Intergovernmental Panel on Climate Change* (IPCC) had predicted that the mean surface temperature on earth will be rising from 2.4 to 6.4 °C by 2100. As a consequence, some regions located at the equator may become unsuitable for cropping and some tropical grasslands could become increasingly arid. Various other impacts have been projected; however, agriculture certainly will be adversely affected by yield loss of two major staple crops (i.e., rice and maize) as well as altered post-harvest quality of horticultural products. An obvious conclusion is that the global temperature rise will deliberately challenge the status *quo* of food-and nutrition-security meaning, in broader terms, that the vicious circle involving poverty-hunger, infectious disease, and even mortality will be strengthened (Schmidhuber and Tubiello 2007).

The current existing plant genetic resources particularly of under-utilized crops such as amaranth (*Amaranthus* spp.) cassava leaves (*Manihot esculenta*) and sweet potato leaves (*Ipomea batatas*) bear great potential to be exploited and targeted as potential parental lines containing specific genes against various biotic and abiotic stresses (Rubaihayo 2002). As such, amaranth emerges as a suitable candidate to be the "future crop" able to face the direct consequences of climate change. Currently, this is the only crop suitable for cultivation in the hot summer months in India, known as "*kharif*", a period in which temperatures range from 20 to 45 °C (Shukla et al. 2010). Thus, breeding novel cultivars with improved adaptations to abiotic challenges via conventional breeding methods, e.g., backcrosses, outcrossing, or inbreeding processes or advance molecular breeding or gene-introgression (Brenner et al. 2000) is highly recommended for reducing the impacts of climate change on global food security (Schmidhuber and Tubiello 2007).

9.2.2 Potential of Amaranth

Amaranthus spp., widely known as amaranth and meaning "ever-lasting" in *Greek*, is a genus belonging to the Amaranthaceae that includes 60–75 species. Historically, it was originated in Central and South America about 5,000–6,000 years ago and evolved on the basis of morphology and geography by leaning on two postulates proposed by Sauer (1967 and 1976) in (Brenner et al. 2000) and most recently by (Mallory et al. 2008). These will be further explained in Sect. 9.3.

Depending on geographical preferences, its utilization may be emphasized for grain or vegetable cultivation, where the sub-continental and temperate regions tend to utilize a group of amaranths (*A. caudatus*, *A. cruentus*, and *A. hypochondriacus*) as "*pseudocereals*" for grain production. *A. cruentus* and *A. hypochondriacus* are mostly cultivated in Mexico, while *A. caudatus* is preferentially grown in South America. In its center of origin, grain amaranths were once served as one of the staple crops of the ancient civilization of Incas, Aztecs, and Mayas, after corn and beans, before the discovery of America in the fifteenth century. For many centuries after, amaranth cultivation was strongly prohibited by the Spanish conquistadors as it was strongly embedded in the native *ceremonies* of the ancient civilizations.

After its disappearance, it was re-discovered in the late 1970s due to its extraordinary protein (13–18%) and lysine content, which exceeds two and three times that of wheat and maize, respectively. Meanwhile, the sulfur amino acid concentration is in the range of 2.6 to 5.5%, which is higher than that of the most important legumes (e.g., peas, beans, and soybeans) that are attributed with protein content in the range of 1.4% (Gorinstein et al. 1998). The grain can be further processed into flour, bran, and oils as cereals or further processed as traditional sweets called "*allegría*" (Zheleznov, Solonenko and Zheleznova 1997). Most of grain amaranths require temperatures ranging between 16 and 35 °C, for seed germination, and between 21 and 27 °C, for optimal growth. Moreover, an annual precipitation of 300 to 2,000 mm is also needed. They can grow well in various altitudes, from sea level to 3,000 m above sea level (*a.s.l.*) (Barba de la Rosa et al. 2009). Grain amaranths are being proposed as a potential crop for the twenty-first century that could provide a well-balanced nutrition. Research, nowadays, is mostly intended to find ways to use them as functional foods considering the anti-cancer, anti-inflammatory, anti-obesity, and hypocholesterolemic effect found in diverse grain protein fractions (Barba de la Rosa et al. 2009). Moreover, grain amaranth is currently a *niche* market crop intended for those in need of a gluten-free diet for celiac disease that is sourced from health food stores in northern part of America or Europe (Gélinas and Seguin 2008).

Amaranth is remarkably known for its photosynthetic efficiency as a C_4 plant, wide adaptability to different eco-geographical regions ranging from 23 1/2° North (N) to South (S). More than half of the species within the family are still classified as natural weeds, e.g., *A. graecizans*, *A. palmeri*, *A. powellii*, *A. spinosus*, *A. thunbergii*, and *A. tuberculatus* (Das 2012; Gerrano, van Rensburg and Adebola 2015). Many of the latter are regarded as invasive weeds in the United States, a trait that has been associated with resistance to herbicides targeting Acetolactate Synthase (ALS) that has raised as a result of interspecific hybridization (Trucco et al. 2005). As a matter of fact, grain and vegetable amaranths are supposed to be two distinct groups that evolved from their respective weed progenitors. Both groups have different centers of origin and have experienced unique processes of domestication (Das 2012). Despite the remarkable morphological features, some dual types such as *A. hypochondriacus*, *A. caudatus* and *A. cruentus* exist, whose young leaves may also be utilized as leafy vegetables in Asian and African countries although they are mostly used as grain species. Meanwhile, *A. dubius*, *A. blitum*, and *A. viridis* are referred to as *feral* species, meaning that these may proliferate without need for cultivation (Andini et al. 2013a, b).

In most of Asia and Africa, vegetable amaranths are attributed with a different morphology characterized by a much shorter stature, with more leaves and branches, a quite smaller inflorescence. Leaf amaranths are perceived as vegetables resembling the European spinach (*Spinacea oleracea*). Although there has been no official record of how they were firstly introduced into Asia and Africa, it was proposed that the sea voyagers in the fifteenth to sixteenth centuries played a great role in the distribution of amaranth plants to the other parts of the world (Brenner et al. 2000). Their gradual, seemingly worldwide adaptation was based on its robustness, superior adaptability to adverse growing conditions, and extraordinary content and quality of the seed proteins. Moreover, it is also known to require less agricultural inputs, thereby constituting a reliable livelihood for many rural households. Thus, amaranths deserve to be considered as potential plants able to challenge the impact of climate change. The utilization of some amaranth species as green leafy vegetables is further explained in Sect. 9.4.1.

9.3 Amaranth: Hybridization Versus Evolutionary Process

In the nature, new combinations of genes resulting from hybridization and introgression between wild relatives and their crop cultivars have played an important role in the evolution of domestic crop species. Moreover, natural introgression between wild relatives and their crop cultivars has been acknowledged as a natural way to increase the genetic diversity of modern crops.

Nowadays, the enlargement of the useful crop gene pools continues to be the subject of active research. As a background, it is known that hybridization and subsequent introgression require that premating and post-mating barriers are overcome. Thus, a range of methods based on cytological or molecular strategies has been reported in order to document how natural hybridization occurs in many important crops such as maize (*Zea mays*), wheat (*Triticum aestivum*), quinoa (*Chenopodium quinoa* Willd.), hemp (*Cannabis sativa*), potato (*Solanum tuberosum*), cassava (*M. esculenta*), radish (*Raphanus sativus*), and cabbage (*Brassica* spp.).

The hybrids, referred to as the product or result of hybridization, are usually morphologically intermediate, uniformly less fit than their parents, and attributed with a large number of novel characters derived from the new combinations of genes produced from hybridization and introgression (Jarvis and Hodgkin 1999).

Most of *Amaranthus* species are dioecious, meaning that separate male and female plants are exist, such as *A. hybridus*. Nevertheless, due to the proximal arrangement of their flowers, it is mostly recognized as a self-pollinating plant, with numerous intricate flowers that sometimes makes it difficult to carry out crosses (Stetter et al. 2016). However, some amaranths are also able to do outcrossing via wind-dependent pollination. An outcrossing rate ranging from 5 to 30% has been reported, which should enable hybridization in amaranth, distinctively among weedy amaranths. Hybridization is hypothesized to increase genetic variation, permit a more rapid adaptation to cropping systems and increase morphological variations. It may also be a way to acquire resistance to various herbicides (Trucco et al. 2005). Theoretically, plants live in open fields or abandoned places, such as along roadsides, may face genetic polymorphisms within those populations (Murren and Pigliucci 2005). In such habitats, some special conditions may prevail that select for a given combination of traits found in hybrid plants (Coons 1982). Increasing genetic variation translated at the molecular level can be acquired either through a number of biochemical/ enzymatic processes via cytogenetic-based activities, e.g., meiotic recombination. Furthermore, a species' genetic variability could reflect its ability to alter allelic forms via introgression (Trucco et al. 2005).

Higher genetic variability is more commonly found among the weedy amaranth species, followed by those cultivated as vegetables. Table 9.1 summarizes the main species representing grain, vegetable, and weed amaranth relatives, or also referred to here as the weedy types. Wild relatives, namely *A. hybridus*, *A. powellii*, *A. quitensis*, *A. retroflexus,* usually represent the secondary gene pools of grain amaranths (Das 2012).

Hybridization between different amaranths, mostly between the weedy and cultivated types, has been amply investigated and reported and has usually been described as a relative straightforward procedure (Brenner et al. 2000; Pandey 1999), although some fertility barriers were also discussed (Gupta and Gudu 1991). Wild *Amaranthus species*, including the weedy ones, usually possess agriculturally desirable traits such as drought-, salt-, and pathogen tolerance, high protein content, or other agromorphological related features, e.g., large seed size and higher biomass.

Interestingly, hybridization among weedy amaranths is hypothesized to increase genetic variation and enable their possibility to adapt more quickly to cropping systems. Recent studies had reported that hybridization occurring among some *Amaranthus* species may be a window opportunity for the acquisition of resistance to various herbicides. In spite of the potential adaptive advantage of hybridization, its evolutionary impact is dependent on its frequency of occurrence, with an expected rate of mutation of -10^{-6} for the specific gene conferring herbicide resistance, later identified as Acetolactate Synthase (ALS) (Trucco et al. 2005). The same workers reported mean hybridization frequencies from 0.4 to 2.3%, which were much higher than expected values. Such a phenomenon was explained by adducing the closer evolutionary proximity existing between the weedy amaranth species used in the experiments (Trucco et al. 2005).

Phylogenetic relationships among the three major grain amaranths' species require further investigation in order to provide useful information for breeding programs and germplasm conservation (Gupta and Gudu 1991). Furthermore, (Chan and Sun 1997; Xu and Sun 2001) highlighted the so-called "*A. hybridus* species complex" covering the evolutionary basis and the relatedness among the three-grain amaranths on the basis of morphology and geography. Their main proposals were (i) the single origin (*monophyletic*) or primary domestication event of *A. cruentus* from *A. hybridus* that occurred presumably in Central America, with the other two domesticated *A. hypochondriacus* and *A. caudatus* species, evolving secondarily through repeated crossing of *A. cruentus* with weedy *A. powellii* and *A. quitensis* in the northern and southern hemisphere, respectively. These two domesticated species subsequently spread into their respective territories, and (ii) the independent domestication event that originated from three putative "wild" progenitors, specifically *A. hybridus*, *A. powellii*, and *A. quitensis*. According to this alternative scenario, *A. caudatus* was mostly likely domesticated from *A. quitensis* in South America, *A. cruentus* from *A. hybridus* in Central America, and *A. hypochondriacus* from *A. powellii* in Mexico (Fig. 9.1, numbers 1 and 2).

Table 9.1 Major, commonly known, *Amaranthus* species

No	Species	Type[a]	Remarks
1	*A. blitum* (2n = 32)	WD, VG	Known as "horse-tooth" and still classified as a weedy amaranth. Grows majorly in most Asian countries. In the highlands of Takengon, in Sumatra, they are mostly referred to as famine foods. Leaves are thicker and wider compared to *A. tricolor* (Andini et al. 2013a, b)
2	*A. caudatus* (2*n* = 32)	GR	The inflorescence-bearing seeds have a characteristic shape resembling a waterfall. They are harvested and consumed as cereal-like grains (Coons 1982)
3	*A. cruentus* (2*n* = 34)	GR, VG	The dual-type grain amaranth from which young leaves can be also harvested and consumed as green vegetables. Most prominent in western Africa. In Indonesia, it is usually grown at higher altitudes (Coons 1982)
1	*A. dubius* (2*n* = 32)	WD	Also known as spleen amaranth. It is still classified as a weedy amaranth having an extraordinary content of foliar protein, up to 32%. Its morphology closely resembles that of *A. tricolor*, except for a more robust appearance, larger leaf area and thicker leaves (Andini et al. 2013a, b)
5	*A. hybridus* (2*n* = 32)	VG	It is majorly perceived as the putative origin of grain amaranths (Mallory et al. 2008)
6	*A. hypochondriacus* (2*n* = 32)	GR	One of the three most cultivated grain amaranths. Proposed as an alternative crop for the twenty-first century due mostly to its nutraceutical qualities (Barba de la Rosa et al. 2009)
7	*A. spinosus* (2*n* = 34)	WD	It is a pantropical widespread species and the only one within the genus having spines, a characteristic from which it derives its name, meaning thorn or prickly amaranth. This is an invasive weed in many crop fields, although it is believed to have medicinal properties (Pal 1972)
8	*A. tricolor* (2*n* = 32)	VG	The most commercial vegetable species. It resembles spinach (*Spinacea oleracea*) and is mostly grown in Southeastern Asian countries, including Indonesia (Andini et al. 2013b)

[a]Type: GR = Grain; VG = Vegetable; WD = Weed

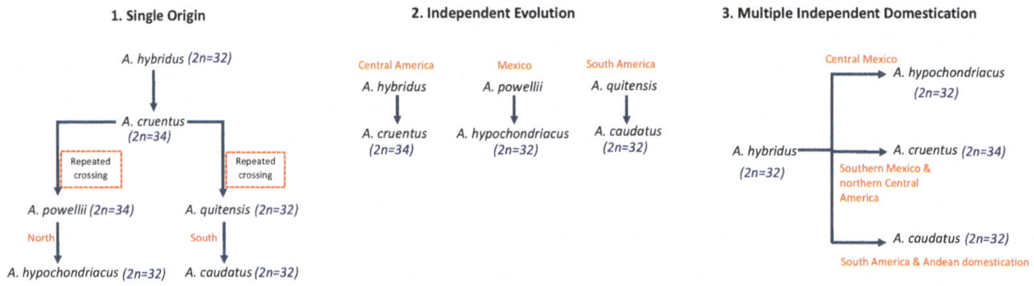

Fig. 9.1 The three theories regarding the evolutionary process of grain amaranths (interpreted from (Mallory et al. 2008; Chan and Sun 1997))

More recently, (Mallory et al. 2008) proposed the polyphyletic placement or the multiple, independent domestication events in three different eco-geographical regions: central Mexico corresponding to *A. hypochondriacus*, southern Mexico-Northern Central America, corresponding to *A. cruentus*, and lastly, the South American-Andean, corresponding to *A.caudatus*.

Their theory increased the information regarding the *A. hybridus* complex, although it still supported the single putative progenitor status of *A. hybridus* (Fig. 9.1, number 3). Despite the several studies conducted to elucidate the evolutionary relationships between grain amaranths, no conclusive evidence to support the final hypothesis has been yet generated.

Whatever their source of origin, the available evidences indicate that (i) *A. hybridus* is the putative origin and most closely associated species with the wild taxon compared to other wild species; (ii) interspecific hybridization among the three major species occurred. These two served as the basis of evolutionary studies. Nevertheless, hybrid development from crosses among the three-grain amaranth species was not so easily obtained. (Pal 1972), previously, had made several interspecific cross attempts between *A. caudatus* × *A. hypochondriacus* and reported that the hybrids produced generated undesired morphological features including deformed leaves showing a virus-like syndrome with tumorous stems and roots, deformed morphology of mature F1-plants having, for example, peculiar twinning of stems and inflorescence axes and deformed anthers, and sterility problems marked by low pollen fertility (3.45%), sterile ovules and no seed production. These factors impeded any gene exchange between these two species (Gupta and Gudu 1991).

Other hybrids as a result of a cross between *A. hypochondriacus* × *A. cruentus* generated partially or completely sterile hybrids, meanwhile, while a reciprocal crossing was also difficult to obtain as an excessive amount of pollen had to be applied in order to produce just a few hybrid seeds. Nevertheless, seeds appeared to be normal without any significant deformities and germinated normally; however, most of them were not able to survive. Four of them were able to reach the maturity stage but were also sterile. Based on the experiment's outcome, they came to three conclusions: (i) *A. hypochondriacus* and *A. caudatus* seemed to be closely related and not reproductively isolated despite the existence of gene exchange restriction; (ii) *A. cruentus* and *A. hypochondriacus* were also distantly related but their genetic differentiation was lower, and (iii) the highest genetic isolation barriers were found between *A. cruentus* and *A. caudatus*, due to the far genetic distance separating them (Gupta and Gudu 1991).

Hence, it can be concluded that natural hybridization often occurs in many crop plants as a natural method to increase genetic variation, permitting better adaptation to cropping systems, increased morphological variations and tolerance to (a)biotic stresses. Moreover, unique or exotic genes present only in many weedy plants may be transferred via natural hybridization (Pickersgill 1981). Interspecific hybridization is believed by many scientists to occur in *Amaranthus*, especially as an apt explanation for the phylogenetic origin of grain amaranths from *A. hybridus*, its putative progenitor. Three theories have been proposed regarding the evolutionary process that led to the present grain amaranths species, known as the "*A. hybridus complex*". Breeding barriers in grain amaranths are accentuated by their extremely small flower size, in addition to their closely-knit arrangement in the glomerule. However, experimental crosses between grain amaranth species have resulted in unfertile hybrids and significant deformation in several important morpho-physiological traits (in stem, anther, and leaves). The process is hindered by breeding barriers that lead to a no-gene exchange status.

9.4 Polyploidy in Amaranths

Polyploidy means the heritable condition of having more than two complete sets of chromosomes. Polyploidy is counted as an important evolutionary process in many living organisms, including plants and animals (Comai 2005). A state of ploidy causes alterations in morphology, such as (i) heterosis and gene redundancy; (ii) bigger cells and more vigorous plant stature; (iii) increased biomass production such as found in wheat, e.g., wider leaves, bigger seeds & stem; and (iv) generates novel variation. In nature, improved performance of many important crop species that provide food, fodder and fiber, e.g., *Brassica* and *Raphanus* spp. (Lü et al. 2008), Triticeae (Feuillet et al. 2008) and *Zea mays* are closely associated with polyploidy. Therefore, polyploidy is recognized as a pinned interest to many plant biologists due to its inextricable link with many important crops. There are two types of polyploids: natural and artificial. Natural polyploids are generated in response to changes in the environment, while artificial polyploids can be generated via a colchicine treatment capable of doubling the chromosome number during plant hybridization (Ánchez et al. 2008; Liu and Wendel 2003).

Ploidy variations usually cause alterations in morphology and may last up to ten generations. A previous study conducted by (Andini et al. 2013b) reported a high variability of morphological changes and high nutritional qualities in the Indonesian weedy accessions, particularly *A. dubius* that exhibited distinct morphological characteristics, e.g., larger plants with many side branches, shorter and thicker stems than those in diploid plants, shorter inflorescences, and increased seed size, up to 42% greater than the diploid seeds (Andini et al. 2018). Thus, the polyploidy types found within *Amaranthus* spp. is assumed to have resulted from an initial hybridization between *A. spinosus* and *A. hybridus*, which has been proposed to be a natural tetraploid (Pal 1972).

In the past, a conventional and simple method, e.g., a chromosome study, was applied to detect the number of chromosomes in amaranths. Generally, the basic chromosome number (referred here as x) in Amaranthaceae is $x = 8$ or $x = 9$. However, in the genus *Amaranthus* it appears to be in allotetraploid form, with counts of $n = 16$ or $n = 17$, except in *A. dubius* with $2n = 4x = 32$, which is representative of an allopolyploid. Table 9.2 presents the polyploidy levels in *Amaranthus* which is based on flow cytometer. They are compared to other available references. From this information, it may be concluded that the majority of amaranths exists as diploids ($2n = 32$ or $2n = 34$).

The flow cytometer results of 87 *Amaranthus* accessions confirmed that the variation of chromosome number within the genus *Amaranthus* exists. Remarkably, no polyploids were detected albeit the remarkable morphological variations observed among the 87 accessions examined, particularly in the *A. dubius* collection. From this result, it may be implied that amaranths show a great number of morphological diversity due to a high genetic diversity, which is strongly affected by genotypic and environmental factors (Das 2012). The high genetic diversity might have resulted in (i) variation in chromosome number, (ii) hybridization derived from frequent outcrossing, (iii) phenotypic plasticity employed as a unique adaptation process or due to genotype and environment interactions, and (iv) geographical isolation during the domestication process, which might have contributed to the genetic variability in several important crop species, including the Indonesian amaranths (Andini et al. 2018).

9.4.1 Indonesian Amaranths

Many indigenous vegetables are characterized by a high nutritional value compared with imported vegetables like cabbage, carrot, and tomato. The succulent leaves of vegetable amaranths are widely consumed in the humid tropical belt countries, e.g., sub-Saharan Africa, major Asian countries (i.e., China and India), and parts of Southeast Asia, including Indonesia, the Philippines, and Malaysia. In fact, grain, vegetable, and weedy amaranths are considered as one of the most nutritious groups of plants in the world, as their leaves, but mostly their seeds, have a desirable spectrum of amino acids for human nutrition, especially a high lysine content ranging from 5 to 7 g 100 g^{-1} DW protein containing in the three types of amaranths.

Lysine, for instance, required to ensure proper growth of small children, is often lacking in common grains. Moreover, the protein content present in both amaranth seeds and leaves ranges between 27.8 and 48.6%. They are also an excellent source of unsaturated oil (45% linoleic acids). Due to these distinct characteristics, its daily consumption is highly recommended by the

Table 9.2 Natural polyploidy levels, based on flow cytometer and chromosome number (2*n*), in *Amaranthus* [modified from (Andini et al. 2018)]

No	Origin	Accession	Place of origin	Type	Species	Polyploidy level	No. of chromosomes (2*n*)
1	WW	USDA 01/Ames 5315/IND	Maharastra–India	VG	*A. blitum*	Diploid	34
2	WW	USDA 02/PI 610,262/IND	Gujarat–India	VG	*A. blitum*	Diploid	34
3	WW	USDA 03/PI 490,298/KEN	Kenya	VG	*A. blitum*	Diploid	34
4	WW	USDA 04/PI 606,281/BGD	Bangladesh	VG	*A. blitum*	Diploid	34
5	WW	USDA 05/PI 606,282/BGD	Bangladesh	VG	*A. blitum*	Diploid	34
6	WW	USDA 06/PI 482,049/ZWE	Zimbabwe	GR	*A. cruentus*	Diploid	34
7	WW	USDA 07/PI 482,051/ZWE	Zimbabwe	GR	*A. cruentus*	Diploid	34
8	WW	USDA 08/PI 490,662/BEN	Benin	GR	*A. cruentus*	Diploid	34
9	WW	USDA 09/PI 494,777/ZMB	Zambia	GR	*A. cruentus*	Diploid	34
10	WW	USDA 10/PI 500,267/ZMB	Zambia	GR	*A. cruentus*	Diploid	34
11	WW	USDA 11/PI 538,319/USDA	United States	GR	*A. cruentus*	Diploid	34
12	WW	USDA 12/PI 566,897/IND	India	GR	*A. cruentus*	Diploid	34
13	WW	USDA 13/PI 604,666/USDA	United States	GR	*A. cruentus*	Diploid	34
14	WW	USDA 15/PI 605,352/JAM	Jamaica	VG	*A. dubius*	Diploid	32
15	WW	USDA 16/PI 642,737/PRI	Puerto Rico	VG	*A. dubius*	Diploid	32
16	WW	USDA 17/PI 608,661/IND	India	VG	*A.graecizans*	Diploid	34
17	WW	USDA 19/PI 500,249/ZMB	Zambia	GR	*A. hybridus*	Diploid	32
18	WW	USDA 20/PI 605,351/GRC	Greece	GR	*A. hybridus*	Diploid	32
19	WW	USDA 21/PI 604,577/MEX	Puebla–Mexico	GR	*A. hypochondriacus*	Diploid	34
20	WW	USDA 22/ PI 604,796/ N.A	Unknown	GR	*A. hypochondriacus*	Diploid	34
21	WW	USDA 23/PI 607,447/JAM	Jamaica	WD	*A. retroflexus*	Diploid	34
22	WW	USDA 24/Ames 5134/USDA	United States	VG	*A. tricolor*	Diploid	34

(continued)

Table 9.2 (continued)

No	Origin	Accession	Place of origin	Type	Species	Polyploidy level	No. of chromosomes (2*n*)
23	WW	USDA 25/PI 349,553/PNG	Papua N Guinea	VG	*A. tricolor*	Diploid	34
24	WW	USDA 26/PI 477,918/N.A	Unknown	VG	*A. tricolor*	Diploid	34
25	WW	USDA 27/PI 566,899/IND	India	VG	*A. tricolor*	Diploid	34
26	WW	USDA 28/PI 604,669/TWN	Taiwan	VG	*A. tricolor*	Diploid	34
27	WW	USDA 29/PI 608,761/IND	India	VG	*A. tricolor*	Diploid	34
28	IDN	USDA 30/PI 540,445/IDN/Java	Java-Indonesia	WD	*A. viridis*	Diploid	34
29	**WW**	**USDA 31/PI 586,680/USA**	**United States**	**OR**	***C. argentea***	**Diploid**	**36**
30	**WW**	**USDA 32/PI 482,244/ZWE**	**Zimbabwe**	**OR**	***C.trigyna***	**Diploid**	**18**
31	**WW**	**NPL 01/KTM**	**Kathmandu-Nepal**	**OR**	***C. argentea***	**Diploid**	**36**
32	**WW**	**NPL 02/KTM**	**Kathmandu-Nepal**	**OR**	***C. argentea***	**n.a**	**n.a**
33	**WW**	**NPL 03/KTM**	**Kathmandu-Nepal**	**OR**	***C. argentea***	**n.a**	**n.a**
34	**WW**	**NPL 04/KTM**	**Kathmandu-Nepal**	**OR**	***C. argentea***	**n.a**	**n.a**
35	**WW**	**NPL 05/KTM**	**Kathmandu-Nepal**	**OR**	***C. argentea***	**Diploid**	**36**
36	IDN	IDN 01/Mongal	Mongal-Takengon	WD	*A. dubius*	Diploid	32
37	IDN	IDN 02/Daling	Kp.Daling-Takengon	WD	*A. dubius*	Diploid	32
38	IDN	IDN 03/Daling	Kp.Daling-Takengon	WD	*A. dubius*	Diploid	32
39	IDN	IDN 04/Bur biah	Kp.Daling-Takengon	WD	*A. spinosus*	Diploid	34
40	IDN	IDN 07/Ulu Nuih	Kp.Burbiah-Takengon	WD	*A. spinosus*	Diploid	34
41	IDN	IDN 08/Asir-asir	Hilir-Takengon	WD	*A. dubius*	Diploid	32
42	IDN	IDN 09/Asir-asir	Kp.Asir-Takengon	WD	*A. dubius*	Diploid	32
43	IDN	IDN 10/P. One–one	One–one-Takengon	WD	*A. dubius*	Diploid	32
44	IDN	IDN 11/Tn.Depet	Tn.Debet-Takengon	WD	*A. dubius*	Diploid	32
45	IDN	IDN 13/Mandua	Mandua-Takengon	WD	*A. dubius*	n.a	n.a

(continued)

Table 9.2 (continued)

No	Origin	Accession	Place of origin	Type	Species	Polyploidy level	No. of chromosomes (2*n*)
46	IDN	IDN 14/Mandua	Mandua-Takengon	WD	*A. dubius*	Diploid	32
47	IDN	IDN 15/Mandua	Mandua-Takengon	WD	*A. dubius*	Diploid	32
48	IDN	IDN 17/Medan	Medan-North Sumatera	WD	*A. blitum*	Diploid	34
49	IDN	IDN 18/Medan	Medan-North Sumatera	WD	*A. blitum*	Diploid	34
50	IDN	IDN 19/Mandua	Mandua-Takengon	WD	*A. blitum*	Diploid	34
51	IDN	IDN 20/Yogya	Yogyakarta	WD	*A. dubius*	Diploid	32
52	IDN	IDN 22/Yogya	Yogyakarta	WD	*A. dubius*	Diploid	32
53	IDN	IDN 23/Yogya	Yogyakarta	WD	*A. dubius*	Diploid	32
54	IDN	IDN 24/JKT	Jakarta	WD	*A. dubius*	Diploid	32
55	IDN	IDN 25/JKT	Jakarta	WD	*A. dubius*	Diploid	32
56	IDN	IDN 26/JKT	Jakarta	WD	*A. dubius*	Diploid	32
57	IDN	IDN 27/JKT	Jakarta	WD	*A. dubius*	Diploid	32
58	IDN	IDN 28/Kl.Urang	Kl.Urg-Yogyakarta	VG	*A. tricolor*	Diploid	34
59	IDN	IDN 29/Kl.urang	Kl.Urg-Yogyakarta	VG	*A. tricolor*	Diploid	34
60	IDN	IDN 30/Kl.Urang	Kl.Urg-Yogyakarta	VG	*A. tricolor*	Diploid	34
61	IDN	IDN 32/Kl.Urang	Kl.Urg-Yogyakarta	VG	*A. tricolor*	Diploid	34
62	IDN	IDN 33/Lembang	West Java	VG	*A. caudatus*	Diploid	32
63	IDN	IDN 34/Lembang	West Java	VG	*A. caudatus*	Diploid	32
64	IDN	IDN 35/BNA	B.Aceh-Sumatera	VG	*A. tricolor*	Diploid	34
65	IDN	IDN 36/BNA	B.Aceh-Sumatera	VG	*A. tricolor*	Diploid	34
66	IDN	IDN 37/BNA	B.Aceh-Sumatera	VG	*A. tricolor*	Diploid	34
67	IDN	IDN 38/BNA	B.Aceh-Sumatera	VG	*A. tricolor*	Diploid	34
68	**IDN**	**IDN 39/BNA**	**B.Aceh-Sumatera**	**OR**	***C. argentea***	**Diploid**	**36**
69	IDN	IDN 40/BNA	B.Aceh-Sumatera	VG	*A. tricolor*	Diploid	34
70	IDN	IDN 41/Marelan	Medan-North Sumatera	VG	*A. tricolor*	Diploid	34
71	IDN	IDN 42/Marelan	Medan-North Sumatera	VG	*A. tricolor*	Diploid	34

(continued)

Table 9.2 (continued)

No	Origin	Accession	Place of origin	Type	Species	Polyploidy level	No. of chromosomes (2*n*)
72	IDN	IDN 43/Marelan	Medan-North Sumatera	VG	*A. tricolor*	Diploid	34
73	IDN	IDN 44/Marelan	Medan-North Sumatera	VG	*A. tricolor*	Diploid	34
74	IDN	IDN 45/Marelan	Medan-North Sumatera	VG	*A. tricolor*	Diploid	34
75	IDN	IDN 46/Kresek	Kresek-Jakarta	VG	*A. tricolor*	Diploid	34
76	IDN	IDN 47/Kresek	Kresek-Jakarta	VG	*A. tricolor*	Diploid	34
77	IDN	IDN 48/Kresek	Kresek-Jakarta	VG	*A. tricolor*	Diploid	34
78	**IDN**	**IDN 49/Yates**	**Commercial-Jakarta**	**OR**	***C. argentea***	**Diploid**	**36**
79	IDN	IDN 50/P.Merah	Commercial-Jakarta	VG	*A. tricolor*	Diploid	34
80	IDN	IDN 51/P.Merah	Commercial-Jakarta	VG	*A. tricolor*	Diploid	34
81	IDN	IDN 52/SHS	Commercial-Jakarta	VG	*A. tricolor*	Diploid	34
82	IDN	IDN 53/SHS	Commercial-Jakarta	VG	*A. tricolor*	Diploid	34
83	IDN	IDN 54/Tanindo	Commercial-Jakarta	VG	*A. tricolor*	Diploid	34
84	IDN	IDN 55/P.Kb	West Sumatra	WD	*A. hybridus*	Diploid	32
85	IDN	IDN 56/P.Kb	West Sumatra	WD	*A. spinosus*	Diploid	34
86	IDN	IDN 57G/P.Kb	West Sumatra	WD	*A. hybridus*	Diploid	32
87	IDN	IDN 57R/P.Kb	West Sumatra	WD	*A. hybridus*	Diploid	32

USDA: United States of Department of Agriculture (US-Genebank), IDN: Indonesian (field) collection, VG: Vegetable, WD: Weedy, OR: Ornamental, GR: Grain, **bold**: diploid *Celosia* spp. as ornamental plants but still included in the genus *Amaranthus*

World Health Organization (UN-WHO). Thus, such potential excellence could be utilized as an effective but affordable tool for alleviating malnutrition and protein deficiency among young children in developing countries (Andini et al. 2013a, b).

Despite a wide range of vegetable amaranth genetic resources available in tropical countries, such as in Indonesia, *A. tricolor* is, by far, the most commercial species (Andini et al. 2013a, b). In Indonesia, vegetable amaranths are ranked as the third most popular green leafy vegetable and are frequently traded in Indonesian traditional markets, after *Ipomoea aquatica* Forskkal and *M. esculenta*. Unfortunately, the scale of the amaranth trade is not registered formally, as the vegetables are usually harvested early in the morning and have been mostly sold-out by the afternoon. Interestingly, from the public health point of view, amaranth has been effectively applied as an effective mean to combat the "dark-blindness" disease among young children living in rural areas, whose parents are persistently unable to purchase meat-based proteins to balance their daily diets. Such initiative was formally made by the Indonesian government in the beginning of the 1980s and was promoted with a slogan broadcasted as "*4 sehat-5 sempurna*" or 4

Fig. 9.2 The three color variants of the most representative vegetable amaranth (*A. tricolor*) in Indonesia (*Source* personal collection)

is healthy-5 is perfect (translated in *English*) in terms of food sources that provided a complete- and perfect-nutrition spectrum for young children. The program was strongly dependent on grass-root health services or community health centers, which are available in each village even in the most remote areas in the eastern part of Indonesia, known as ***Pos Pelayanan Terpadu*** or abbreviated as *Posyandu* in *bahasa*.

There are, in general, three species that are widely cultivated, *A. tricolor* known as the uprooted type, is the most economically representative species, with three color variants: whitish green (Fig. 9.2a), red or deep purple (Fig. 9.2c) and variegated (Fig. 9.2b). *A. dubius*, generally known as the picked-up type, has distinct morphological features (i.e., more robust, wider leaf area, bigger leaves, and dark green leaf-color), and *A. cruentus*, including both the picked-up or uprooted types, possessing distinct features like a robustness, wide leaf area and greyish green pigmented leaves that can be harvested within three weeks. This species is more prominently cultivated at higher altitudes in Asia.[1] The highly pigmented, or red amaranths, have been reported to have superior antioxidants properties linked to total polyphenols and betalain pigments (i.e., betacyanins and betaxanthins) and are considered to have nutraceuticals attributes similar to those present in green amaranths (Ali et al. 2009).

Their statistics regarding amaranth leaf production are not specifically known, as the official Indonesian statistic website presents all kinds of vegetables in one single bulk category designated as "*horticulture*".[2] However, approximately 152,334 tons of *A. tricolor* were produced in 2010. A vast area of cultivation is found on the island of Java, consisting of private farmer fields, fallow lands, riverbanks, abandoned spaces in many villages or in housing settlements in urban cities with a total area up to 700 square meters, etc. Leaves are commonly sold in bundles of 10–20 plants and can be harvested within 3 weeks, with a yield ranging from 5 to 10 tons ha^{-1} per harvest season depending on the available inputs. Thus, they are considered as economically important to many Indonesian smallholder farmers, considering their rapid income generation.

Despite their popularity and importance in Indonesia one of the three major cultivated vegetable species, there have been no concentrated efforts to improve the nutritional value and productivity of vegetable amaranths. Based on a direct field interview, most of the farmers that grow leafy amaranths on their scattered plot that are smaller than one hectare, do it for subsistence purposes. They tend to use their own seeds, selected from the previous planting seasons, and very seldom buy new or hybrid seeds for each planting season. On the other hand, Indonesia,

[1]Internet source: https://hortikultura.litbang.pertanian.go.id/teknologi-detail-44.html; accessed on 3rd Sept. 2020.

[2]Internet source: https://www.bps.go.id/indicator/55/61/1/produksi-tanaman-sayuran.html; accessed on 6 Sept. 2020.

encompassing the widest archipelago in the world, is recognized as a biodiversity hotspots in the tropics. This suggests that there is a potential to increase the nutritional value of the existing amaranth genetic variants, particularly the conventional or commercialized type (*A. tricolor*). Some cultivars belonging to the *A. blitum* weedy type were reported to have as much as 38% foliar protein, compared to average protein level of 12%, and showed a broad range of variation in favorable crop morphologies, e.g., high leaf number and copious branching. In the remote highland areas of Takengon on central Aceh, in Sumatra, local people have utilized some weedy types that closely resemble the vegetable amaranths. As mentioned above, these are known as "famine food", while other weedy types have been used in traditional medicine. The potential superiority of the weedy amaranths, e.g., *A. blitum*, *A. dubius* and *A. viridis* in terms of protein content over the conventional *A. tricolor*, was confirmed by data reported by (Andini et al. 2013a, b).

9.5 Amaranth Crop Improvement

Plant breeders have to deal constantly with new challenges. At the same time, agricultural practices keep changing. Amaranth, which is already established as a grain crop in temperate regions, and as a vegetable source in most of the tropical regions of the world, should be further improved. Several breeding targets for the grain amaranths have been recognized, in order to achieve important agronomical objectives such as seed shattering reduction at the maturity stage and larger seed sizes.

Nutritional or nutraceutical perspectives, could involve increased protein and/or antioxidant content in seeds, taking into account that they are a valuable source of rutin (quercetin-3-*O*-rutinoside) and quercetin (the precursor of rutin) that have been found to have the potential to be used for the prevention and treatment of colon cancer. Moreover, quercetin could be applied as a constituent of chemotherapeutic drugs for the treatment in many cancer-related diseases (Kalinova and Dadakova 2008).

On the other hand, breeding targets in vegetable amaranths could be directed to increase abiotic stress resistance, such as to excess heat, without altering their low agricultural input. Other interesting breeding objectives could be an increment of leaf nutritional values, especially the protein content. This could be increased up to 32% as found in weedy *A. dubius*. In general, the average protein content found in *A. tricolor*, the most commercial type in most of the Southeast Asian countries, is still low, namely in the range of 12–14%. Such target is not impossible, as the gene resources do exist, as exhibited by the weedy types. Various methods could be applied for this purpose, including the possible interspecific hybridization and cytoplasmic male sterility for producing hybrids, such as previously done in Brassicaceae plants (Gudu and Gupta 1988; Sodhi et al. 2006) and *Zea mays* (Duvick 2001). By doing so, the attractiveness of vegetable amaranths as a cheap source of protein for many rural people living in developing countries who can't afford animal-based protein could be guaranteed for many generations.

9.6 Conclusion

Amaranth (*Amaranthus* spp.) encompasses valuable dicotyledonous plants used as grain crops, vegetables, ornamentals, and medicinal herbs attributed to their various nutraceuticals properties. It had been already cultivated by the ancient civilization in Central and South America about 5,000–6,000 years ago and reached its paramount as one of their most important staple crops. Although grain amaranths exhibit morphological features that differ from the vegetable types, *A. hybridus* is presumed as their putative progenitor. Three theories had been proposed to elucidate the phylogeny and evolution of amaranths, though none of them has been confirmed. Amaranth has a high genetic variability, characterized by varied morphological features that have resulted from frequent outcrossing, with an outcrossing rate up to 30% being reported. Due to this ability, interspecific hybridization between the wild and cultivated amaranths could be possible in Nature. Polyploidy, which is important for the evolution and speciation of plants has been also discussed here. Generally, the basic chromosome number (referred here as x) in Amaranthaceae is $x = 8$ or $x = 9$. However, the genus Amaranthus appears to be an allotetraploid form, with counts of $n = 16$ or $n = 17$, except in *A. dubius* with an allopolyploid $2n = 4x = 32$.

References

Ali B, Khandaker L, Oba S (2009) Comparative study on functional components, antioxidant activity and color parameters of selected colored leafy vegetables as affected by photoperiods. J Food Agric Environ 7:392–398

Ánchez CSI, Osa APBADELAR, Alván MFLEÓN et al (2008) Bioactive peptides in Amaranth (*Amaranthus hypochondriacus*) seed. J Agric Food Chem 56:1233–1240

Andini R, Sulaiman MI, Ohsawa R (2018) Natural polyploidy in Amaranths (*Amaranthus* spp.). In: AIP conference proceedings

Andini R, Yoshida S, Ohsawa R (2013a) Variation in protein content and amino acids in the leaves of grain, vegetable and weedy types of Amaranths. Agronomy 3:391–403. https://doi.org/10.3390/agronomy3020391

Andini R, Yoshida S, Yoshida Y, Ohsawa R (2013b) *Amaranthus* genetic resources in Indonesia: morphological and protein content assessment in comparison with worldwide amaranths. Genet Resour Crop Evol 60:2115–2128. https://doi.org/10.1007/s10722-013-9979-y

Barba de la Rosa AP, Fomsgaard IS, Laursen B, Mortensen AG, Olvera-Martínez L, Silva-Sánchez C, Mendoza-Herrera A, Gonzáles-Castañeda J, De León-Rodríguez A (2009) Amaranth (*Amaranthus hypochondriacus*) as an alternative crop for sustainable food production: phenolic acids and flavonoids with potential impact on its nutraceutical quality. J Cereal Sci 49:117–121. https://doi.org/10.1016/j.jcs.2008.07.012

Brenner D, Baltensperger D, Kulakow P, Lehmann JW, Myers RL, Slabbert MM, Sleugh BB (2000) Genetic resources and breeding of *Amaranthus*. In: Janick J (ed) Plant breeding reviews. Wiley & Sons, Inc., pp 227–285

Chan KF, Sun M (1997) Genetic diversity and relationships detected by isozyme and RAPD analysis of crop and wild species of *Amaranthus*. Theor Appl Genet 95:865–873. https://doi.org/10.1007/s001220050637

Comai L (2005) The advantages and disadvantages of being polyploid. Nat Rev Genet 6:836–846. https://doi.org/10.1038/nrg1711

Coons MP (1982) Relationships of *Amaranthus caudatus*. Econ Bot 36:129–146. https://doi.org/10.1007/BF02858709

Das S (2012) Systematics and taxonomic delimitation of vegetable, grain and weed amaranths: a morphological and biochemical approach. Genet Resour Crop Evol 59:289–303. https://doi.org/10.1007/s10722-011-9684-7

Duvick DN (2001) Biotechnology in the 1930s: the development of hybrid maize. Nat Rev Genet 2:69–74. https://doi.org/10.1038/35047587

Ebert AW (2014) Potential of underutilized traditional vegetables and legume crops to contribute to food and nutritional security, income and more sustainable production systems. Sustainability 6:319–335. https://doi.org/10.3390/su6010319

Feuillet C, Langridge P, Waugh R (2008) Cereal breeding takes a walk on the wild side. Trends Genet 24:24–32. https://doi.org/10.1016/j.tig.2007.11.001

Gélinas B, Seguin P (2008) Evaluation of management practices for grain Amaranth production in Eastern Canada. Agron J 100:344–351. https://doi.org/10.2134/agronj2007.0189

Gerrano AS, van Rensburg WSJ, Adebola PO (2015) Genetic diversity of *Amaranthus* species in South Africa. South African J Plant Soil 32:39–46. https://doi.org/10.1080/02571862.2014.973069

Gorinstein S, Zemser M, Fliess A, Fliess A, Shnittman I, Paredez-Lopez O, Yamamoto K, Kobayashi S, Taniguchi H (1998) Computational analysis of the amino acid residue sequences of amaranth and some other proteins. Biosci Biotechnol Biochem 62:1845–1851. https://doi.org/10.1271/bbb.62.1845

Gudu S, Gupta VK (1988) Male-sterility in the grain amaranth (*Amaranthus hypochondriacus* ex-Nepal) variety Jumla. Euphytica 37:23–26. https://doi.org/10.1007/BF00037218

Gupta VK, Gudu S (1991) Interspecific hybrids and possible phylogenetic relations in grain amaranths. Euphytica 52:33–38. https://doi.org/10.1007/BF00037854

Jarvis A, Lane A, Hijmans RJ (2008) The effect of climate change on crop wild relatives. Agric Ecosyst Environ 126:13–23. https://doi.org/10.1016/j.agee.2008.01.013

Jarvis DI, Hodgkin T (1999) Wild relatives and crop cultivars: detecting natural introgression and farmer selection of new genetic combinations in agroecosystems. Mol Ecol 8:S159–S173. https://doi.org/10.1046/j.1365-294X.1999.00799.x

Kalinova J, Dadakova E (2008) Rutin and total quercetin content in Amaranth (*Amaranthus* spp.). Plant Foods Hum Nutr 64:68–74. https://doi.org/10.1007/s11130-008-0104-x

Liu B, Wendel JF (2003) Epigenetic phenomena and the evolution of plant allopolyploids. Mol Phylogenet

Evol 29:365–379. https://doi.org/10.1016/S1055-7903(03)00213-6

Lü N, Yamane K, Ohnishi O (2008) Genetic diversity of cultivated and wild radish and phylogenetic relationships among *Raphanus* and *Brassica* species revealed by the analysis of trnK/matK sequence. Breed Sci 58:15–22. https://doi.org/10.1270/jsbbs.58.15

Mallory MA, Hall R V, McNabb AR, Pratt DB, Jellen EN, Maughan PJ (2008) Development and characterization of microsatellite markers for the grain Amaranths. Crop Sci 48:1098–1106. https://doi.org/10.2135/cropsci2007.08.0457

Murren CJ, Pigliucci M (2005) Morphological responses to stimulated wind in the genus Brassica (*Brassicaceae*): allopolyploids and their parental species. Am J Bot 92(5):810–818

Pal M (1972) Evolution and improvement of cultivated amaranths. III. *Amaranthus spinosus—dubius* complex. Genetica 43:106–118

Pandey RM (1999) Evolution and improvement of cultivated amaranths with reference to genome relationship among *A. cruentus, A. powellii* and *A. retroflexus*. Genet Resour Crop Evol 46:219–224. https://doi.org/10.1023/A:1008664611893

Pickersgill B (1981) Biosystematics of crop-weed complexes. Kulturpflanze 377–388

Rubaihayo E (2002) Uganda-the contribution of indigenous V egetables to household food security. Washington, D.C.

Schmidhuber J, Tubiello FN (2007) Climate change and food security special feature: global food security under climate change. In: Proceeding of the national academy of sciences, pp 19703–19708

Shukla S, Bhargava A, Chatterjee A, Pandey AC, Mishra BK (2010) Diversity in phenotypic and nutritional traits in vegetable amaranth (*Amaranthus tricolor*), a nutritionally underutilised crop. J Sci Food Agric 90:139–144. https://doi.org/10.1002/jsfa.3797

Sodhi YS, Chandra A, Verma JK, Arumugam N, Mukhopadhyay A, Gupta V, Pental D, Pradhan AK (2006) A new cytoplasmic male sterility system for hybrid seed production in Indian oilseed mustard *Brassica juncea*. Theor Appl Genet 114:93. https://doi.org/10.1007/s00122-006-0413-0

Stetter MG, Zeitler L, Steinhaus A, Kroener K, Biljecki M, Schmid KJ (2016) Crossing methods and cultivation conditions for rapid production of segregating populations in three grain Amaranth species. Front Plant Sci 7:816. https://doi.org/10.3389/fpls.2016.00816

Trucco F, Jeschke MR, Rayburn AL, Tranel PJ (2005) *Amaranthus hybridus* can be pollinated frequently by *A. tuberculatus* under field conditions. Heredity (Edinb) 94:64–70. https://doi.org/10.1038/sj.hdy.6800563

Xu F, Sun M (2001) Comparative analysis of phylogenetic relationships of grain amaranths and their wild relatives (*Amaranthus*; Amaranthaceae) using internal transcribed spacer, amplified fragment length polymorphism, and double-primer fluorescent intersimple sequence repeat. Mol Phylogenet Evol 21:372–387. https://doi.org/10.1006/mpev.2001.1016

Zheleznov AV, Solonenko LP, Zheleznova NB (1997) Seed proteins of the wild and the cultivated *Amaranthus* species. Euphytica 97:177–182. https://doi.org/10.1023/A:1003073804203

Amaranth Transcription Factors in Response to Biotic and Abiotic Stresses

10

Paola A. Palmeros-Suárez, Kena Casarrubias-Castillo, and Julio A. Massange-Sánchez

Abstract

Amaranth species have inherent tolerance to diverse environmental stresses such as drought, saline soil, extreme temperature, defoliation, pathogens and herbivory attacks. Transcriptomic analysis of the *Amaranthus* genus has been a powerful tool to identify genes induced in response to biotic and abiotic stress conditions. Various amaranth genes of unknown function and transcription factors (TFs) have been functionally characterized in recent years. TFs are of main interest due to their ability to modulate the expression of several responsive genes, which is a crucial part of the plant response to abiotic and biotic stresses. In this chapter, we identified from different amaranth studies the most promising TFs families in response to biotic and abiotic stresses. Also, we compiled functional studies of amaranth TFs conferring stress tolerance, which are potential candidates to be used in plant engineering to increase stress tolerance in commercial crops.

10.1 Introduction

Plants are adversely affected by a broad range of stressful conditions, including abiotic stress such as drought, salinity, extreme temperatures and intense illumination. Plants are also subject to biotic stresses including pathogen and insect attacks (Ohta et al. 2006; Dodds and Rathjen 2010). Stress can have devastating effects on growth and productivity, especially when multiple stresses occur in combination. To survive, plants have evolved complex, well-coordinated adaptive responses that operate at molecular and physiological levels (Ahuja et al. 2010). At the molecular level, gene expression is modified under stress; depending on the stress type/intensity, and plant species (Shinozaki and Yamaguchi-Shinozaki 2007). Stress inducible genes comprise genes involved in direct protection from stress, i.e. osmoprotectants and detoxifying enzymes, and genes that encode regulatory proteins such as transcription factors, protein kinases and phosphatases (Krasensky and Jonak 2012).

Transcription factors (TFs) are DNA-binding proteins that play roles in the modulation of gene expression by binding to specific DNA sequences called *cis*-elements in the gene promoters. TFs act as positive (on) or negative (off)

P. A. Palmeros-Suárez · K. Casarrubias-Castillo
Centro Universitario de Ciencias Biológicas y Agropecuarias, Universidad de Guadalajara, Camino Ramón Padilla Sánchez 2100, Las Agujas, 44600 Zapopan, Jalisco, Mexico

J. A. Massange-Sánchez (✉)
Unidad de Biotecnología Vegetal, Centro de Investigación y de Asistencia en Tecnología y Diseño del Estado de Jalisco A.C. (CIATEJ), Av. Normalistas 800, Colinas de la Normal, 44270 Zapopan, Jalisco, Mexico
e-mail: jmassange@ciatej.mx

D. Adhikary et al. (eds.), *The Amaranth Genome*, Compendium of Plant Genomes,
https://doi.org/10.1007/978-3-030-72365-1_10

regulators and they are involved in plant development from the germinating seed through to senescence and seed production, as well as the response mechanisms to biotic and abiotic stresses (Udvardi et al. 2007). In higher plants, 58 different TF families are known, comprising 6% to 10% of the plant genome, which is indicative both of their importance and the complexity of the transcriptional regulation process (Jin et al. 2017).

In the last decades, a great effort has been devoted to elucidating the function and regulation of several TF families using different approaches and model plants, including the large-scale analysis of omics data available in public databases (Singh et al. 2002; Seo and Choi 2015; Wang et al. 2016; Jin et al. 2017). Several gene function studies have been focused on Arabidopsis because it is easy to genetically modify, it is ephemeral, is of small size and it was the first plant genome sequenced (Arabidopsis and Initiative 2000). Often, agronomic crops have been used to study gene function in an attempt to directly improve quality characters, e.g. tomato, maize, rice, barley and wheat, but these crops require more effort and resources (Eckardt 2011; Tomato and Consortium 2012; (IWGSC) 2014; Mascher et al. 2017; Jiao et al. 2017). Additionally, extremophytes or resilient plants such as glasswort and amaranth, respectively, have been used to find novel gene functions and alternative regulatory mechanisms to increase environmental stress resistance in crops (Oh et al. 2012).

Amaranths (*Amaranthus* spp.) are C4 plants that belong to the Amaranthaceae family and the Caryophyllales order, which based on their utility can be divided into grain amaranth, vegetable amaranth, ornamental, and weedy amaranth (Brenner et al. 2000). Amaranths are known to be tolerant to adverse environmental conditions, including drought, poor and/or saline soils, intense illumination, severe defoliation and insect herbivory (Brenner et al. 2000; Castrillón-Arbeláez et al. 2012; Rastogi and Shukla 2013), even glyphosate resistance (Gaines et al. 2010). In this chapter, we focused on recent studies of amaranth TFs that play roles in abiotic and biotic stress tolerance. In the omics era, there has been major progress identifying amaranth genes induced under several stress conditions. We enlist a group of the most interesting induced TFs in grain amaranth transcriptome associated with stress tolerance. Also, we discuss specific TF members; AhERF-VII, AhDOF-AI and AhNF-YC increasing tolerance to water and salt stress by ABA-dependent and ABA-independent pathways, when they are overexpressed in Arabidopsis.

10.2 Identification of Amaranth TFs Involved in Stresses Tolerance

Amaranth is an excellent food source owing to the high nutritional value of its leaves and seeds, which have been reported to have unique nutraceutical properties (Rastogi and Shukla 2013). Further, numerous research approaches and resources have been developed in recent years, e.g. development of genetic markers (Transue et al. 1994), simple-sequence repeats (SSRs) (Mallory et al. 2008), bacterial artificial chromosome (BAC) library (Maughan et al. 2009) and genome sequencing (Sunil et al. 2014; Clouse et al. 2016). Amaranth is also valued for its ability to tolerate drought and salt conditions. This tolerance has been attributed to various physiological properties of amaranths, including C4 photosynthesis (Miller et al. 1984), a high water use efficiency (WUE) (Johnson and Henderson 2002; Omami et al. 2006), an indeterminate flowering habit, and the ability to develop long tap-roots and an extensive web of lateral roots (Miller et al. 1984; Johnson and Henderson 2002; Grobelnik-Mlakar et al. 2012). In addition, several studies have attempted to elucidate which molecular mechanisms make amaranth resistant to adverse environmental conditions. These investigations have been carried out mainly in grain amaranths (*A. hypochondriacus, A. cruentus* and *A. caudatus,*), and recently in weed amaranths (*A. palmeri and A. tuberculatus*), the latter because of resistance against the herbicide glyphosate (Salas-Perez et al. 2018; Kohlhase et al. 2019).

We show a list of amaranth studies related to stress tolerance that have been classified as biotic, abiotic, (a)biotic or chemical (Table 10.1). Three types of studies are frequently used to look for amaranth-induced genes under stress conditions; transcriptomic, proteomic and gene expression with enzyme activity. Aguilar-Hernández et al. 2011 performed the first attempt to identify calcium stress-induced genes in amaranth by using the 'old fashioned' suppression subtractive hybridization (SSH) technique. In this study, screening of the libraries showed 420 upregulated genes, which were grouped into six functional categories including stress response, signal transduction, transcription factors, gene regulation, and genes with unknown or hypothetical function. Among the TFs that were upregulated, they focus on CID9 (RNA binding) and Zinc finger (C3HC4-type RING finger), both related in response to drought and salt stress (Ma et al. 2006).

Table 10.1 Addressed studies to investigate the response mechanism to biotic and abiotic stresses in grain and weed amaranths

	Type of study	Species	Tissues	Type of stress	References
Abiotic	Gene expression and enzyme activity	*A. hypochondriacus, A. cruentus, A. caudatus, and A. hybridus,*	Leaves and roots	Water deficit stress	(González-Rodríguez et al. 2019)
	Proteomic	*A. cruentus*	Roots	Salt stress	(Huerta-ocampo et al. 2014)
	Proteomic	*A. hypochondriacus*	Roots	Water stress	(Huerta-Ocampo et al. 2011)
	Proteomic	*A. hypochondriacus*	Leaves	Drought	(Huerta-Ocampo et al. 2009)
Biotic	Gene expression and enzyme activity	*A. cruentus*	Leaves, stem and roots	Mechanical injury and insect herbivory	(Castrillón-Arbeláez et al. 2012)
	Gene expression and enzyme activity	*A. cruentus*	Leaves	Bacterial infection	(Casarrubias-Castillo et al. 2014)
(A)biotic	Gene expression and enzyme activity	*A. hypochondriacus*	Leaves, stem and roots	Salt, drought and insect herbivory	(Casique-Arroyo et al. 2014)
	Transcriptomic	*A. hypochondriacus*	Leaves and stems	Water stress, salt stress, herbivory and bacterial infection	(Délano-Frier et al. 2011)
Chemical	SSH Library	*A. hypochondriacus*	Leaves	Calcium stress	(Aguilar-hernández et al. 2011)
	Transcriptomic	*A. palmeri*	Leaves	Herbicide resistance	(Salas-Perez et al. 2018)
	Transcriptomic	*A. tuberculatus*	Leaves	Herbicide resistance	(Kohlhase et al. 2019)

Most studies are addressed to understand amaranth response to abiotic stress (drought and salt), in part for the current problem of global warming causing changes in rainfall and water availability to crops. Huerta-Ocampo et al. (2009); (2011), detected amaranth proteins in response to drought stress both in roots and leaves by using two-dimensional electrophoresis and LC/ESI–MS/MS. They identified 17 and 15 induced proteins in leaves and roots, respectively (Huerta-Ocampo et al. 2009, 2011). In roots, DOF1 and MIF1 TFs were associated with drought stress together with osmolyte accumulation and heat shock proteins, contrary to leaves, where TFs were not found. They also identified 77 amaranth proteins in response to salt stress. These proteins were grouped into different categories of biological processes according to Gene Ontology (GO), but once again no TFs were detected (Huerta-ocampo et al. 2014). Recently, Gonzalez-Rodriguez et al. (2019), found that water deficit stress (WDS) tolerance in *A. hypochondriacus*, *A. cruentus*, *A. caudatus* and *A. hybridus* is multifactorial and have a different tolerance range between species. WDS tolerance was correlated with an enhanced osmotic adjustment, a stronger expression of abscisic acid marker genes and a more robust sugar starvation response (González-Rodríguez et al. 2019).

Otherwise, biotic stress is caused by living organisms, especially viruses, bacteria, fungi, and insects (Frost et al. 2008; Kessler 2010). Plant defence results in upregulation of genes associated with the production of metabolites such as proteinase inhibitors and toxins, but the induced response is complex because each pathogen or insect elicit different types of defence reaction (Bruce and Pickett 2007). However, only a limited number of defence-related genes have been reported in amaranth (Délano-Frier et al. 2004; Sánchez-Hernández et al. 2004). In Table 10.1, we included two studies addressed to biotic stress in *A. cruentus* (Castrillón-Arbeláez et al. 2012; Casarrubias-Castillo et al. 2014). In the first, amaranth plants show exceptional tolerance to severe defoliation, which was associated with an efficient utilization of high levels of stored carbon reserves in stem and roots (Castrillón-Arbeláez et al. 2012). In the second, amaranth plants treated with benzothiadiazole (BTH), methyl jasmonate (MeJA) showed differences mechanism to induce systemic acquired resistance against bacterial pathogens (Casarrubias-Castillo et al. 2014). Both studies were addressed to metabolic pathways, non-structural carbohydrates and systemic acquired resistance, respectively, thus TFs were not analyzed.

Transcriptomic analysis of grain amaranth plants subjected to various (a)biotic stresses (Délano-Frier et al. 2011) is the most significant study performed to identify induced genes in amaranth. This revealed that more than 1900 genes were induced in at least one out four stress conditions; water stress (WS), salt stress (SS), herbivory (Hb) and bacterial infection (BI). Some of these genes have been functional characterized and associated with growth and stress tolerance, e.g. Ah24, AhDGR2 and NF-YC. We did a search of induced genes in grain amaranth transcriptome of the most relevant TFs families associated with stress tolerance, to look for new strong candidates that may coordinate the response mechanism to biotic and abiotic stresses (Table 10.2). In the section below, we describe and discuss three main families: MYB (derived from the avian **my**elo**b**lastosis), bHLH (**b**asic **h**elix-**l**oop-**h**elix) and WRKY (called by the invariant **WRKY** amino acid sequence at the N-terminus) (Table 10.2; Fig. 10.1). In addition, we added other TF candidates to be associated in response with biotic and abiotic stresses such as C2H2, NAC and bZIP.

Notably, all identified TFs showed lower relative abundance of transcripts in the transcriptomic analysis. This can be explained because cDNA library for pyrosequencing was a mix of two tissues (leaves and stem), four type of stress with specific time points; WS (6 day), SS (6 day), Hb (9, 12 and 24 h) and BI (6 day). This idea is supported by the published quantitative expression of *AhERF-VII*, *AhDofAI* and *AhNF-YC* genes (Palmeros-Suárez et al. 2015; Massange-Sánchez et al. 2016), where detailed analysis of tissue-specific and stress-induced expression patterns was performed. For instance, *AhERF-VII* was $\approx$9

Table 10.2 Strong amaranth TFs candidates to be associated in response mechanism with biotic and abiotic stress tolerance

Transcription factor family	No. of genes in the genome	No. of induced gene	Isotig/ Gene	Relative abundance			
				SS	WS	Hb	BI
MYB	259	2	8,132	0.1	0	0	0.4
			15,743	0.7	0.7	0.2	0
WRKY	60	2	15,239	1.8	1.1	0.6	3
			8,051	0.1	0.2	0.1	0.1
bHLH	111	1	15,725	0.2	0.2	0.2	1.2
AP2/ERF	100	6	17,105	0	0	0	0.4
			11,384	0.2	1.4	1	2.9
			AhERF-VII	**0.4**	**0.1**	**0.2**	**1.4**
			00,416	1.2	1.1	1.6	4.6
			3,639	2.8	1.1	1.7	7.4
			11,758	1.5	0.1	1.1	4.4
Dof	33	2	***AhDof-AI***	**1.6**	**1.1**	**0.6**	**3**
			1,871	0.2	0.7	0.2	0
NFY	25	1	***AhNFY-C***	**2.8**	**1.7**	**0.8**	**2.9**
C2H2	61	1	9,391	0.1	0.1	0	0.5
NAC	55	1	17,652	1.9	0.9	0.2	1.3
bZIP	55	1	2,197	0.3	0.2	0.2	2.1

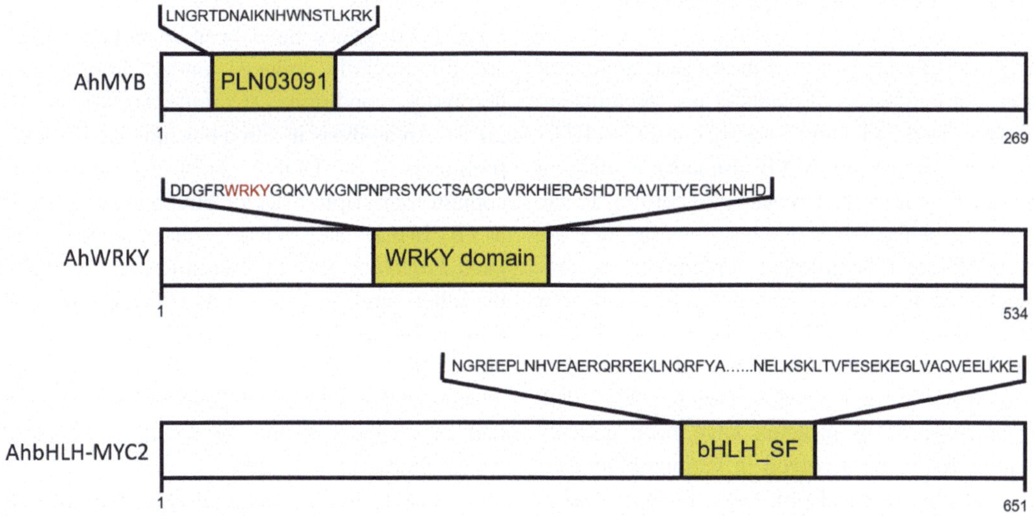

Fig. 10.1 Scheme and domain aminoacidic sequences of putative AhWRKY, AhbHLH-MYC2 and AhMYB TF proteins

Table 10.3 Functional studies of Amaranth TFs (AhERF-VII, Dof-A1 and NF-YC) conferring abiotic stress tolerance

Transcription factor	Isotig	Increase stress tolerance to:	Mechanism of regulation	Reference
AhERF-VII	12,988	Water stress	ABA-independent	(Massange-Sánchez et al. 2016)
AhDof-AI	03,395	Salt stress	ABA-dependent and ABA-independent	(Massange-Sánchez et al. 2016)
AhNFY-C	20,395	Water stress	Partly ABA-dependent	(Palmeros-Suárez et al. 2015)

and ≈twofold higher expressed in leaves of amaranth plants subjected to WS and SS at 6 day, respectively. This is contrasting with the 0.4 and 0.1 relative abundance found in WS and SS, respectively, in the transcriptome. Thus, relative abundance in grain amaranth transcriptome can only work as a guideline for selection of potential genes. Another two transcriptomic analyses performed in weedy amaranth were addressed to glyphosate resistance mechanism, but we here focus only on induced genes from grain amaranth transcriptome.

10.2.1 MYB Transcription Factor

The MYB family is one of the largest families of transcription factors in plants. These proteins are characterized by having a highly conserved DNA binding domain (or MYB domain) containing one to four imperfect repeats. According to the structure of the DNA binding domain, they are grouped into four different subfamilies, according to the number of adjacent repeats in their MYB domains: 1R-MYB (one repeat), R2R3-MYB (two repeats), 3R-MYB (three repeats), and 4R-MYB (four repeats) (Dubos et al. 2010). Until now, MYB proteins have been actively studied for being involved in various biochemical and physiological processes, including regulation of primary and secondary metabolism, control of development and cell cycle, participation in defence and response to biotic and abiotic stresses, flavonoid biosynthesis, hormonal synthesis and signal transduction (Chen et al. 2014). The MYB TFs that have been identified have been extensively studied through functional characterization, especially by regulating biotic and abiotic stresses (Dubos et al. 2010).

Most of the MYB genes in the plant kingdom encode R2R3-MYB class proteins, which contain two repeats, and therefore most studies have focused on the functions of R2R3-type MYB proteins. In beet, there are 70 genes of the R2R3-MYB type, analysis of genome revealed the identification of a specific group of betalain regulators, and the functional study indicated that the *R2R3-MYB Bv_iogq* gene encodes a flavonol regulator *BvMYB12* (Stracke et al. 2014). In *A. hypochondriacus*, *AhDODA*-1, *AhDODA*-2 and *CYP76AD1* genes have been identified, which are differentially induced under drought and herbivory conditions (Casique-Arroyo et al. 2014). An analysis in silico has indicated that the promoters of the DODA genes of *P. americana* contain transcription factor binding sites of type MYB, bHLH, and WRKY that could be regulated *PaDODA gene* (Takahashi et al. 2009). On the other hand, in beet, an MYB-type transcription factor has been identified for leaf pigmentation (Hatlestad et al. 2015) and in grain amaranth, the *AmMYBL1* gene has been identified as a candidate for the colour of the seed cover (Stetter et al. 2020).

A search in the amaranth genome (https://phytozome.jgi.doe.gov/pz/portal.html) with the MYB keyword exhibited 259 genes. From these, only two sequences (Isotigs; 08,132 and 15,743) encoding for MYB-type transcription factors were identified to be induced in the grain amaranth transcriptome. Isotig 08,132 is approximately

80% similar to species of *Beta vulgaris*, *Spinacia oleracea* and *Chenopodium quinoa*, whilst Isotig 15,743 is matched to the species of *Salvia miltiorrhiza* and *Gossypium hirsutum* around 59%. MYB domain (PLN0309) aminoacidic sequence of amaranth proteins is represented in Fig. 10.1. However, Isotig 08,132 was curiously induced both SS (0.1) and BI (0.4), and Isotig 15,743 was mainly induced by abiotic stress, with relative abundance of 0.7 both SS and WS (Table 10.2). These make Isotig 15,743 attractive for future studies related to increasing abiotic stress tolerance and Isotig 08,132 interesting to investigate a putative regulator involved in crosslink response to SS and BI.

10.2.2 *bHLH Transcription Factor*

The superfamily named bHLH (basic helix-loop-helix) is the second-largest plant transcription factor (TF) family, and it is distributed among all eukaryotes. The bHLH TFs are characterized by a highly conserved bHLH domain of 50–60 amino acids that form two distinctive regions, the basic region (approximately 15 amino acids in length) with six basic residues for DNA binding in a palindromic hexanucleotide region E-box (CANNTG), such as the G-box (CACGTG), in promoters of the target genes and HLH (helix-loop-helix) region for protein–protein interaction, which forms two amphipathic helices with a linking loop, and acts as a region of homo- and heterodimerization (Jones 2004; Saibo et al. 2009; Pires and Dolan 2009; Strygina and Khlestkina 2019). It has been reported that bHLHs are key regulatory components in transcription networks that control a number of biological processes in plants, including growth and development as well as response to various stresses, such as injury, drought and salinity stress (Kavas et al. 2016; Mao et al. 2017). A recent genome-wide analysis of bHLH TFs from Arabidopsis, poplar, rice, moss and algae classified the bHLH TFs into 32 subfamilies (Carretero-Paulet et al. 2010); within which there are some special members, named myelocytomatosis oncogenes (MYC). These members are characterized by a bHLH_MYC_N region in the N-terminal and a bHLH region in the C-terminal (Ledent and Vervoort 2001; Kazan and Manners 2013).

MYC genes play important roles in response to abiotic and biotic stresses. MYC TFs participate in the ABA-dependent pathway of stress signalling for the upregulation of the abiotic stress-induced genes; it was demonstrated that AtMYC2 function as transcriptional activators in the dehydration- and ABA-inducible *rd22* expression (Yadav et al. 2013). Other research showed that AtMYC2 is induced by drought and salt stresses, and overexpressing this gene improves osmotic stress tolerance. Abe et al. (2003) showed that increased expression levels of AtMYC2 enhance ABA sensitivity and improve osmotic tolerance, and overexpression of 35S:AtMYC2 in Arabidopsis induced ABA-responsive stress genes and showed an ABA-hypersensitive phenotype with increased osmotic stress tolerance. Microarray studies in transgenic plants overexpressing these TFs revealed that not only were ABA-related stress genes differentially regulated, but also jasmonic acid-related genes, thus indicating a crosstalk pathway between abiotic and biotic stress responses (Abe et al. 2003; Saibo et al. 2009). AtMYC2 likely plays roles in JA-mediated tolerance to the herbivore *Helicoverpa armigera* and jasmonate associated MYC2-like (JAM) 1, 2 and 3 were identified as targets of MYC2; which negatively regulate other targets of MYC2, such as genes related to defence against herbivores during JA responses (Dombrecht et al. 2007; Sasaki-Sekimoto et al. 2013). In other plant species like *Nicotiana attenuata,* NaMYC2 regulates plant resistance against herbivores (Woldemariam et al. 2013), and rice OsMYC2 negatively regulates JA-mediated resistance to a necrotrophic pathogen in bacterial blight (Uji et al. 2016).

Approximately 111 and 90 bHLH elements were found in grain amaranth genome (https://phytozome.jgi.doe.gov/pz/portal.html) and transcriptome, respectively, but only one (Isotig 15,725) was induced under biotic and abiotic stress conditions. Isotig 15,725 is approximately 79% similar to *Beta vulgaris*, followed by 78%

and 72% of identity with *C. quinoa and S. oleracea,* respectively. bHLH domain aminoacid sequence of AhbHLH-MYC2 proteins is represented in Fig. 10.1. Notable, Isotig 15,725 was higher induced under BI (1.2) conditions, but lower and steadily expressed in SS (0.2), WS (0.2) and Hb (0.2), which make it an attractive candidate to study tolerance both biotic and abiotic stresses.

10.2.3 WRKY Transcription Factor

The WRKY family is characterized by the presence of WRKY domains containing a WRKYGQK motif highly conserved in the N-terminal region and a zinc finger motif Cys(2)-His(2) or Cys(2)-HisCys in the C-terminal region (Bakshi and Oelmüller 2014). Both elements are essential for WRKY proteins can bind to TTGAC(C/T) W-box *cis*-element in the promoter of their target genes, which is frequently present in the promoters of many defence-related genes (Yu et al. 2001; Turck et al. 2004). Plant stress involves hormone signalling, therefore, WRKYs are closely related in hormone signalling net (Bakshi and Oelmüller 2014). This TF′s family plays a major role in plant defence to abiotic and biotic stresses and has been extensively investigated for functional studies in many agronomic and horticulture crops (Seo and Choi 2015; Chen et al. 2017).

Amaranth genome (https://phytozome.jgi.doe.gov/pz/portal.html) has a reported 60 genes annotated with the WRKY keyword (Table 10.2). Two sequences (Isotigs; 08,051 and 15,239) encoding for WRKY proteins were induced in the grain amaranth transcriptome. Isotig 08,051 is approximately 67% similar to *C. quinoa*, follow by $\approx$ 65% of identity both *S. oleracea* and *B. vulgaris*, and Isotig 15,239 is around 88% similar to *C. quinoa* and *B. vulgaris*. WRKY domain aminoacid sequence of amaranth proteins is represented in Fig. 10.1. Isotig 15,239 was relatively higher induced in four stress conditions; SS (1.8), WS (1.1), Hb (0.6) and BI (3), with highest relative abundance in BI (Table 10.2). Isotig 08,051 also was induced in four stress conditions SS (0.1), WS (0.2), Hb (0.1) and BI (0.1), but with lower relative abundance in comparison to the other WRKY TF (Isotig 15,239). These make it more attractive to investigate Isotig 15,239 than Isotig 08,051 for future studies related to increasing biotic and abiotic stresses tolerance. Nonetheless, it will be interesting to test whether both WRKY TFs have a putative redundant function or are more specific to any stress condition.

10.3 Functional Studies of Amaranth TFs Conferring Stress Tolerance

Transcriptomic analysis of grain amaranth plants subjected to various (a)biotic stresses (Délano-Frier et al. 2011) started an ongoing characterization of a selected number of these genes. Already TFs and unknown genes have been functionally characterized and associated with both growth and stress tolerance. In this section, we focus on the three TFs; AhERF-VII, Dof-A1 and NF-YC overexpressed in transgenic Arabidopsis plants. Also, we show a scheme and domain aminoacid sequence of AhERF-VII, AhDof-A1 and AhNY-YC TF proteins (Fig. 10.2). In addition, we searched for other members of ERF and Dof TF families in the grain amaranth transcriptome, which are strong candidates to be studied in response mechanism to biotic and abiotic stresses (Table 10.2).

10.3.1 AhERF-TF Conferred Tolerance to Water Stress by ABA-Independent Mechanisms

The ERF TF family belongs to the AP2/ERF super family, which also contains the AP2 and RAV families (Riechmann et al. 2000). AP2/ERF TFs are characterized by the presence of at least one AP2/DNA binding domain, consisting of 60–70 amino acid (aa) residues that directly interact with GCC box and/or dehydration responsive element (DRE)/C-repeat element

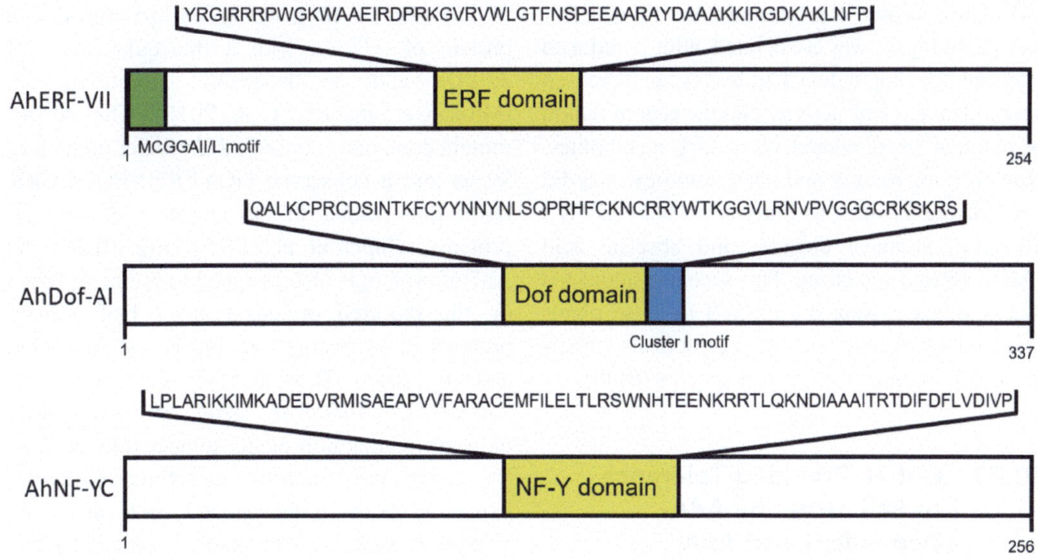

Fig. 10.2 Scheme and domain aminoacidic sequences of AhERF-VII, AhDof-A1 and AhNY-YC TF proteins

(CRT) *cis*-acting elements (Allen et al. 1998) (Fig. 10.2). ERF family proteins have a single AP2/ERF domain, AP2 family proteins possess two repeated AP2/ERF domains, and RAV family proteins contain a single AP2/ERF domain with the addition of a B3 domain that is a DNA-binding domain in another plant transcription family (Nakano et al. 2006). A blast search in the amaranth genome (https://phytozome.jgi.doe.gov/pz/portal.html) using the AhERF-VII protein query of 254 aa length (Massange-Sánchez et al. 2016), identified a total of 100 members of the AP2/ERF family proteins. Meanwhile, transcriptomic analysis of grain amaranth (*A. hypochondriacus*) subjected to various (a)biotic stresses revealed that at least four AP2/ERF TFs were induced in one or more stress conditions (Délano-Frier et al. 2011). One out of these four TFs (*AhERF-VII*) was functionally analyzed by overexpression in Arabidopsis plants.

The *AhERF-VII* gene has a length of 1022 bp and encodes for a protein of 254 aa with a molecular weight of 28.6 kDa and an isoelectric point of 5.08. The AhERF protein contains a conserved AP2 /ERF domain of 58 aa and two conserved N-terminal motifs (MCGGAII and CMVII-1) (Fig. 10.2). A phylogenetic analysis clustered it together with group VII ERF Arabidopsis genes, this is the cause is referred to as AhERF-VII. *AhERF-VII* gene was induced in amaranth plants under water stress (WS) and avirulent (Avr) bacterial infection. Specifically, *AhERF-VII* is highly expressed both in leaves (≈9 and ≈fivefold higher) and roots (≈3 and ≈ninefold higher) of amaranth plants subjected to WS for 6 and 8 days, respectively. Whilst, in amaranth plants infected with *Pseudommonas syringae* pathovar (Avr), *AhERF-VII* was consistently expressed about three-fold higher at 4- and 6-days post-infection (dpi), with highest expression (eightfold higher) at 8 days dpi. Also, *AhERF-VII* was rapidly and strongly induced by exposure to jasmonic acid and mechanical defoliation, respectively.

In agreement with *AhERF-VII* expression level, *AhERF-VII* overexpressing (OE-AhERF-VII) Arabidopsis plants increased tolerance to WS and suggest that AhERF-VII plays a role regulating plant responses to wounding and/or biotic stressors. This last hypothesis is supported, in part, by microarray data showing the upregulated expression of several defence-related genes as thionin 2.1 and thaumatin-like protein 1, and by the observed enrichment of gene ontology categories associated with biotic stress in OE-AhERF-

VII plants. While WS tolerance in OE-AhERF-VII Arabidopsis was associated with a reduced stomatal opening in detached leaves, an increased antioxidant enzyme activity and the accumulation of putative stress-related secondary metabolites. However, microarray and gene ontology data did not indicate an obvious correlation between WS tolerance, stomatal closure, and abscisic acid (ABA)-related signalling. This scenario suggested that stomatal closure during WS in these plants involved ABA-independent mechanisms, possibly involving reactive oxygen species (ROS).

10.3.2 Dof TF Provided Tolerance to Salt Stress by ABA-Dependent and ABA-Independent Pathways

The DNA binding with one finger (Dof) family proteins are a group of plant-specific TFs that contain a 50-to-52 aa conserved domain in the N-terminal region (Fig. 10.2). This Dof domain corresponds to a C_2C_2 configured zinc finger that binds specifically to the 5′-T/AAAAG-3′ sequence motif in the promoters. The C-terminal part of the protein forms a highly variable structure that may contain specific protein–protein interaction domains and other regulatory elements (Yanagisawa 2002; Gupta et al. 2015). Dof TF family is widely distributed in the plant kingdom. In silico analysis indicates that the number of Dof genes in different plant species can vary from 1 to 8 members in algae, mosses ferns and gymnosperms, to 54 and 78 members in monocots and dicots, respectively (Gupta et al. 2015). Arabidopsis and rice have been predicted to possess 30 and 36 Dof genes, respectively. They were further classified into four major clusters, i.e. A-D, of orthologous genes (Lijavetzky et al. 2003), a system that was also applicable to the classification of the 34 Dof TF genes predicted in tomato (Corrales et al. 2014). Amarant genome has predicted 31 Dof genes (https://phytozome.jgi.doe.gov/pz/portal.html), from these two Dof genes were found to be mainly induced under abiotic stress (Délano-Frier et al. 2011), but only one, the AhDof-AI, was chosen for functional studies.

The *AhDof-AI* gene (1734 bp) encodes a protein of 337 aa with a molecular mass of 37 kDa and an isoelectric point of 5.8 (Massange-Sánchez et al. 2016). The AhDof protein contains a conserved Zn finger domain of 29 aa and a conserved GGVLRNVPGGGCRK motif that belong to the cluster I of Dof TF proteins (Gupta et al. 2015) (Fig. 10.2). The AhDof protein is also assigned to group A, based on the reported organization of Dof domain proteins in soybean, Arabidopsis and rice (Guo and Qiu 2013). Thus, this TF is referred to as AhDof-AI. *AhDof-AI* gene is temporarily induced in amaranth plants subjected to WS and SS. Under WS, *AhDof-AI* is ≈threefold higher expressed both in leaves and roots at 2 and 8 days, respectively. In comparison, under SS, *AhDof-AI* is mainly expressed in roots, about ≈2 and ≈threefold higher expressed at 6 and 8 days. In contrast to *AhERF*, AhDof-AI did not respond to AJ treatment, but is induced locally and systemically by both ABA and BTH.

AhDof-AI overexpressing (OE-AhDof-A1) Arabidopsis plants increased tolerance to SS, but not WS. SS tolerance was associated with significantly higher SOD activity levels and Glu accumulation. However, SS tolerance did not involve modified stomatal closure, and it was supported by a down-regulation of GO categories related to stomatal complex morphogenesis and development. Likewise, genes associated with stomata movement were mostly down-regulated as well. Otherwise, GO analysis indicated that SS tolerance in OE-Ah- Dof-AI plants coincided with an enrichment of categories related with response to xenobiotic/ biotic stimulus and to flooding and related nitric oxide metabolic processes. Further, SS tolerance conferred by AhDof-AI in Arabidopsis involves ABA-dependent and ABA-independent stress amelioration mechanisms.

10.3.3 NF-YC Increase Water Stress Resistance by Partly ABA-Dependent

Nuclear Factor Y (NF-Y) family also called heme-activated proteins (HAPs) or CCAAT binding factors (CBFs) are TFs ubiquitous in higher eukaryotes. NF-Y is a heterotrimeric complex comprised of three distinct subunits: NF-YA (HAP2 or CBF-B), NF-YB (HAP3 or CBF-A), and NF-YC (HAP5 or CBF-C) (Mantovani 1999; Romier et al. 2003; Nardini et al. 2013). Each subunit is required for DNA binding, subunit association and transcriptional regulation in animals and plants (Hackenberg et al. 2012; Calvenzani et al. 2012). First, NF-YB and NF-YC initially form a dimer complex in the cytoplasm and subsequently translocate to the cell nucleus where they interact with NF-YA to form a mature heterotrimeric complex and bind CCAAT sites, to activate or repress transcription of genes (25–30% of the mammalian genes have this site in their promoters) (Steidl et al. 2004; FitzGerald et al. 2004; Testa et al. 2005). In animals and yeast, there is a single gene encoding each NF-Y subunit with multiple splicing isoforms. In contrast plant genomes have multigene families encoding each NF-Y subunits. For example, Arabidopsis has 30 members of the NF-Y family, 10 from each family, *O. sativa* harbours at least 30 NF-Y genes and 37 in *Triticum aestivum* (Stephenson et al. 2007; Thirumurugan et al. 2008; Cao et al. 2011; Chen et al. 2014). The amaranth genome is predicted to have 25 NF-Y genes (Table 10.2; https://phytozome.jgi.doe.gov/pz/portal.html), where only one NF-YC TF was induced by biotic and abiotic stresses (Table 10.2).

NF-YC subunit from *Amaranthus hypochondriacus* has been reported to be a gene that is upregulated under stress conditions, and Palmeros-Suárez et al., 2015 showed that Ah NF-YC was induced in leaves and roots of grain amaranth plants subjected to water stress. AhNF-YC is a 1,257 bp long coding DNA sequence of length, including a 768 bp ORF which have 255 amino acids in length and a conserved histone domain (Fig. 10.2), belonging to the H2A superfamily (Mantovani 1999) to allow the interaction with NF-YB subunits, which can then interact with NF-YA to conform the active DNA-binding hetero-trimer with the two amino acids arginine (R52) and aspartate (D59) necessary in stabilizing the structural conformations of NF-YB and NF-YC through the HFMs (Hackenberg et al. 2012). Additionally, the protein also shows a glutamine-rich hydrophobic domain in the C-terminus, considered to be necessary for transcription activation.

The overexpression of AhNF-YC in Arabidopsis strongly supported its active role in development, in accordance with the versatile participation of NF-Y genes in plant growth and development by modifying the root length and root hair abundance delays flowering and increased vegetative tissue and modified the sensitivity to ABA by decreasing seed germination and increasing 3-day-old seedling growth, according with the increased sensitivity to ABA observed in transgenic rice OE an NF-YC transcription factor from *Bermuda grass* (Chen et al. 2014). Additionally, it was observed that AhNF-YC conferred less damage compared with wild-type plants, with more turgor on leaves and less wilting under drought stress conditions (Palmeros-Suárez et al. 2015).

10.4 Conclusions

Amaranth, specifically grain amaranth, transcriptomes have been an excellent tool to identify induced genes under one or more stress conditions. Functional characterization of selected amaranth genes using a transgenic approach in Arabidopsis confirmed their role in response mechanism to adverse environmental conditions. Specifically, amaranth TFs (AhERF-VII, Dof-A1 and NF-YC) have been shown to increase either salt stress or water stress tolerance by both ABA-dependent and ABA-independent mechanisms and they are available to be used in plant engineering to increase stress tolerance in commercial crops. We also identified strong TF candidates (MYB, WRKY and bHLH) associated with biotic abiotic stress tolerance, making them attractive to study in specific or nonspecific types of stress tolerance.

Acknowledgements We thank Consejo Nacional de Ciencia y Tecnología (CONACYT) and Laboratorio Nacional PlanTECC, México for the facilities and support given to write this chapter.

References

Abe H, Urao T, Ito T et al (2003) Arabidopsis AtMYC2 (bHLH) and AtMYB2 (MYB) function as transcriptional activators in abscisic acid signaling. Plant Cell 15:63–78. https://doi.org/10.1105/tpc.006130

Aguilar-hernández HS, Santos L, León-galván F et al (2011) Identification of calcium stress induced genes in amaranth leaves through suppression subtractive hybridization 168:2102–2109. https://doi.org/10.1016/j.jplph.2011.06.006

Ahuja I, de Vos RCH, Bones AM, Hall RD (2010) Plant molecular stress responses face climate change. Trends Plant Sci 15:664–674. https://doi.org/10.1016/j.tplants.2010.08.002

Allen MD, Yamasaki K, Ohme-Takagi M et al (1998) A novel mode of DNA recognition by a β-sheet revealed by the solution structure of the GCC-box binding domain in complex with DNA. EMBO J 17:5484–5496. https://doi.org/10.1093/emboj/17.18.5484

Arabidopsis T, Initiative G (2000) Analysis of the genome sequence of the—owering plant Arabidopsis thaliana 408

Bakshi M, Oelmüller R (2014) WRKY transcription factors: Jack of many trades in plants. Plant Signal Behav 9:e27700. https://doi.org/10.4161/psb.27700

Brenner DM, Baltensperger DD, Kulakow PA et al (2000) Genetic resources and breeding of amaranthus. In: Plant breeding reviews. Wiley, Ltd., pp 227–285

Bruce TJA, Pickett JA (2007) Plant defence signalling induced by biotic attacks 387–392. https://doi.org/10.1016/j.pbi.2007.05.002

Calvenzani V, Testoni B, Gusmaroli G et al (2012) Interactions and CCAAT-binding of *Arabidopsis thaliana* NF-Y subunits. PLoS ONE 7:e42902

Cao S, Kumimoto RW, Siriwardana CL et al (2011) Identification and characterization of NF-Y transcription factor families in the monocot model plant Brachypodium distachyon. PLoS ONE 6:e21805. https://doi.org/10.1371/journal.pone.0021805

Carretero-Paulet L, Galstyan A, Roig-Villanova I et al (2010) Genome-wide classification and evolutionary analysis of the bHLH family of transcription factors in arabidopsis, poplar, rice, moss, and algae. Plant Physiol 153:1398. https://doi.org/10.1104/pp.110.153593

Casarrubias-Castillo K, Martínez-Gallardo NA, Délano-Frier JP (2014) Treatment of Amaranthus cruentus with chemical and biological inducers of resistance has contrasting effects on fitness and protection against compatible Gram positive and Gram negative bacterial pathogens. J Plant Physiol 171:927–939. https://doi.org/10.1016/j.jplph.2014.02.004

Casique-Arroyo G, Martínez-Gallardo N, De La Vara LG, Délano-Frier JP (2014) Betacyanin biosynthetic genes and enzymes are differentially induced by (a)biotic stress in Amaranthus hypochondriacus. PLoS ONE 9:19–22. https://doi.org/10.1371/journal.pone.0099012

Castrillón-Arbeláez PA, Martínez-gallardo N, Arnaut HA et al (2012) Metabolic and enzymatic changes associated with carbon mobilization, utilization and replenishment triggered in grain amaranth (Amaranthus cruentus) in response to partial defoliation by mechanical injury or insect herbivory

Chen F, Hu Y, Vannozzi A et al (2017) The WRKY transcription factor family in model plants and crops. CRC Crit Rev Plant Sci 36:311–335. https://doi.org/10.1080/07352689.2018.1441103

Chen N, Yang Q, Pan L et al (2014) Identification of 30 MYB transcription factor genes and analysis of their expression during abiotic stress in peanut (Arachis hypogaea L.). Gene 533:332–345. https://doi.org/10.1016/j.gene.2013.08.092

Clouse JW, Adhikary D, Page JT et al (2016) The amaranth genome : genome, transcriptome, and physical map. Assembly. https://doi.org/10.3835/plantgenome2015.07.0062

Corrales AR, Nebauer SG, Carrillo L et al (2014) Characterization of tomato cycling Dof factors reveals conserved and new functions in the control of floweri. J Exp Bot 65:995–1012. https://doi.org/10.1093/jxb/ert451

Délano-Frier JP, Avilés-Arnaut H, Casarrubias-Castillo K, et al (2011) Transcriptomic analysis of grain amaranth (Amaranthus hypochondriacus) using 454 pyrosequencing: Comparison with A. tuberculatus, expression profiling in stems and in response to biotic and abiotic stress. BMC Genomics 12. https://doi.org/10.1186/1471-2164-12-363

Délano-Frier JP, Martínez-Gallardo NA, Martínez-de La Vega O et al (2004) The effect of exogenous jasmonic acid on induced resistance and productivity in Amaranth (Amaranthus hypochondriacus) is influenced by environmental conditions. J Chem Ecol 30:1001–1034. https://doi.org/10.1023/B:JOEC.0000028464.36353.bb

Dodds PN, Rathjen JP (2010) Plant immunity: towards an integrated view of plant–pathogen interactions. Nat Publ Gr 11:539–548. https://doi.org/10.1038/nrg2812

Dombrecht B, Xue GP, Sprague SJ et al (2007) MYC2 differentially modulates diverse jasmonate-dependent functions in *Arabidopsis*. Plant Cell 19:2225. https://doi.org/10.1105/tpc.106.048017

Dubos C, Stracke R, Grotewold E et al (2010) MYB transcription factors in Arabidopsis. Trends Plant Sci 15:573–581. https://doi.org/10.1016/j.tplants.2010.06.005

>Eckardt NA (2011) Sequencing the rice genome, 2011–2017

FitzGerald PC, Shlyakhtenko A, Mir AA, Vinson C (2004) Clustering of DNA sequences in human promoters. Genome Res 14:1562–1574. https://doi.org/10.1101/gr.1953904

Frost CJ, Mescher MC, Carlson JE et al (2008) Plant defense priming against herbivores: getting ready for a different battle 1(146):818–824. https://doi.org/10.1104/pp.107.113027

Gaines TA, Zhang W, Wang D et al (2010) Gene amplification confers glyphosate resistance in Amaranthus palmeri 107. https://doi.org/10.1073/pnas.0906649107

González-Rodríguez T, Cisneros-Hernández I, Acosta Bayona J et al (2019) Identification of Factors Linked to Higher Water-Deficit Stress tolerance in amaranthus hypochondriacus compared to other grain Amaranths and A. hybridus, their shared ancestor. Plants (Basel, Switzerland) 8. https://doi.org/10.3390/plants8070239

Grobelnik-Mlakar S, Bavec M, Jakop M, Bavec F (2012) The effect of drought occurring at different growth stages on productivity of grain Amaranth Amaranthus cruentus G6. J Life Sci 6:283–286

Guo Y, Qiu L-J (2013) Genome-wide analysis of the Dof transcription factor gene family reveals soybean-specific duplicable and functional characteristics. PLoS ONE 8. https://doi.org/10.1371/journal.pone.0076809

Gupta S, Malviya N, Kushwaha H et al (2015) Insights into structural and functional diversity of Dof (DNA binding with one finger) transcription factor. Planta 241:549–562. https://doi.org/10.1007/s00425-014-2239-3

Hackenberg D, Keetman U, Grimm B (2012) Homologous NF-YC2 Subunit from arabidopsis and tobacco is activated by photooxidative stress and induces flowering. Int J Mol Sci 13:3458–3477. https://doi.org/10.3390/ijms13033458

Hatlestad GJ, Akhavan NA, Sunnadeniya RM et al (2015) The beet Y locus encodes an anthocyanin MYB-like protein that activates the betalain red pigment pathway. Nat Genet 47:92–96. https://doi.org/10.1038/ng.3163

Huerta-ocampo JA, Barrera-pacheco A, Mendoza-herna CS et al (2014) Salt stress-induced alterations in the root proteome of Amaranthus cruentus L

Huerta-Ocampo JÁ, Briones-Cerecero EP, Mendoza-Hernández G et al (2009) Proteomic analysis of Amaranth (Amaranthus hypochondriacus L.) leaves under drought stress. Int J Plant Sci 170:990–998. https://doi.org/10.1086/605119

Huerta-Ocampo JA, León-Galván MF, Ortega-Cruz LB et al (2011) Water stress induces up-regulation of DOF1 and MIF1 transcription factors and down-regulation of proteins involved in secondary metabolism in amaranth roots (Amaranthus hypochondriacus L.). Plant Biol 13:472–482. https://doi.org/10.1111/j.1438-8677.2010.00391.x

(IWGSC) TIWGSC (2014) A chromosome-based draft sequence of the hexaploid bread wheat (Triticum aestivum) genome Ancient hybridizations among the ancestral genomes of bread wheat Genome interplay in the grain transcriptome of hexaploid bread wheat structural and functional pa. Science 80(345):1251788. https://doi.org/10.1126/science.1251788

Jiao Y, Peluso P, Shi J et al (2017) Improved maize reference genome with single-molecule technologies. Nature 546:524–527. https://doi.org/10.1038/nature22971

Jin J, Tian F, Yang DC et al (2017) PlantTFDB 4.0: toward a central hub for transcription factors and regulatory interactions in plants. Nucleic Acids Res 45:D1040–D1045. https://doi.org/10.1093/nar/gkw982

Johnson BL, Henderson TL (2002) Water use patterns of grain Amaranth in the northern great plains. Agron J 94:1437–1443. https://doi.org/10.2134/agronj2002.1437

Jones S (2004) An overview of the basic helix-loop-helix proteins. Genome Biol 5:226. https://doi.org/10.1186/gb-2004-5-6-226

Kavas M, Baloğlu MC, Atabay ES et al (2016) Genome-wide characterization and expression analysis of common bean bHLH transcription factors in response to excess salt concentration. Mol Genet Genomics 291:129–143. https://doi.org/10.1007/s00438-015-1095-6

Kazan K, Manners JM (2013) MYC2: the master in action. Mol Plant 6:686–703. https://doi.org/10.1093/mp/sss128

Kessler A (2010) Silencing the jasmonate cascade: induced plant defenses and insect populations 665. https://doi.org/10.1126/science.1096931

Kohlhase DR, O'Rourke JA, Owen MDK, Graham MA (2019) Using RNA-seq to characterize responses to 4-hydroxyphenylpyruvate dioxygenase (HPPD) inhibitor herbicide resistance in waterhemp (Amaranthus tuberculatus). BMC Plant Biol 19:182. https://doi.org/10.1186/s12870-019-1795-x

Krasensky J, Jonak C (2012) Drought, salt, and temperature stress-induced metabolic rearrangements and regulatory networks. J Exp Bot 63:1593–1608. https://doi.org/10.1093/jxb/err460

Ledent V, Vervoort M (2001) The basic helix-loop-helix protein family: comparative genomics and phylogenetic analysis. Genome Res 11:754–770. https://doi.org/10.1101/gr.177001

Lijavetzky D, Carbonero P, Vicente-Carbajosa J (2003) Genome-wide comparative phylogenetic analysis of the rice and Arabidopsis Dof gene families. BMC Evol Biol 3:1–11. https://doi.org/10.1186/1471-2148-3-17

Ma S, Gong Q, Bohnert HJ (2006) Dissecting salt stress pathways. J Exp Bot 57:1097–1107. https://doi.org/10.1093/jxb/erj098

Mallory MA, Hall RV, McNabb AR et al (2008) Development and characterization of microsatellite markers for the grain Amaranths. Crop Sci 48:1098–1106. https://doi.org/10.2135/cropsci2007.08.0457

Mantovani R (1999) The molecular biology of the CCAAT-binding factor NF-Y. Gene 239:15–27. https://doi.org/10.1016/S0378-1119(99)00368-6

Mao K, Dong Q, Li C et al (2017) Genome wide identification and characterization of apple bHLH transcription factors and expression analysis in response to drought and salt stress. Front Plant Sci 8:480

Mascher M, Gundlach H, Himmelbach A et al (2017) A chromosome conformation capture ordered sequence of the barley genome. Nature 544:427–433. https://doi.org/10.1038/nature22043

Massange-Sánchez JA, Palmeros-Suárez PA, Espitia-Rangel E et al (2016) Overexpression of grain amaranth (Amaranthus Hypochondriacus) AhERF or AhDOF transcription factors in Arabidopsis Thaliana increases water deficit-and salt-stress tolerance, respectively, via contrasting stress-amelioration mechanisms. PLoS ONE 11. https://doi.org/10.1371/journal.pone.0164280

Maughan PJ, Yourstone SM, Jellen EN, Udall JA (2009) SNP discovery via genomic reduction, barcoding, and 454-pyrosequencing in Amaranth. Plant Genome 2:260–270. https://doi.org/10.3835/plantgenome2009.08.0022

Miller TE, Wing JS, Huete AR (1984) The agricultural potential of selected C4 plants in arid environments [carbon]. J Arid Environ 7:275–286

Nakano T, Suzuki K, Fujimura T, Shinshi H (2006) Genome-Wide Analysis of the ERF Gene Family. Plant Physiol 140:411–432. https://doi.org/10.1104/pp.105.073783.currently

Nardini M, Gnesutta N, Donati G et al (2013) Sequence-specific transcription factor NF-Y displays histone-like DNA binding and H2B-like ubiquitination. Cell 152:132–143. https://doi.org/10.1016/j.cell.2012.11.047

Oh D, Dassanayake M, Bohnert HJ, Cheeseman JM (2012) Life at the extreme : lessons from the genome 1–9

Ohta M, Schumaker KS, Zhu JK (2006) Mechanisms underlying plant tolerance to abiotic stresses. Multigenic Induc Syst Resist Plants 360–385. https://doi.org/10.1007/0-387-23266-4_16

Omami EN, Hammes PS, Robbertse PJ (2006) Differences in salinity tolerance for growth and water-use efficiency in some amaranth (Amaranthus spp.) genotypes. New Zeal J Crop Hortic Sci 34:11–22. https://doi.org/10.1080/01140671.2006.9514382

Palmeros-Suárez PA, Massange-Sánchez JA, Martínez-Gallardo NA et al (2015) The overexpression of an Amaranthus hypochondriacus NF-YC gene modifies growth and confers water deficit stress resistance in Arabidopsis. Plant Sci 240. https://doi.org/10.1016/j.plantsci.2015.08.010

Pires N, Dolan L (2009) Origin and Diversification of basic-helix-loop-helix proteins in plants. Mol Biol Evol 27:862–874. https://doi.org/10.1093/molbev/msp288

Rastogi A, Shukla S (2013) Amaranth: a new millennium crop of nutraceutical values. Crit Rev Food Sci Nutr 53:109–125. https://doi.org/10.1080/10408398.2010.517876

Riechmann JL, J. Heard GM, Reuber L et al (2000) Arabidopsis transcription factors: genome-wide comparative analysis among eukaryotes. Science 290 (80):2105–2110. https://doi.org/10.1126/science.290.5499.2105

Romier C, Cocchiarella F, Mantovani R, Moras D (2003) The NF-YB/NF-YC structure gives insight into DNA binding and transcription regulation by CCAAT factor NF-Y. J Biol Chem 278:1336–1345. https://doi.org/10.1074/jbc.M209635200

Saibo NJM, Lourenço T, Oliveira MM (2009) Transcription factors and regulation of photosynthetic and related metabolism under environmental stresses. Ann Bot 103:609–623

Salas-Perez RA, Saski CA, Noorai RE et al (2018) RNA-Seq transcriptome analysis of Amaranthus palmeri with differential tolerance to glufosinate herbicide. PLoS ONE 13:e0195488. https://doi.org/10.1371/journal.pone.0195488

Sánchez-Hernández C, Martínez-Gallardo N, Guerrero-Rangel A et al (2004) Trypsin and α-amylase inhibitors are differentially induced in leaves of amaranth (Amaranthus hypochondriacus) in response to biotic and abiotic stress. Physiol Plant 122:254–264. https://doi.org/10.1111/j.0031-9317.2004.00398.x

Sasaki-Sekimoto Y, Jikumaru Y, Obayashi T et al (2013) Basic helix-loop-helix transcription factors JASMONATE-ASSOCIATED MYC2-LIKE1 (JAM1), JAM2, and JAM3 are negative regulators of jasmonate responses in arabidopsis. Plant Physiol 163:291. https://doi.org/10.1104/pp.113.220129

Seo E, Choi D (2015) Functional studies of transcription factors involved in plant defenses in the genomics era. Brief Funct Genomics 14:260–267. https://doi.org/10.1093/bfgp/elv011

Shinozaki K, Yamaguchi-Shinozaki K (2007) Gene networks involved in drought stress response and tolerance. J Exp Bot 58:221–227. https://doi.org/10.1093/jxb/erl164

Singh KB, Foley RC, Oñate-Sánchez L (2002) Transcription factors in plant defense and stress responses. Curr Opin Plant Biol 5:430–436. https://doi.org/10.1016/S1369-5266(02)00289-3

Steidl S, Tüncher A, Goda H et al (2004) A single subunit of a heterotrimeric CCAAT-binding complex carries a nuclear localization signal: piggy back transport of the pre-assembled complex to the nucleus. J Mol Biol 342:515–524. https://doi.org/10.1016/j.jmb.2004.07.011

Stephenson TJ, McIntyre CL, Collet C, Xue G-P (2007) Genome-wide identification and expression analysis of the NF-Y family of transcription factors in Triticum aestivum. Plant Mol Biol 65:77–92. https://doi.org/10.1007/s11103-007-9200-9

Stetter MG, Vidal-Villarejo M, Schmid KJ (2020) Parallel seed color adaptation during multiple domestication attempts of an ancient new world grain. Mol Biol Evol 37:1407–1419. https://doi.org/10.1093/molbev/msz304

Stracke R, Holtgräwe D, Schneider J et al (2014) Genome-wide identification and characterisation of R2R3-MYB genes in sugar beet (Beta vulgaris). BMC Plant Biol 14:249. https://doi.org/10.1186/s12870-014-0249-8

Strygina KV, Khlestkina EK (2019) Myc-like transcriptional factors in wheat: structural and functional organization of the subfamily i members. BMC Plant Biol 19. https://doi.org/10.1186/s12870-019-1639-8

Sunil M, Hariharan AK, Nayak S et al (2014) The draft genome and transcriptome of Amaranthus hypochondriacus: a C4 dicot producing high-lysine edible pseudo-cereal. DNA Res an Int J Rapid Publ Reports Genes Genomes 21:585–602. https://doi.org/10.1093/dnares/dsu021

Takahashi K, Takamura E, Sakuta M (2009) Isolation and expression analysis of two DOPA dioxygenases in Phytolacca americana. Z Naturforsch C 64:564–573. https://doi.org/10.1515/znc-2009-7-816

Testa A, Donati G, Yan P et al (2005) Chromatin immunoprecipitation (ChIP) on chip experiments uncover a widespread distribution of NF-Y binding CCAAT sites outside of core promoters. J Biol Chem 280:13606–13615. https://doi.org/10.1074/jbc.M414039200

Thirumurugan T, Ito Y, Kubo T et al (2008) Identification, characterization and interaction of HAP family genes in rice. Mol Genet Genomics 279:279–289. https://doi.org/10.1007/s00438-007-0312-3

Tomato T, Consortium G (2012) The tomato genome sequence provides insights into fleshy fruit evolution. https://doi.org/10.1038/nature11119

Transue DK, Fairbanks DJ, Robison LR, Andersen WR (1994) Species Identification by RAPD analysis of grain amaranth genetic resources. Crop Sci 34.https://doi.org/10.2135/cropsci1994.0011183X003400050044x

Turck F, Zhou A, Somssich IE (2004) Stimulus-dependent, promoter-specific binding of transcription factor WRKY1 to Its native promoter and the defense-related gene PcPR1-1 in parsley. Plant Cell 16:2573–2585. https://doi.org/10.1105/tpc.104.024810

Udvardi MK, Kakar K, Wandrey M et al (2007) Legume transcription factors: global regulators of plant development and response to the environment. Plant Physiol 144:538–549. https://doi.org/10.1104/pp.107.098061

Uji Y, Taniguchi S, Tamaoki D et al (2016) Overexpression of OsMYC2 results in the up-regulation of early JA-rresponsive genes and bacterial blight resistance in rice. Plant Cell Physiol 57:1814–1827. https://doi.org/10.1093/pcp/pcw101

Wang H, Wang H, Shao H, Tang X (2016) Recent advances in utilizing transcription factors to improve plant abiotic stress tolerance by transgenic technology. Front Plant Sci 7:1–13. https://doi.org/10.3389/fpls.2016.00067

Woldemariam MG, Dinh ST, Oh Y et al (2013) NaMYC2 transcription factor regulates a subset of plant defense responses in Nicotiana attenuata. BMC Plant Biol 13:73. https://doi.org/10.1186/1471-2229-13-73

Yadav NR, Taunk J, Rani A, et al (2013) Role of transcription factors in abiotic stress tolerance in crop plants. In: Climate change and plant abiotic stress tolerance. Wiley, Ltd., pp 605–640

Yanagisawa S (2002) C2C2-Dof Family4.pdf. 7:555–560

Yu D, Chen C, Chen Z (2001) Evidence for an important role of WRKY DNA binding proteins in the regulation of NPR1 gene expression. Plant Cell 13:1527–1540. https://doi.org/10.1105/tpc.010115

The manufacturer's authorised representative in the EU is Springer Nature Customer Service Centre GmbH, Europaplatz 3, 69115 Heidelberg, Germany. If you have any concerns regarding our products, please contact ProductSafety@springernature.com

Printed and bound by CPI Group (UK) Ltd, Croydon, CR0 4YY
30/09/2026
02225691-0013